RECENT ADVANCES IN LASER PROCESSING OF MATERIALS

Elsevier Internet Homepage - http://www.elsevier.com
Consult the Elsevier homepage for full catalogue information on all books, journals and electronic products and services.

Elsevier Titles of Related Interest

Laser Processing of Engineering Materials
John Ion
2005, ISBN 0750660791
www.elsevier.com/locate/isbn/0750660791

Related Journals
The following journals related to laser processing can all be found at
http://www.sciencedirect.com

Applied Surface Science
Journal of Materials Processing Technology
Materials Science and Engineering: A Structural Materials: Properties, Microstructure
and Processing
Materials Science and Engineering: B Solid State Materials for Advanced Technology
Science and Technology of Advanced Materials
Thin Solid Films

To Contact the Publisher
Elsevier welcomes enquiries concerning publishing proposals: books, journal special issues, conference proceedings, etc. All formats and media can be considered. Should you have a publishing proposal you wish to discuss, please contact, without obligation, the commissioning editor responsible for Elsevier's materials science books publishing programme:

Amanda Weaver
Publisher, Materials Science
Elsevier Limited
The Boulevard, Langford Lane   Tel.:   +44 1865 84 3634
Kidlington, Oxford   Fax: +44 1865 84 3920
OX5 1GB, UK   E-mail: a.weaver@elsevier.com

General enquiries including placing orders, should be directed to Elsevier's Regional Sales Offices – please access the Elsevier internet homepage for full contact details.

RECENT ADVANCES IN LASER PROCESSING OF MATERIALS

EDITORS

JACQUES PERRIÈRE
INSP, Université Paris 6 and CNRS UMR 7588
Paris, France

ERIC MILLON
LSMCL, Université Paul Verlaine, Metz, France

ERIC FOGARASSY
InESS, ULP-CNRS, Strasbourg, France

Published in Association with

E·MRS

EUROPEAN
MATERIALS RESEARCH SOCIETY

ELSEVIER

Amsterdam – Boston – Heidelberg – London – New York – Oxford
Paris – San Diego – San Francisco – Singapore – Sydney – Tokyo

ELSEVIER B.V.
Radarweg 29
P.O. Box 211, 1000 AE
Amsterdam, The Netherlands

ELSEVIER Inc.
525 B Street, Suite 1900
San Diego,
CA 92101-4495, USA

**ELSEVIER Ltd
The Boulevard, Langford Lane
Kidlington, Oxford OX5 1GB UK**

ELSEVIER Ltd
84 Theobald's Road
London, WCIX 8RR UK

First edition 2006

ISBN-13: 978-0-08044-727-8
ISBN-10: 0-080-44727-9

Printed in Great Britain.

06 07 08 09 10   10 9 8 7 6 5 4 3 2 1

Contents

Introduction to the Laser Processing of Materials

J. Perrière[a], E. Millon[b] and E. Fogarassy[c]

[a]*INSP, Université Paris 6 et CNRS, Campus Boucicaut, 140 rue de Lourmel, 75015 Paris, France;* [b]*LSMCL, Université Paul Verlaine de Metz, 1 bd Arago, 57078 Metz Cedex 3, France;* [c]*InESS, CNRS, rue du Loess, 67037 Strasbourg Cedex 2, France*

During the last 20 years new technological needs in various key areas of materials processing such as microelectronics, nanoscience or biology have motivated the development of novel alternative techniques which are able to respond to the new technological requests for more precision, higher resolution and better surface and volume localization. This development would certainly not be possible without the strong development of new laser sources. Indeed, in addition to the intrinsic properties of lasers, i.e. monochromaticity, spatial and temporal coherency, divergence as low as 10^{-4} rad, and very high power density up to 10^{15} W/cm^2 are now easily obtained with ultrashort laser pulses. This represents therefore a powerful tool to induce structural or morphological modifications in the near surface region of solids, as well as structural ordering and phase transformation of materials.

The importance of laser processing of materials clearly appears through the number of publications in this field, i.e. between 1200 and 1500 papers are regularly published each year on the various aspects of the use of lasers to modify the near surface region of solids, i.e. the composition, the morphology or the structure of the materials. Moreover, the main international conferences covering this domain (SPIE, MRS, COLA, etc.) always have a great number of attendees, and among all these meetings, the classical symposium on laser processing of the European Materials Research Society (EMRS) is now the annual 'rendez-vous' of scientists from both academic institutions and industries, working in this field. Indeed, nucleating the very first of EMRS Spring Meetings, the symposium on

Laser Materials Interaction and Processing has established a long standing and highly successful tradition in the materials science community.

Moreover this symposium allows to display the latest state of the art and to anticipate new trends in the field. Every year, more than 200 communications are presented in this regular symposium by around hundreds of attendees from more than 30 countries all over the world. Taking into account this point and the noticeable recent progresses in this field, both in the scientific and technological aspects of laser matter interaction covering various emerging materials science domains, a book on 'Recent Advances in Laser Processing of Materials' could be useful for all the laser community as it appeared through discussions between the main experts of the field during the EMRS Conference in Strasbourg. In addition, such book may be a powerful and invaluable tool for scientists and engineers strongly interested in laser-assisted processes in a wide range of disciplines: fundamental and applied physics, organic and inorganic chemistry, electronics, optics and biology.

Laser processing of materials is a multidisciplinary field of research, which cannot be exhaustively treated in a single book. Indeed, due to the large number of works in this domain, a continuous evolution of thematic occurs, new developments appear, and thus it is not possible to cover all the numerous aspects of laser processing of materials. We have therefore decided to focus this book on the very new trends involving laser treatments which are now developed. First the formation of nanomaterials by laser-assisted methods, the modifications induced by laser irradiations in the near surface region of organic or biomaterials, the various processes used for the micro-and nanostructuration of solid surfaces, and then due to a renewing interest, the classical aspects of laser crystallization, doping or cleaning are also treated in this book. This book is therefore divided into 13 chapters which have been written by the most relevant researchers recognized for their work in their respective field of interest. The chapters are gathered in 4 main parts corresponding to aspects of laser processing toward principles and applications of lasers in research and development as well as in industrial surrounding: i) laser-assisted formation of nanomaterials, ii) organic and biomaterial modifications by laser, iii) laser processing for micro- and nanostructuration of solids and iv) laser crystallization, laser doping and laser cleaning.

Laser-Assisted Formation of Nanomaterials

Nanoparticles and materials consisting of nanometer-sized building blocks present a lot of promising properties which could lead to potential applications in various fields of science and technology, i.e. optics, electronics, etc... As a matter of fact, nanometer size semiconductors proved to be very interesting materials since quantum confinement in such systems modifies the band structure and leads to unique physical properties applicable in optoelectronics. In the same way, particle size also plays an important role in other domains like in the ferroelectric to paraelectric phase transition and dielectric properties of ferroelectric materials. These findings lead to an extraordinary big effort on the synthesis of nanoparticles and nanostructures to study their physical properties.

The possibility to form nanoparticles by laser-assisted methods has been of course rapidly checked, and the laser-induced pyrolysis of a gaseous phase is now a well-known method to produce nanoparticles. Indeed, by carefully choosing the precursors, the composition of the gas mixture, the pressure in the reaction chamber, the laser power and flowrate of the gaseous precursors, nanocrystallites of various materials are synthesized. Oxides, metals or semiconducting nanoparticles are therefore obtained, with generally a small distribution in size leading to a large domain of applications. Besides this classical approach, new uses of laser to assist the formation of nanoparticles have been considered, and the laser ablation of a target under a gas pressure (in the 5 to 50 mbar range) has been found as an efficient way to produce nanocrystallites of a broad range of materials. The state of the art in this domain will be presented by A.V. Kabashin and M. Meunier from the École Polytechnique in Montréal (Canada) in Chapter 1 dealing with the 'Laser ablation-based synthesis of nanomaterials'. This chapter describes first the general aspects of nanocluster formation according to the regimes involved in laser ablation. The synthesis of nanomaterials in a gaseous environment by laser ablation of a solid target may lead to nanostructured films deposited on a substrate. The properties of nanostructured semiconductor-based films are particularly discussed in this chapter. Another way to produce high quality nanoparticles consists in pulsed-laser ablation in a liquid which is expected to be advantageous in the preparation of metastable nanocrystals. This method needs a state of high pressure and high temperature for their synthesis. As a result it has been theoretically predicted that nano-diamonds could be obtained by pulsed-laser ablation of a graphite target in water. This new and exciting field and more generally the chemical and physical aspects of laser ablation-based colloidal nanoparticles are also discussed in this chapter. A special emphasis is given on the development of chemical methods to control nanofabrication during laser ablation in liquids, for example by the use of surfactants.

The insertion of nanocrystallites of a given compound in a matrix of a different material can be used to enhance some physical properties of the matrix or to induce new properties. As an example, the insertion of silicon nanocrystals in an Er-doped silica matrix which allows to increase the yield of luminescence of the dopant. Oxide nanoparticles inserted in a silica matrix can also be the basis of nanocomposite presenting new and original properties, for example maghemite (γ-Fe_2O_3) nanocrystals embedded in SiO_2 can be used to develop new magneto-optical planar devices. In the same way, metal nanoparticles inserted in a dielectric matrix shows very attractive optical, magnetic or thermal properties which can be used in various application domains.

The formation of nanocomposite thin films is thus an important challenge, and laser ablation seems to be particularly well suited to grow such nanostructures with a perfect control of the size and density of nanoparticles. The state of the art in this domain is presented in Chapter 2 titled 'Metal–dielectric nanocomposites produced by pulsed-laser deposition: A route for new functional materials' written by C.N. Afonso, J. Gonzalo, R. Serna and J. Solís from CSIC in Madrid (Spain). This contribution shows that the use of pulsed-laser deposition (PLD) with a multitarget system allows one to control both the morphology of the nanoparticles (size and shape), and their distribution within the dielectric matrix host. This is a key point since these nanoparticles characteristics determine the final functional properties of the nanocomposites. The role of PLD parameters on the nanostructuring of the composites is evidenced through the precise analysis of the influence of laser pulse number and fluences, gas nature and pressure on the nucleation, size and shape of the metallic nanoparticles. The main properties and applications of such metal–dielectrics nanocomposites are reviewed. As a matter of fact, the appearance of optical resonance or enhanced absorption at the surface plasmon resonance wavelength induces significant variation in the refractive index of the nanocomposite which can be used in optical applications. Moreover, these nanocomposite films show large third-order optical non-linearity which makes them particularly attractive for the development of non-linear photonic devices, like all optical switches. The insertion of ferromagnetic nanoparticles in a dielectric matrix leads to special magnetic properties which are related to the magnetic anisotropy of the nanoparticles and to their magnetic interactions. The size and shape of the nanoparticles, their distribution and the magnetic properties of the embedding host play important roles in the final functional properties of the nanocomposites. Finally, the thermal properties of the metal dielectric nanocomposite are examined. Such systems appear very well adapted to the fundamental studies on the thermodynamical properties of materials in the nanometric scale. For example, the effect of nanoparticle size on the solid–liquid transition can be studied on nanoparticles independent of each other without any contamination. Moreover, these studies demonstrate the potential interest for the development of thermally driven optical switches based on the reflectivity change of the nanocomposite at the solid–liquid phase transition.

Though carbon has been known since using fire, new facets of this fascinating element are still discovered. The various existing types of pure carbon materials including: diamond, graphite, amorphous carbon, diamond-like carbon (DLC), fullerenes and nanotubes differ in bond configuration and bond hybridization.

Differences in their chemical structures materialize in the diversity of physical and chemical properties. This diversity in properties broadens even further when considering the millions of organic compounds and carbon-based materials in general. Over the last decades these different forms of carbon-based materials, which were produced in thin films using both chemical (CVD) and physical (PVD) vapor deposition techniques, were demonstrated to present a wide range of potential technological applications. Pulsed-laser deposition proved to be one of the most successful preparation techniques, being able to fabricate all members of this class. Chapter 3 'Carbon-based materials by pulsed-laser deposition: From thin films to nanostructures' by T. Szörényi from the Hungarian Academy of Sciences in Szeged (Hungary) reviews the peculiarities of the PLD for the growth of controlled thin films of carbon-based materials presenting a wide range of physico-chemical properties. The contribution of the laser parameters in tailoring the properties of the material produced and the benefits of ablation in different neutral and reactive atmospheres will be discussed in detail. All these results will confirm in particular the unique interest of the PLD technique for preparing nanostructured carbon-based materials.

Organic and Biomaterial Modifications by Laser

The fabrication of high quality micro-optical components for diffractive or refractive lens may be now achieved by laser-induced processing. The chapter written by G. Kopitkovas, L. Urech and T.K. Lippert from Paul Scherrer Institut in Villigen (Switzerland) entitled 'Fabrication of micro-optics in polymers and in UV transparent materials' (Chapter 4) deals with this field of application. After a brief description of basic mechanisms of laser ablation of polymer and UV transparent material, this chapter summarizes the various methods for fabrication of micro-optical elements in polymers and dielectrics: lithography combined with reactive ion etching, laser writing, laser ablation combined with the projection of a gray tone mask and laser-assisted wet etching. Fresnel micro-lens arrays in polymers (i.e. triazene polymer, polyimide) are therefore fabricated by laser ablation associated with the projection of the so-called diffractive gray tone phase mask (DGTPM). Micro-optical elements with dimension from 250 μm to several millimeter are obtained by this process. The projection of DGTPM combined with a laser-induced backside wet etching (LIBWE) process is also used for the one-step fabrication of plano-convex lens in quartz. These processes allow the laser structuration of transparent material with fluences weaker than the damage threshold of materials which induces a surface roughness low enough to create optical elements as beam homogenizer for high power Nd:YAG lasers.

A full understanding of the complex mechanisms of ultraviolet laser ablation of polymers is the target of efforts of the scientific community due to its increasing importance in both academic and application fields. S. Lazare and V. Tokarev from University of Bordeaux (France) review in Chapter 5 'Ultraviolet laser ablation of polymers and the role of liquid formation and expulsion', new recent drilling experiments and theoretical works on the mechanisms of material removal in micro-drilling processes. In this chapter, the basics of laser ablation of polymers are first recalled, then the specific phenomena related to transient liquid flow when laser irradiation melt the target and produce liquid matter are described and modelled.

In fact, in the microdrilling process, a poor knowledge is available on mechanisms of material transport to the outside when the ablation front is deep in the material. This point is analyzed in this chapter, and it is shown that upon laser irradiation of polymers, a thin layer of liquid can be formed at the surface, and depending upon its physical properties (mainly viscosity), its expulsion behavior can be strongly different. In some cases, a substantial amount of liquid is expulsed under the form of droplets, while in other cases, laser ablation leads to the expulsion of long nanofibers with a 150 to 200 nm diameter. These two cases are described and related to the physical phenomena taking place during laser ablation of polymers.

The chapter on 'Nanoscale laser processing and micromachining of biomaterials and biological components' (Chapter 6) proposed by D.B. Chrisey, S. Qadri, R. Modi, D.M. Bubb, A. Doraiswamy, T. Patz and R. Narayan from Naval Research Laboratory in Washington and Georgia Institute of Technology of Atlanta (USA) summarizes several successful examples of the laser processing of materials dealing with life sciences. New methods have been developed to exploit the laser material interactions in a way to process the fragile biological materials into films and patterns for numerous electronic, optical and sensing applications. For example matrix-assisted pulsed laser evaporation (MAPLE) is able to give biomaterial films of complex carbohydrates, proteins, hydrogels, etc. with a perfect control of thickness, roughness, homogeneity and reliability. Moreover, MAPLE has been used for the formation of bioresorbable nanoscale composite films which can provide continuous and constant release of fragile drug in the body while avoiding any inflammatory response.

In this field of drug delivery coatings, resonant infrared pulsed laser deposition has been developed to avoid irreversible photochemical modifications of the fragile biomaterial by electronic transitions caused by UV lasers. Resonant infrared PLD modifies the laser–material interaction by tuning the laser to vibrational modes of the material. The combination of these two approaches (MAPLE and resonant infrared PLD) eliminates the possibility of any photochemical modifications. This is particularly important in the case of large organic molecules such as DNA which can be active elements in sensors.

The most recent developments of laser processing of biomaterials are related to tissue engineering and regenerative medicine and are based on the use of MAPLE direct write to the fabrication of three dimensional tissue constructs. The concept of transfer of live cells combined with inorganic materials is a novel approach in developing bioactive, bioresorbable scaffolds and is presented in this chapter. It is thus shown that the versatility of such a process offers good potentials for engineering functionalized tissue constructs.

Laser Processing for Micro- and Nanostructuration of Solids

A promising technique using lasers for fabricating submicronic patterns of organic, inorganic and biological materials consists in the forward transfer of a compound from a transparent support onto a suitable substrate. This method named laser-induced forward transfer (LIFT) and its related application is discussed in Chapter 7 'Direct transfer and microprinting of functional materials by laser induced forward transfer' by K.D. Kyrkis, A.A. Andreadaki, D.G. Papazoglou and I. Zergioti from National Technical University of Athens and from Foundation for Research and Technology of Heraklion (Greece). Originally, it is used to transfer metals for producing conductive lines in microelectronics, the LIFT process has been extended to metal oxides, semiconductors, superconductors, and, more recently, biomaterials and organics. This chapter overviews the viabilty of LIFT for the fabrication of microdevices for electronics and optoelectronics such as capacitors, resistors, chemosensors, biosensors, organic and inorganic thin film transistors (TFT) and organic light-emitting diode (OLED). As an example, the LIFT process is able to fabricate Li-ion microbatteries consisting of printed electrodes around 50 μm thick with a 4×4 μm². The obtained cells are proved to exhibit similar capacity to sputtered thin-film Li-ion cells having active surface ten times greater. Regarding light-emitting polymer material, the corresponding properties (i.e. light-emitting efficiency) is comparable to those of spin-coated devices. Finally, the most promising future of this technique may be found in the field of biomolecules or biopolymers for biosensor and medical sensor devices and microarrays for which the chemical and structural properties of molecules must be conserved during the fabrication. This aspect is also developed in this chapter as well as the mechanisms and the physics involved in the LIFT process.

The recent development of femtosecond lasers has opened many exciting possibilities for high resolution material processing. They are virtually able to process all types of materials with a submicron resolution. The main features of material processing with ultrashort pulses are: (i) an efficient, fast and localized energy deposition, (ii) a well-defined deformation and ablation thresholds and (iii) a minimal thermal and mechanical damage of the substrate material. The overwhelming majority of the research groups working in this field are using Ti:Sapphire femtosecond laser systems working at 1 kHz repetition rate. Such systems are pumped by a visible laser radiation, which is usually provided by frequency-doubled diode-pumped solid state lasers. As shown in Chapters 8 and 9, nearly arbitrary shaped 2D and 3D structures can be produced by direct photofabrication techniques using femtosecond lasers. Chapter 8 titled 'Recent progress in direct write 2D and 3D photofabrication technique with femtosecond laser pulses' by J. Koch, T. Bauer, C. Reinhardt and B.N. Chichkov from Laser Zentrum of Hannover (Germany) is presently a brief review of recent progress in using these techniques for high precision drilling, cutting, surface structuring, and 3D shaping. In Chapter 9 'Self-organized surface nanostructuring by femtosecond laser processing' presented by J. Reif, F. Costache and M. Bestehorn from Brandenburgische Technische Universitat of Cottbus (Germany), have more specifically discussed the fundamental mechanisms to understand the generation of periodic nanostructures at solid surfaces upon femtosecond laser ablation. In the ultrashort pulse duration regime, the nanostructures are believed to be the result of a self-assembly from a surface instability created by the ablation regime. This chapter also reviews the main potential applications of such phenomena in several emerging fields, e.g. sensor devices in micro- and nanoelectronics, field emitter matrix for integrated display technology and magnetic applications.

The very high intensities ($>10^{13}$ W/cm^2) currently available with femtosecond laser pulses are likely to induce a strong multiphoton absorption process in transparent material (wide band gap insulators) even in the case of near-infrared light. As a result, permanent structural modifications can be obtained inside the bulk of transparent materials by focused femtosecond laser pulses. These phenomena are precisely described by Watanabe and Itoh from the Osaka University (Japan) in Chapter 10 on 'Three-dimensional micromachining with femtosecond laser pulses'. Such three-dimensional micromachining is possible when intense femtosecond laser pulses are focused inside the bulk of a transparent material, since the intensity in the focal volume becomes high enough to cause non-linear absorption, leading to localized modifications in the focal volume, while leaving the surface unaffected. The influence of laser parameters (wavelength, pulse duration energy, repetition rate) focusing conditions and materials on the type of tracks resulting from the femtosecond laser pulses irradiation is analyzed in details, with special emphasis on refractive index change, formation of voids, color centers, microchannels, etc.

The potential applications of such 3D micromachining are reviewed in this chapter. More precisely, the fabrication of optical elements like waveguides embedded in glass, couplers with the potential of spectroscopic couplers, Bragg gratings in silica glass, diffractive lens, binary data storage, etc. The most recent and spectacular result of this approach is related to the fabrication of microchannels in glass which present the potential for use in microphotonics, microelectronics, microchemistry and biology. These microchannels are obtained through an original experimental process: drilling by in-water ablation with femtosecond laser pulses. This method leads, for example, to the realization of 3D microchannels of constant diameter with complex shapes like square wave or spiral microchannels.

Laser Crystallization, Laser Doping and Laser Cleaning

In the pioneering experiments on laser–matter interactions during the 1970s, laser annealing of semiconductors, such as silicon, germanium and silicon carbide were largely studied mainly for scientific reasons. However, in the 1990s, these investigations took a great technological importance for industrial applications. In the same way, laser doping of silicon and silicon carbide is now attractive for device processing and cleaning of various surfaces by using laser beam appears to be mature enough to provide an industrial processing technology in respect with environmental considerations. These three aspects highlighting the use of laser sources toward industrial problems are described in the last part of this book.

In Chapter 11, A.T. Voutsas from Sharp Laboratory (USA) discusses about the 'Laser crystallization of Si thin films for flat panel displays: From glass to plastic substrates'. Polysilicon thin-film-transistors (TFT) are key building blocks for active-matrix-driven flat panel displays (FPDs). Many studies have demonstrated the ability of poly-Si-based transistors to support a variety of functions beyond pixel switching, which has been the traditional role of TFTs in FPD applications. High quality poly-Si microstructure is needed for the fabrication of high quality poly-Si TFTs. The crystallization process is a very critical step of the TFT fabrication process, as it needs to satisfy conflicting requirements on material quality and cost and, at the same time, comply with the thermal–budget constraints imposed by the display substrate. Historically, solid phase crystallization (SPC) was the first technology to produce poly-Si films for display applications, followed by the development of laser-annealing crystallization (LAC). Both the technologies evolved significantly over the past 20 years with a variety of spin-offs that aimed at improving different features of the poly-Si crystallization process and/or the poly-Si microstructure. This chapter reviews the motivation behind the evolution process in the crystallization technology. The different aspects of various laser crystallization techniques developed during the last decade are also discussed in detail. Finally, it is demonstrated why the laser crystallization approach seems to possess today the best collection of features to enable the formation of very high quality poly-Si films compatible with the fabrication of state of the art, ultra-high performance poly-Si TFT devices.

In the last 20 years, laser processing of semiconductors has been regarded as one of the most promising technique for manufacturing microelectronic devices both in silicon (Si) and silicon carbide (SiC) according to the international technology roadmap of semiconductors (ITRS) requirements. In silicon, the ultra shallow junctions (USJ) formation for the sub-65 nm CMOS node is a major challenge. As an alternative to rapid thermal processing (RTP), various nanosecond-pulsed laser doping techniques, including laser annealing of ion implanted dopants and laser-induced diffusion of dopants from gas and solid sources were investigated recently in order to achieve such requirements. These approaches offer the possibility to suppress transient enhanced diffusion (TED) phenomena observed in RTP by applying ramp up and down times in the order of hundreds of nanoseconds, which enables USJ (less than 20 nm in depth) formation, very abrupt profiles and a solubility limit higher than that usually encountered with any conventional rapid thermal process.

Silicon carbide is a wide-gap semiconductor which presents unique material properties especially suitable for high temperature, high power and high frequency applications. However, SiC device fabrication has to face various technological difficulties. Among them, one of the more crucial appears to be the doping step. The high melting point and limited diffusion of impurities into SiC have greatly restricted the use of ion implantation and furnace annealing commonly employed in the silicon microelectronics industry to incorporate and activate the dopants. As for Si, excimer laser processing was demonstrated to be suitable for the doping of SiC in various experimental conditions. All these aspects are developed in Chapter 12 'Laser doping of semiconductors: Application to silicon and silicon carbide' written by E. Fogarassy and J. Venturini from laboratoire InESS in Strasbourg and SOPRA company in Bois-Colombes (France), respectively. This part will review the most advanced aspects of the laser doping technology both in silicon and silicon carbide. It is also demonstrated why the laser approach will provide the best way for the fabrication of such future devices.

The last chapter of this book (Chapter 13) is devoted to the cleaning of materials or compounds such as steel, paints, art objects, semiconductors. In this chapter titled 'Laser cleaning: state of the art', Ph. Delaporte from University of Marseille and R. Oltra from Université of Dijon (France) demonstrate that the use of laser as a source for surface cleaning problems may advantageously be an alternative method allowing waste reduction or organic solvents non-employment. Most of the industrial fields (microelectronics, optics, surface coating, medical, etc.) require surface free of chemical pollution and with a controlled surface morphology. Among the classical, chemical or mechanical cleaning treatment, new technology involving cryogenic and photonic (plasma, laser) assisted process are now emerging. This chapter describes the physic and the mechanisms of laser removal of particles in dry (DLC) or steam (SLC) environments as well as cleaning induced by shock-wave. As a result, near UV laser source has successfully been used to achieve efficient cleaning of field emitters arrays (FEA) for microelectronic devices. Another outstanding laser cleaning development is related to the removal of radioactive oxide from metallic surfaces (aluminum, steel, copper, lead) in nuclear facilities. The efficiency of such process is close to 100% and allows a strong reduction of radioactive wastes compared to other techniques. Laser cleaning is also becoming a major process for restoration of artworks, antiquities and historic buildings. Its advantages are linked to a selective removal of the pollution layers without affecting the original surface. Other applications are also discussed by Ph. Delaporte and R. Oltra: The surface preparation of polymer and metals for adhesion improvement, and a patterned surface pre-treatment which combines a conventional thermal spraying process with a pulsed-laser surface preparation. If the laser cleaning tends to be more and more conventionally used, the process efficiency is strongly dependant on the laser parameters (wavelength, pulse duration, fluence) which have to be carefully chosen as a function of the contaminant and the substrate to be treated.

Chapter 1

Laser Ablation-Based Synthesis of Nanomaterials

A.V. Kabashin and M. Meunier

Department of Engineering Physics, École Polytechnique de Montréal, Case Postale 6079, succ. Centre-ville, Montréal (Québec), Canada, H3C 3A7

Abstract

This chapter reviews progress in the development of nanofabrication methods using laser ablation. We present general aspects of the cluster production in the laser ablation process and review experimental data on the laser-assisted synthesis of nanostructures in both gaseous and liquid environment. In particular, we discuss the production of nanostructured thin films and colloidal nanoparticles in solutions, which are of importance for both photonics and biological sensing/imaging applications.

Keywords: Colloidal nanoparticles; Laser-assisted nanofabrication; Nanostructured films; Pulsed-laser ablation.

1.1. Introduction

Properties of material are known to be strongly dependant on the chemical nature and the structure of its constituents [1]. In particular, due to the overlapping of atomic or molecular orbitals, bulk materials, composed of a large number of atoms, are characterized by the presence of energy bands, which are responsible for most of physical and chemical properties of solids. However, for nanomaterials with dimension between few to 50 nanometers, the number of atoms becomes so small that the electronic energy bands are significantly modified, strongly affecting almost all physical properties of the materials (see, e.g. Refs [2–4]). In particular, nanostructured semiconductors such as Si and Ge are known to exhibit a strong photoluminescence (PL) emission in the visible [4], while in the bulk state

Recent Advances in Laser Processing of Materials
J. Perrière, E. Millon and E. Fogarassy (Editors)

these elements have small and indirect band gaps. Although the origin of the visible PL is still disputable, the PL signals can depend on the size of the nanostructures, suggesting an involvement of the quantum confinement mechanism [4,5]. Nanostructured metals are also known to exhibit novel interesting properties. In particular, ensembles of metal nanoparticles show spectacular changes of color, as well as changes in transmitted and reflected spectra, when the size and the shape of nanoparticles or the average distance between them change [6,7]. These color effects are related to the effective resonant absorption of light by individual metal nanoparticles and the excitation of localized surface plasmons, coupled to collective oscillations of electrons in metals [6,8]. It is expected that the employment of new properties of semiconductor or metal-based nanomaterials can lead to significant progress in many areas including optoelectronics, biosensing and imaging, magnetic and optical memory, nanoelectronics, hard coatings, chemical–mechanical polishing, etc.

Laser ablation has shown itself as one of the most efficient physical methods for nanofabrication. The method consists in the ablation of a target (mostly solid) by an intense laser radiation, yielding to an ejection of its constituents and to the formation of nanoclusters and nanostructures. When the target is ablated in vacuum or in a residual gas, the nanoclusters can be deposited on a substrate, placed at some distance from the target, leading to the formation of a nanostructured film. In contrast, the ablation in a liquid ambience (e.g. in aqueous solutions) causes a release of nanoclusters to the environment and the formation of a colloidal nanoparticle solution. In both cases, properties of synthesized nanostructures can be efficiently controlled by parameters of laser ablation and properties of the environment.

In this chapter, we will review works on the laser ablation-based nanofabrication in gaseous and liquid environment. We will limit our review to methods involving direct ablation of material from a target and further nanofabrication of species in the ablated state. To describe all possibilities of laser-assisted nanofabrication, the proposed review must be completed by methods implying a laser-assisted treatment of surfaces and a formation of sub-micron structures on them (see, e.g. [9–22]).

1.2. General Aspects of Laser Ablation and Nanocluster Formation

When focused on the surface of a solid target, pulsed-laser radiation can be absorbed through various energy transfer mechanisms, leading to thermal and non-thermal heating, melting, and finally ablation of the target. Here, independently of the mechanism and rate of the radiation energy deposition, the material is mostly ablated in the form of atoms and nanoclusters [23–28].

However, parameters of ablated species strongly depend on characteristics of laser radiation (intensity, pulse length, wavelength). The laser intensity, measured in W/cm^2, determines the intensity of radiation-matter interaction. For typical laser ablation experiments considered here, the intensity can vary from 10^8 to 10^{13} W/cm^2, depending on radiation parameters. Under the same laser intensity, the mass of ablated species depends on the total radiation energy absorbed by the surface. Micro- and nanosecond radiation have much higher pulse energy compared to picosecond (ps) and femtosecond (fs) radiation that makes long pulse ablation more efficient to produce a larger mass of the ablated material. Nevertheless, the ablation by ultrashort pulses (ps, fs) is normally characterized by the ejection of much more energetic species (see, e.g. [29]) due to a higher radiation intensity. We will later show that in some conditions this property can give some important advantages for nanofabrication processes. Note that the laser ablation process can be accompanied by a production of larger microscale fragments or droplets. The droplets are generally due to a melting and detachment of microspikes, formed on the target surface as a result of a multi-pulse illumination of the same spot [30,31]. This effect, reported in many papers on the nanosecond laser ablation, can be significantly reduced by using a rotating target, excluding the multi-pulse regime of ablation from the same irradiation spot.

Laser ablation from a target can also be accompanied by a production of hot plasma. When formed in front of the target, this plasma can not only reduce the efficiency of laser energy transmission toward the target surface, but also cause a secondary ablation of material through the thermal heating of the target [32] or cavitation phenomena [33]. Normally, the plasma-related effects are much stronger for IR pumping radiation, since this radiation is strongly absorbed by the plasma itself through the inverse Bremsstrahlund mechanism [34]. On the other hand, the use of UV radiation, weakly absorbed by the plasma, enables to minimize the loss of radiation energy on the target surface. The impact of plasma-related phenomena is much enhanced during the ablation in a dense environment, such as a high pressure gas or a liquid. In these conditions, a phenomenon of laser-induced breakdown of surrounding medium can take place. The breakdown is a self-supporting detonation wave, which propagates in the medium toward the focusing lens and absorbs the main radiation energy [34]. Although in some conditions the laser-induced breakdown plasma can be employed, e.g. for an efficient treatment of semiconductors (Si, Ge) and metals (Zn), and their nanostructuring [19–22], this phenomenon is normally considered as detrimental since it leads to a further loss of radiation energy on the target surface and to an undesirable plasma-related ablation of the target material [32].

After the ablation process, the evolution of properties of formed nanostructures is mainly determined by the interaction of the ejected species with the environment.

In fact, in the first approximation of quasi-stationary conditions the formation process can be described by the classical theory of condensation and nucleation in a vapor layer being in thermodynamic equilibrium, which was developed in the 1970s (see, e.g. [35–37]). According to this theory, the rate of the homogeneous nucleation of a condensate is strongly dependent on the saturation ratio. For a given temperature, the latter parameter is a ratio of an actual vapor pressure p_v to the pressure of a saturated vapor p_s:

$$S = \frac{p_v}{p_s} \tag{1}$$

This ratio depends on the system cooling rate and in general, the greater is this rate relatively to the rate of condensation, the higher is the level of saturation. Nuclei are formed in this layer of saturated vapor due to fluctuations. The free energy of a cluster formation follows the equilibrium between forces of atomic cohesions in the condensate phase and the energy barrier caused by forces of a surface tension. In terms of the nucleus radius r, this Gibbs free energy variation can be expressed as:

$$\Delta G(r) = -\frac{4}{3}\pi r^3 (n\Delta\mu) + 4\pi r^2 \sigma \tag{2}$$

where $\Delta\mu = k_B T \ln S$, is the difference of chemical potentials between the condensed and the non-condensed atoms, σ is the surface energy per unit area, and n is the atomic density.

Plotted in Fig. 1, Eq. (2) clearly shows that the growth of nucleus becomes energetically favorable when the system overcomes a potential barrier given by the maximum value of ΔG, i.e. when $\partial\Delta G/\partial r = 0$. When the nucleus becomes larger than the critical radius given by:

$$r_c = \frac{2\sigma}{n\Delta\mu} \tag{3}$$

it continues to grow lowering the condensate energy. However, at the first moments the nucleus growth is compensated by the increase of the surface energy. The probability to produce a nucleus of radius larger than r_c can be obtained from Statistical Physics [37]:

$$\omega \propto \exp\left(-\frac{16\pi\sigma^3 V_n^2 p_v^2}{3k_B^3 T^3 \Delta p^2}\right) \tag{4}$$

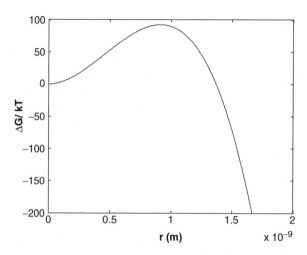

Figure 1: Gibbs free energy variation as a function of the radius of a nucleus.

where k_B is the Boltzmann constant, T the temperature, V_n the volume available for every atom in the gaseous phase and $\Delta p = p_v - p_s \ll p_v$.

If we assume that the environment is a buffer gas of a reduced pressure, the process of condensate formation can be broken into several phases, which correspond to different nucleus growth regimes, as shown in Fig. 2.

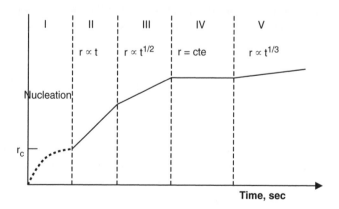

Figure 2: Schematic of various nucleation and growth processes: I – Nucleation, II – Kinetics-controlled growth, III – Diffusion-controlled growth, IV – Transition regime, and V – Ostwald ripening.

1.2.1. Nucleation

The nucleation arises as a statistical phenomenon in a balance of surface energy and the free energy of condensation. To overcome the energetic barrier of nucleation, given by Eq. (2) either thermal energy of atoms or a thermal energy of the buffer gas can be used. The nucleation is mainly controlled by temperature of material and its cooling rate. This phenomenon takes place at $r = r_c$.

1.2.2. Kinetics-Controlled Growth

In this regime, nuclei are small relatively to their mean free path λ_m in gas. In terms of the radius of the nucleus, this means that $r_c < r \ll \lambda_m$. Since the growth in a volume $(d(r^3)/dt)$ is related to a number of collisions with gas atoms, which is proportional to the effective cross section of nucleus (αr^2), we may write:

$$\frac{d(r^3)}{dt} \propto r^2 \Rightarrow r \propto t \qquad (5)$$

1.2.3. Diffusion-Controlled Growth

In this regime, the radius of a condensate is of the same order of magnitude than λ_m and much larger than r_c ($r_c \ll r \cong \lambda_m$). Since the nuclei are almost in equilibrium with the environmental vapor, we can write that the increase of the condensate volume is proportional to the flux arriving to its surface:

$$\frac{d(r^3)}{dt} \propto J \propto r^2 D_i [N(\infty) - N_s(r)]/r \qquad (6)$$

where D_i is the diffusion coefficient, $N(\infty)$ is the actual vapor density on the surface, and $N_s(r)$ is the density of the saturated vapor over the surface. Using Laplace expression [37], we can finally obtain:

$$\frac{d(r^3)}{dt} \propto (r - r_c) \Rightarrow r \propto t^{1/2} \qquad (7)$$

1.2.4. Transition Phase

In this regime, the cluster radius remains approximately constant during a certain period, depending on the initial temperature of the buffer gas. The slowing of the growth is usually related to a competition between nuclei of the same size.

1.2.5. Ostwald Ripening

This phase is very important, since it strongly affects the final size of particles produced. Basically, it is characterized by an increase of the condensate size due to a coalescence of two particles. Here, the coalescence means a fusion of two spherical particles to generate a single spherical one. The total surface energy decreases during this process and the excess energy is released in the form of heat. Notice also that the formation of chemical links during the coalescence process can additionally lead to some energy release. In both cases, heat transfer through conduction involving colder atoms of the buffer gas or radiation losses are the main processes to dissipate the energy. As followed from detailed studies of the coalescence process (see, e.g. [38]), the growth of the particle radius in this regime is determined as:

$$r \propto t^{1/3} \tag{8}$$

It should be noted that in general case the coalescence is in competition with the agglomeration process, which consists in the aggregation of two particles as a result of their collision without any surface modification. These phenomena have the following characteristic times [36]:

– Coalescence: $\tau_c \propto \exp(C/t)$
– Agglomération: $\tau_a \propto 1/t^{0.5}$

Therefore, if collisions are faster than coalescence process, agglomerations are formed, while the opposite case leads to a formation of larger particles.

Although the classical theory of condensation ignores direct interactions between clusters and is valid only in quasi-stationary conditions, it can be applied to describe the nanoparticle growth in the laser ablation process under relatively moderate radiation intensities and large pulse lengths when the dynamics of laser plume development is quasi-stationary. More precise description of the phenomenon of nuclei formation during laser-matter interaction requires a consideration of many other additional parameters. First, the size of ablated species, the ablation rate and the dynamics of the laser plume can be significantly affected by parameters of pumping radiation (intensity, intensity distribution, etc.), as well as by properties of a surrounding medium. Second, the radiation can directly affect the nucleation process if the pumping laser pulse is long enough. In particular, the energy of photons can be sufficient to produce nucleation centers, change the dynamics of the nuclei growth, and modify the diffusion of species in the vapor phase. In fact, a simulation of the nanocluster growth during the ablation process requires the involvement of many additional data such as the dynamics of plume expansion, initial size, energy distributions, material and density of ablated species, as well as properties of the environment where the nanoparticles are ejected. These data can be obtained uniquely from experiments. Next section gives a review of experimental results on the laser ablation-based growth of nanoclusters.

1.3. Synthesis of Nanomaterials in Gaseous Environment

Methods of laser ablation in residual gases have been used for a cluster production and the deposition of nanostructured thin films from mid-1990s. Although films of nanostructured metals and other materials can have attractive applications such as e.g. substrates for Surface-Enhanced Raman Scattering (SERS), most studies on the laser ablation-based nanofabrication were devoted to nanostructuring of silicon (Si). Such interest to nanostructured Si is mainly explained by very promising optical properties of this material. Indeed, both nanoporous silicon [4] and different nanostructured Si-based thin films (see e.g. [5,39–42]) exhibit strong visible PL, whereas in the case of bulk Si this visible PL is essentially absent. The appearance of PL is usually connected to the presence of nanoscale crystallites in the deposit, while different mechanisms such as quantum confinement effect [4], surface radiative states [39], defects in SiO_2 structure [42], etc. were proposed to explain PL characteristics. Taking into account that Si is the leading semiconductor in modern microelectronics industry, the PL property of nanostructured Si opens up novel opportunities for the creation of a novel generation of inexpensive Si-based optoelectronics devices. Note that these applications require dry methods of silicon nanostructuring, compatible with industrial silicon processing technology. This condition is well satisfied in the case of pulsed-laser ablation (PLA), which has become one of most promising techniques for the production of nanostructured Si-based films. In this section, we will mainly concentrate on the laser ablation-based silicon nanocluster growth and the deposition of Si-based nanostructured films. Since laser ablation is not strongly selective to the ablated material, the described approach may be generalized to many other materials.

1.3.1. Configuration of Conventional Pulsed-Laser Deposition for Production of Nanomaterials

When ablated from a solid target in gaseous environment, the material can be collected in the form of either a powder or a deposit on a substrate. Although some applications require the production of nanostructured powders, as e.g. in the case of fuel cell applications of carbon nanotubes and nanohorns [43–46], most applications require the deposition of thin nanostructured films [47–64]. Conventional pulsed-laser deposition (PLD) technique, which does not imply any synthetic cluster size selection, was found to be well suited for these tasks. In PLD nanofabrication experiments, the ablation of material is generally performed in a scheme, shown in Fig. 3, by a nanosecond excimer laser (KrF, ArF), operating at UV wavelengths (248 and 193 nm, respectively), or by a second harmonic of Nd:YAG laser (532 nm). These wavelengths are weakly absorbed by the plasma, which minimizes plasma influence on the ablation process. The radiation is focused on the surface of a rotated target (e.g. Si) at the incident angle of approximately 45°. The radiation intensity is usually about 10^8 to 10^9 W/cm^2. The laser-induced plasma plume expands perpendicularly to the target surface. The substrates are placed on a rotated substrate holder at some distance (generally few centimeters) from the target. The substrate is either kept at room temperature or heated to improve the adhesion of deposited material. The deposition is carried out in the presence of residual inert (He, Ar) or reactive (O_2, N_2) gases, maintained at a reduced

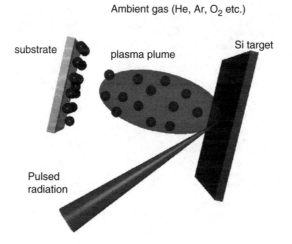

Figure 3: Schematic of PLD experiment for deposition of Si-based nanostructured films.

pressure (basically 0.01–20 Torr). The chamber is pumped down to high vacuum prior to filling with gases.

1.3.2. Studies of Laser Ablation-Based Nanocluster Growth in a Residual Buffer Gas

The first important parameter is the size of nanoclusters produced by the laser ablation process. The size distribution of laser-ablated species was examined in many independent studies either by atomic force microscopy (AFM) or by transmission electron microscopy (TEM). For this purpose, several well separated nanoclusters were deposited on the surface of highly oriented pyrolytic graphite (HOPG) or carbon-coated TEM grid, respectively, placed at some distance from the irradiation spot on the target surface. Almost all studies reported the presence of very small nanoclusters with the mean size of about 1–10 nm, although some agglomeration of 10–20 particles could also be seen. A typical example of a Si cluster observed by TEM is shown in Fig. 4. The mean size of nanoclusters depends on a combination of parameters, including laser fluence, target-to-substrate distance, and the pressure of ambient gas. In particular, an increase of the ambient gas pressure resulted in the increase of the mean nanocluster size [49–51,57–59], while light He gas was the best to finely vary the size of ablated clusters [26]. The increase

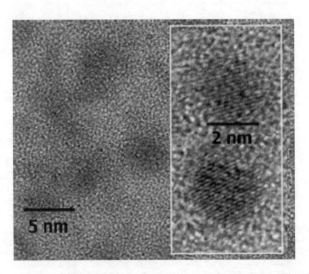

Figure 4: High resolution TEM images of as-deposited Si particles. The data were taken from [61].

of the mean cluster size could be explained by a more efficient condensation of nanoclusters due to faster cooling of ablated species under higher pressure of the buffer He gas. Movtchan *et al.* also reported similar efficient control of the mean nanocluster size by changing the laser fluence [53,54]. Larger nanoclusters for high fluences could also be explained by a higher cooling rate of the more energetic ejected species. It should be noted that many studies reported the production of rare microscale droplets, although their population could be drastically reduced by avoiding multi-pulse ablation from the same spot.

To understand conditions of laser ablation-based nanostructure synthesis, it is very important to know properties of nanoclusters in gas-suspended state, before arriving to the substrate. Several *in situ* measurements of nanocluster growth have been undertaken. In 1994, Chiu *et al.* attempted the first PL measurements of gas-suspended silicon nanoclusters [23]. The nanoclusters were generated in a Nd:YAG laser vaporization cluster source, in which collisions between Si monoatomic particles and the expanding high pressure gas (generally, He at pressures of about 1–10 atm) cool the particles and allow heterogeneous nucleation and growth of the clusters. Despite strong Rayleigh scattering (RS) from the particles, no detectable PL signal could be observed.

Movtchan *et al.* [24] were the first to apply an optical emission spectroscopy in combination with optical time-of-flight measurements (TOF) during conventional pulsed ArF laser ablation of Si targets in He, Ar, and O_2 atmospheres and to detect cluster emission. Plasma emission was captured by an optical system in the direction transverse to the expansion of the plume. Several broad emission bands in ultraviolet, blue, and green-yellow spectral ranges were detected without external excitation and identified as emissions from Si nanoclusters formed at early stage of laser-induced plasma expansion. The emission was mainly observed close to the target surface ($d < 2$ cm) within delays of $\Delta t < 10$ μs after laser ablation. Space- and time-resolved analysis also showed considerably different expansion dynamics for the clusters and monoatomic particles.

Muramoto *et al.* [25] later used a combination of laser-induced fluorescence imaging from atomic Si and Rayleigh scattering (RS) from nanoparticles during the ablation of silicon (Si) in He ambience. In this approach, an external laser source is used to excite PL signals from formed gas-suspended nanoparticles. Experimental data inferred a much later time ($\Delta t > 100$–200 μs) for the onset of dimerization and subsequent nanoparticle growth following Si ablation into 10 Torr He. The growth of the particles was completed in about 20 ms after ablation in a helium (He) ambient pressure of 1.33 kPa and 50 ms at 0.67 kPa. The spatial distribution of the particles was retained for more than one second.

Considerable progress in the understanding of nanocluster growth in the laser ablation process was obtained in the works by Geohegan *et al.* [26,27]. In [26],

time-resolved photoluminescence from gas-suspended nanoparticles was measured and correlated with *ex situ* scanning TEM analyses of individual nanoparticles during laser ablation from a silicon target in He and Ar. A KrF laser was used
to ablate a Si target in a turbopumped vacuum chamber (base pressure 10^{-6} Torr)
maintained at 1–10 Torr of pure He or Ar. To obtain RS and PL from clusters
suspended in the gas phase, a pulse of light from a XeCl-laser illuminated a slice
of the silicon plume from below. Probe laser light scattered by nanoparticles with
estimated diameters larger than 1 nm was imaged by gating the camera during the
XeCl laser pulse. Under the external excitation, three broad bands of PL at ~2.0, 2.6,
and 3.2 eV were measured, as shown in Fig. 5, and correlated with 1–10 nm diameter, spherical particles collected from the gas phase and measured by TEM.

It was also found that an efficient PL signals from gas-suspended species
requires the presence of a certain concentration of residual oxygen in the experimental chamber. This phenomenon was related to the production of optimal stoi-
chiometries of gas-suspended SiO_x species. Later in 1998, Geohegan *et al.* [27]
obtained time-resolved images of the plasma plume during the ablation from a
silicon target in He and Ar. In these experiments, nanoclusters forming the plume
were screened by their own emission at the first moments and by externally excited
PL at later moments after laser-matter interaction. Here, surprisingly different

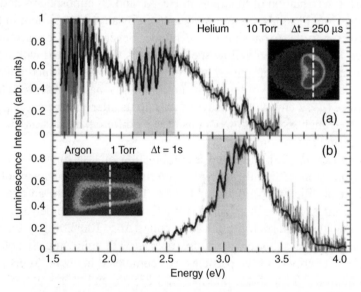

Figure 5: Gated photoluminescence spectra from XeCl-laser-excited, gas-suspended
nanoparticles 2 cm away from a laser-ablated Si target in (a) 10 Torr helium, $\Delta t = 250$ μs
after the ablation laser and (b) 1 Torr argon, $\Delta t = 1$ s. The data were taken from [26].

transport dynamics and the nanoparticle synthesis were obtained for experiments in relatively heavy (Ar) and light (He) residual gases.

Figure 6(a) shows a sequence of images detailing the visible plasma luminescence during the ablation of Si into flowing argon gas at 1 Torr. One can see that within $\Delta t = 20$ μs, Ar stops (the velocity of plume decreases from 2 to 0.01 cm/μs) and reflects the Si plume, resulting in a stationary, uniformly distributed nanoparticle cloud. Here, a backward-propagating flux of material is observed at $\Delta t = 300$ μs after ablation, before a relatively uniform cloud of confined plasma forms and cools over several milliseconds. The first detectable PL is observed in Fig. 6(b) $\Delta t = 3$ ms after ablation, just as the last remnants of plasma luminescence disappear. The PL images of Fig. 6(b) also indicate that silicon nanoparticles form first near the target surface and then occupy the entire plume volume within the next two milliseconds. Over the next hundreds of milliseconds, the stationary nanoparticle cloud becomes denser near the center of the initial spatial distribution. In contrast, for experiments in He plasma, the luminescence drops much rapidly, during first $\Delta t = 400$ μs after ablation, whereas laser-induced PL could be first

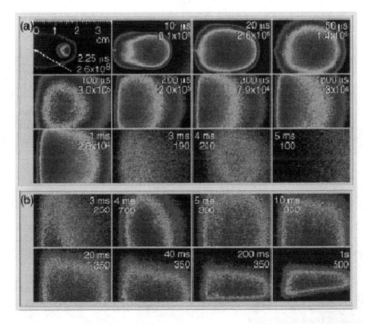

Figure 6: (a) Gated-ICCD (Intensified Charge Coupled Devices) photographs of the nascent visible plasma PL observed when a Si wafer is laser-ablated by KrF laser into 1 Torr Ar ~5 ns to 15 ms exposures. (b) 3 ms exposures of PL form nanoparticles after a XeCl-laser pulse. The data were taken from [27].

discerned at $\Delta t = 150$–200 µs, near the front of the expanding plume. As shown in Fig. 7, Helium (10 Torr) slows the silicon plume, angularly segregating most of the nanoparticles to a turbulent smoke ring which grows in diameter as it propagates forward at 10 m/s, while the PL in the central region rapidly collapses. Nanoparticles prepared in He exhibited more efficient PL compared to Ar. This phenomenon was explained by the turbulent nature of the nanoparticle flow in the case of He. It was proposed that unlike the propagation in static argon, the turbulent mixing and forward propagation of the plume in helium introduced fresh oxygen containing molecules to the condensing nanoparticles.

Makimura *et al.* [28] developed a decomposition method to observe photoluminescent nanoclusters, with sizes intermediate between the dimer and the 10 nm nanoparticles, and those without PL. In this approach, the ablation from a Si target in ambient Ar was performed by the second harmonic of Nd:YAG laser (532 nm), while the nanoparticles were probed by detecting light emission resulting from decomposition of ablated species by a second laser (355 nm). Experiments showed that the main emission from the decomposed plasma is from Si and Ar neutrals. By measuring the emission from these decomposed particles, the authors could estimate the typical growth rate of nanoparticles. The nanoparticles were found to

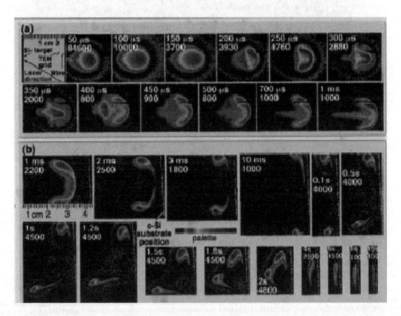

Figure 7: (a) ICCD images of plasma luminescence ($\Delta t < 400$ µs) plus PL ($\Delta t > 200$ µs) from nanoparticles produced by silicon ablation into 10 Torr He 3 µs exposures. (b) later images. The data were taken from [27].

grow in time periods of 1–1.8 ms following ablation at 5 Torr of Ar. The growth mainly occurred above ablation spots slightly apart from the target just after the thermalization of the plume. It was also mentioned that higher fluxes of laser light delayed the growth of nanoparticles.

It should be noted that significant attention is now focused on the investigation of properties of nanoclusters produced by ultrashort (mainly femtosecond) laser radiation. The interest is mainly explained by anticipated different mechanisms and properties of material ablation by ultrashort radiation. Indeed, the use of ultrashort laser pulses (picosecond and less) is characterized by (i) the reduction of the pulse energy which is necessary to induce ablation for fixed laser wavelength and focusing conditions and (ii) the reduction or absence of heat-related effects as a direct consequence of the pulse being shorter than the heat diffusion time, given by the phonon transport [65]. It is expected that such novel conditions of laser-matter interaction can lead to interesting properties of ablated species. In particular, Bulgakov *et al.* [66] used reflection time-of-flight spectrometry to analyze size and velocity distributions of clusters produced after femtosecond laser ablation (80 fs, Ti:sapphire) of Si in vacuum. It was found that this kind of ablation provides a very small neutral silicon clusters Si_n (up to $n = 6$) and their cations Si_n^+ (up to $n = 10$), while their relative yield depends on laser fluence. Using these experimental data, the authors proposed that mechanism of cluster ablation was essentially non-thermal. Amoruso *et al.* [67] recently examined clusters of different materials (Si, Ni, Au, Ag) produced during the femtosecond laser ablation (120 fs, Ti:sapphire) at relatively high radiation intensities 10^{12}–10^{13} W/cm^2 in vacuum. AFM analysis showed that the size of produced nanoclusters was between 5 and 25 nm with very weak size dispersion for most materials. These nanoclusters could be interesting for the deposition of different nanocluster films.

1.3.3. Properties of Nanostructured Si-Based Films Prepared by Pulsed-Laser Deposition

The deposition of gas-suspended nanoclusters leads to a formation of a thin Si-based nanocluster film on a substrate facing the target. Typically, the thickness of the film is 100–1000 nm after several thousand laser shots. When relatively thick (few hundred nanometers and more), the films become colored. They can also manifest distinct multicolor interference fringes due the non-uniform film thickness. Some studies of the film morphology, performed by scanning electron microscopy (SEM), revealed a porous texture, as shown in Fig. 8(a–d) for thin Si-based films deposited at different He pressure [56,58]. Here, the porosity increases

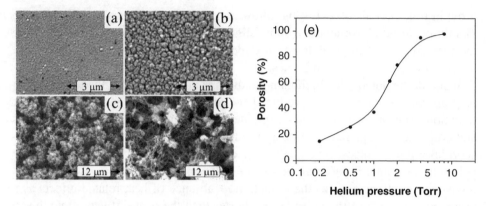

Figure 8: SEM images of the laser-ablated silicon films fabricated under 1 Torr (a), 2 Torr (b), 4 Torr (c), 8 Torr (d), and porosity of the films estimated from SXRR curves as a function of helium pressure during the deposition (e). The data were taken from [58].

as the pressure of the residual gas increases [59]. While the deposition under 1 Torr resulted only in some germs of roughness, the experiment under 2 Torr provided a developed porous structure with pore size of about 50–100 nm. Further pressure increase up to 4 Torr led to a formation of web-like aggregations of particles. The porosity of the laser-ablated films can be quantified by using specular X-ray reflectivity measurement [68]. As shown in Fig. 8e, the porosity of the laser-ablated films gradually increases with the increased of the helium pressure and exceeded 90% for the films deposited at relatively high pressures.

Significant attention has been given on the composition and structural properties of laser-ablated films. It was found from X-ray photoelectron spectroscopy (XPS) and FTIR studies that the layers are composed of unoxidized Si core and SiO_x oxide layers ($1 < x < 2$) just after the fabrication (see, e.g. [48–52,56–58]). A prolonged exposition of the film to ambient air leads to a further increase of x in SiO_x composition, suggesting a formation of an upper SiO_2 layer [58]. However, studies of films by X-ray diffractometry (XRD) showed a presence of all peaks typical for crystalline silicon, suggesting that films were mostly crystalline [49,50,55,58]. In addition, some studies detected the presence of a minor amorphous phase for films deposited at relatively low pressures [55,57]. XRD data were also used in [57] to roughly estimate the size of nanocrystals in the films by the Debye–Scherrer formula, using the fact that the broadness of XRD peaks is mainly determined by the smallest crystals in the deposit [69]. Such estimation gave a size of the order of few nanometers, which is in good agreement with AFM and TEM data on the size of laser-ablated clusters [49–53,58].

Since applications of Si-based nanostructures are mainly related to optoelectronics, luminescent properties of laser-ablated films were intensively studied by many research groups [47–58]. It was found that PL emission takes place only after the exposition of films to ambient air (see, e.g. [57]). This phenomenon was attributed to the necessity to passivate most dangling bonds yielding to the suppression of non-radiative transitions [4]. It should be noted that PL emission of gas-suspended Si-based nanoclusters was also strongly connected to the presence of a certain concentration of oxygen in the experimental chamber [26].

Most studies were focused on the clarification whether quantum confinement mechanism [4] is responsible for the PL signals. Basically, this mechanism must provide size-selective PL spectra. However, results of different groups were rather contradictory. Makimura *et al.* [50,51], Yoshida *et al.* [49] reported the presence of nearly fixed PL peaks at 1.5–1.7 eV, 2.1–2.3 eV, and 2.7 eV from films, produced by the ablation of Si in He. The PL signals were independent of a mean size of nanocrystals, which could be controlled by changing the pressure of He during the deposition. They also found that the intensity of 1.5–1.7 eV, 2.1–2.3 eV emission could be significantly enhanced through a post-deposition thermal annealing of the films. Umezu *et al.* [52] later showed that the annealing in oxygen led to the enhancement of the PL peak at 2.2 eV, whereas the annealing in N_2 contributed to the one at 1.4 eV. Although the appearance of the fixed PL signals was strongly connected to oxidation phenomena, the origin of the observed bands remained unclear. In particular, some properties of such size-independent PL signals enabled to relate them to mechanism, which implies the formation of radiative states present in the interfacial layer between Si core and SiO_2 oxide shell [39], whereas other properties suggest a mechanism related to defects in the SiO_2 structure [42]. In contrast, Patrone *et al.* [53,54] reported size-dependent PL spectra from films, produced by the ablation of Si in helium. The decrease of the mean nanocrystal size, obtained by decreasing the laser fluence, was accompanied by a clear shift of PL spectra to the blue range, as shown in Fig. 9. This result was in good agreement with the quantum confinement mechanism [4]. Such a difference of results was striking, taking into account almost identical conditions of film fabrication in [49–52] and in [53,54].

Later, Kabashin *et al.* [58] showed that not only the nanocluster size, but also the microstructure of deposited films can determine PL properties of the laser-ablated film. By changing the pressure of the ambient He during the ablation, they managed to drastically change the porosity of deposited films, as shown in Fig. 8. Low-porous films, deposited at reduced pressures of helium ($P < 1.5$ Torr), exhibited size-dependent PL, similar to [53,54], while the peak energy could be controlled by changing He pressure. In particular, a pressure decrease from 1.5 to 0.15 Torr during deposition caused a blue shift of the peak from 1.6 to 2.15 eV.

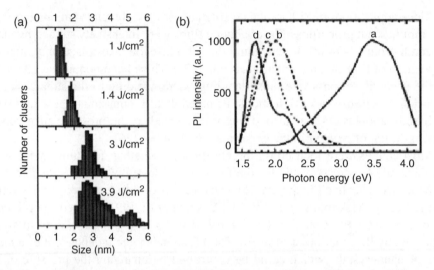

Figure 9: (a) Cluster size distributions for deposits prepared at different laser fluences. (b) The PL spectra of nanostructured films prepared at a – 1 J/cm²; b – 1.4 J/cm²; c – 3 J/cm²; d – 3.9 J/cm². The data were taken from [53].

Meanwhile, films with an enhanced porosity, deposited at P > 1.5 Torr, provided only spectra with fixed peaks around 1.6 and 2.2–2.3 eV, similar to ones observed in [49–52]. Here, the first band (1.6–1.7 eV) could be seen just after the exposition of films to air and the prolonged oxidation of the films led to a significant increase of the integral intensity of such PL signals. The second peak (2.2–2.3 eV) appeared only after the oxidation of samples in humid air or the thermal annealing of all as-deposited films. It was concluded that this was the porosity difference that caused the dramatic distinction of long-term PL properties. Dense and self-coagulated structures of the films fabricated under P<1 Torr minimized the impact of ambient atmosphere on the film properties. For these films, core crystal-based mechanisms such as quantum confinement became predominant, giving rise to a blue shift of the spectra under the decrease of the nanocrystal size. On the other hand, the porosity enhanced the surface area, which was subjected to surface chemistry modifications due to interactions of nanocrystallites with oxygen and other elements or impurities in ambient air. This drastically enhanced the role of oxidation in the formation of PL centers and led to a domination of oxygen-related PL mechanisms. In fact, the observed 2.1–2.2 eV and 1.6–1.7 eV PL correlated well with typical PL signals due to defects in the SiO_2 structure (usually, this mechanism provides PL peaks around 2–2.4 eV [42]) and to the interfacial layer (1.65 eV [39]). It should be noted that these results with the absence and presence of size-dependent spectra for high and low-porous films reproduced well contra-dictory results of [49–52] and [53,54], respectively. Therefore, this was probably the difference of porosities of films in [49–52] and [53,54] that was responsible for quite different PL data.

Main applications of Si-based nanostructured films are related to optoelectronics. In real photonics devices, the luminescence should be excited electrically rather that optically. Yoshida *et al.* [70] were the first to demonstrate the effective electroluminescence using laser-ablated films. They basically observed 1.66 eV emission, which correlates well with one of PL peaks.

1.3.4. Use of Other Laser Ablation-Based Methods for Producing Nanostructures

In this section, we will briefly list laser-assisted nanofabrication methods, which do not use conventional PLD configuration. Few studies are known, which use an external control of plume dynamics and cluster size selection. In particular, El-Shall [62] applied an external temperature gradient to control the plume dynamics (laser vaporization controlled condensation method). Although PL signals were similar to the case of the conventional laser ablation, the method enables to finely control structural properties of the formed films. Seto *et al.* [63,64] used a differential mobility analyzer (DMA) to control the size of the silicon nanoparticles generated by laser ablation. In this case, the nanoparticles generated by laser ablation are transported by a helium gas stream and classified by a DMA in the gas phase. This synthetic method showed a very high efficiency in the size selection of Si clusters, which enables to control PL emission. Sharp PL spectra were observed from the monodispersed size-selected Si-based nanoparticles and the peak shifted to higher energy ~1.34–1.79 eV with decrease in particle size, as shown in Fig. 10. The results are of importance for the fabrication of tunable light emitting nanostructured Si-based on the quantum confinement effect.

Chen *et al.* [71] used laser ablation of a metal-containing target in combination with a heating of the experimental chamber (over a temperature range 910–1120°C) to synthesize silicon nanowires (SiNWs) with different diameters and morphologies. The distribution of the morphology and diameter of SiNWs were very promising compared to nanowires prepared by conventional thermal evaporation of SiO powders. In similar approach, Purezky *et al.* [45,46] used pulsed-laser vaporization method to form carbon nanotubes. In this method, a laser pulse evaporates a solid target of graphite which contains a small amount of metal catalyst atomic (Ni and Co) into a background gas (Ar) which gently flows through a quartz tube inside a high temperature (1000°C) oven. The method shows a high efficiency

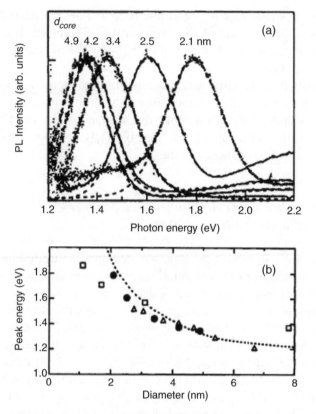

Figure 10: (a) PL spectra of monodispersed nanostructured Si for different sizes of Si crystals. (b) PL peak energy versus the core diameter. The data were taken from [63,64].

in the formation of carbon nanostructures, which are produced in the form of a powder. Note that conventional laser ablation can also be used to form carbon-based nanostructures. Iijima *et al.* [43,44] showed single wall carbon nanotubes (SWCN) can be efficiently formed even during a simple CO_2 laser ablation of graphite at room temperature without any metal catalyst in 760 Torr Ar atmosphere.

1.4. Synthesis of Nanomaterials in Liquid Environment: Production of Colloidal Nanoparticles

At present, there is an explosive growth of interest to the fabrication of colloidal nano-materials in liquid environment. This interest is mainly explained by newly emerged biosensing and bioimaging applications of colloidal nanoparticles. There are two classes of nanomaterials, which are of particular importance. The first class is related to fluorescent inorganic materials, which can be used as markers or labels for biological species. In this case, an attachment of biological species to the surface (biosensing applications) or their presence in a living organism (bioimaging applications) can be screened by a fluorescent signal from the label. It is now actively discussed (see, e.g. [72–76]) that the use of semiconductor-based nanoparticles (quantum dots) can solve most problems of conventional fluorescent labeling-based biodetection with the use of organic fluorophores, which suffer from photobleaching, environmental quenching, broad and asymmetric emission spectra, and the inability to excite more than three colors at a single wavelength. The second class is related to metallic or plasmonics nanostructures. The ability of noble metals to support surface plasmons gives rise to novel effects, such as the appearance of an absorption peak, whose position depends on the size and the shape of nanoparticles, as well as on the average distance between them [6,7]. When used as labels of biomaterials, metal nanoparticles also make possible the control of biointeractions on the surface or in a solution by monitoring absorption spectra characteristics.

Ideally, colloidal nanoparticles are supposed to have the following properties for biosensing applications:

(i)   They must be formed in an aqueous solution;
(ii) They must be small enough (1– 50 nm), with a relatively narrow size dispersion;
(iii) They must be free of toxic impurities on the surface;
(iv) They must contain reactive chemical groups to simplify a further attachment to biomolecules.

In principle, colloidal semiconductor and metal nanoparticles can be routinely produced by chemical reduction methods (micelles, organometallic procedures) (see, e.g. [77–80]). However, conventional wet chemical procedures such as, e.g. the reduction of a precursor salt with a reducing agent, often lead to surface contamination, mainly by residual anions and reducing agent [81]. In addition, the oxidation of some chemically prepared materials in the presence of UV radiation often leads to citoxicity problems [82]. All these problems strongly complicate the use of chemically prepared particles in biosensing and imaging applications [83].

Laser ablation in liquids, which consists of the pulverization of a solid target in liquid ambience, gives a unique opportunity to solve the toxicity problems. In contrast to chemical nanofabrication methods, laser ablation can be performed in a clean, well-controlled environment, such as deionized water, giving rise to the production of ultrapure nanomaterials. The use of these particles decreases toxicity risks, which is especially important in *in vivo* biosensing and imaging applications. In this section, we will review works on the laser ablation-based synthesis of nanomaterials in liquid environment.

1.4.1. Physical Aspects of Laser Ablation in Liquid Environment

First works of laser ablation in a liquid milieu for a fabrication of colloidal nano-structures were undertaken by researchers in chemistry [78–101]. These, mostly fundamental studies examined the potential of laser ablation technique for the production of stable solutions of various nanoparticles. In contrast, little attention was given to the study and the optimization of physical conditions of laser ablation-based nanofabrication.

Systematic investigations of physical aspects of the method started about 5 years ago. Although all collected data are rather empiric, they enable to make first conclusions on conditions and mechanisms of laser ablation in liquids. Summarizing experimental data from different studies, the main difference of laser ablation in liquids compared to the one in residual gases consists of the following:

(i) Much more efficient breakdown of a relatively dense liquid medium, which can absorb a significant amount of radiation energy; an efficient energy transfer from laser plasma to the surrounding liquid and a formation of a cavitation bubble.

The phenomenon of breakdown of liquids was investigated in many works. Vogel *et al.* [33] studied the efficiency of conversion of the radiation energy to a cavitation bubble during focusing of radiations of different wavelengths in water (without a target). As shown in Fig. 11, the conversion efficiency is much

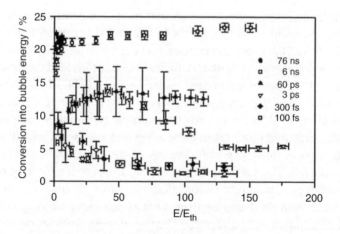

Figure 11: Conversion efficiency of absorbed light energy into cavitation bubble energy as a function of the normalized laser pulse energy E/E_{th} for various laser pulse durations. The data were taken from [33].

higher for nanosecond laser pulses compared to femtosecond ones. Indeed, the decrease of the pulse length from 76 ns to 100 fs leads to a decrease of the efficiency from 22–25 to 1–2%.

Tsuji *et al.* [102] later studied the dynamic of formation of plasma in liquid during nanosecond laser ablation of a target. Figure 12 shows images of the ablation process after different times from the beginning of the laser pulse.

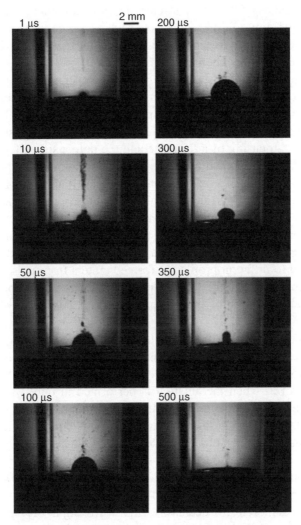

Figure 12: Shadow graph images for nanosecond laser ablation of an Ag target in water. The data were taken from [102].

Here, two phenomena can be clearly distinguished. The first one is the formation of a plasma jet at the first moments after the laser action (before 50 µs). This jet was attributed to radiation-related ablation of material. The second one is the formation of the cavitation bubble, which grows until 150–250 µs and then collapses. The collapse of bubble releases a significant amount of mechanical energy and can become an additional source of the material ablation.

(ii) An absorption of incoming radiation by nanoparticles produced, which are suspended in a solution. If the wavelength of pumping radiation is near some absorption bands of the nanoparticles, the radiation energy can be efficiently absorbed, yielding to different effects such as a secondary ablation of material [92,103–105], fragmentation of colloidal particles [100,106], and even a formation of complex chemical compounds [107–110].

(iii) Appearance of self-focusing effects and the generation of white continuum for ultrashort pulses. In particular, it is known that a phenomenon of filamentation can take place during the propagation of an ultrashort laser pulse in an optical medium. The filamentation appears mainly as a result of a balance between self-focusing of laser pulse and the defocusing effect of the plasma generated at high intensities in the self-focal region. This plasma basically forms as a consequence of multiphoton/tunnel ionization. Liu *et al.* [111] showed conditions of a relative domination of the optical breakdown and filamentation regimes during the femtosecond radiation focusing in water. Nevertheless, despite a presence of filamentation effects, mentioned in many studies, it can hardly influence the ablation process. Indeed, these phenomena should lead to different profiles of formed craters from pulse to pulse due to casual focusing of light to different points, whereas many groups reported the same craters produced by the femtosecond laser ablation [112].

In spite of the presence of these effects, the laser ablation in liquids leads to an efficient production of nanoparticles, which are released to the liquid forming a colloidal solution. Such process is usually accompanied by a visible coloration of the liquid. However, when the ablation is performed in pure water or any other solution in the absence of chemically active components, the size of nanoparticles produced is relatively large, since the coagulation and aggregation processes of hot ablated atoms after the ablation process cannot be easily overcome. In particular, nanosecond ablation, used in most works, generally gives relatively large (10–300 nm) and strongly dispersed (50–300 nm) particles [84–99,104,105,113–115], as shown in Fig. 13. It should be noted that for nanosecond pulses, certain size control can be achieved by decreasing the wavelength of pumping radiation [104,105] or decreasing the pulse width [116]. The size properties can also be somewhat controlled by varying the laser fluence, although the range of size variations was rather moderate in the case of nanosecond pulses (see e.g. [84,98,105]).

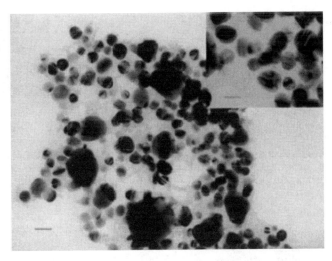

Figure 13: TEM images of Ag nanoparticles produced by nanosecond ablation from an Ag target in water. A scale bar denotes 100 nm. The data were taken from [114].

Much more significant success in the control of size and size dispersion of produced nanoparticles was obtained by using ultrashort laser radiation. Ablating a gold target by femtosecond laser radiation (120 fs, 800 nm) in pure deionized water, Kabashin and Meunier [117] demonstrated a possibility of a very efficient control of nanoparticle size parameters by changing the radiation fluence. In particular, the mean particle size dropped from 120 to 4 nm as F decreased from 1000 to 60 J/cm^2, as shown in Fig. 14. In addition, two populations of nanoparticles were revealed. The first population, with a small mean size and narrow size dispersion, was obtained at relatively low fluences (Fig. 14c), whereas the second one, with a large mean size and a broad dispersion, could be produced at high fluences (Fig. 14a). Intermediate fluences were characterized by the presence of both populations, as shown in Fig. 14b. Notice that similar populations were observed when the radiation intensity on the target surface was changed through a change of the target surface position with respect of the focal point position [118]. The data on two populations suggested an involvement of two different mechanisms of the nanoparticle growth. The production of the first, less dispersed population was characterized by the absence of target melting effects, suggesting a direct radiation-related ablation of material. In contrast, the production of the highly dispersed population was accompanied by strong melting of material inside ablated craters. This melting is usually attributed to plasma effects [32], suggesting a plasma-related origin of the highly dispersed population. Here, the thermal heating of the target by the plasma [32], or its mechanical erosion by the collapse of a plasma-induced cavitation bubble [33,102] were proposed as possible plasma-related ablation mechanisms.

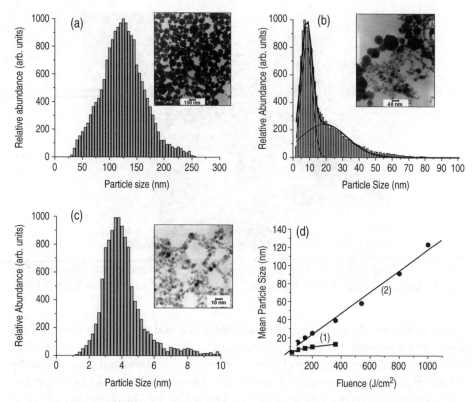

Figure 14: TEM micrograph images and corresponding size distributions of gold nanoparticles prepared by the femtosecond ablation in water at different fluences; (a) 1000 J/cm^2, (b) 160 J/cm^2, (c) 60 J/cm^2, and (d) mean size for the narrowly (1) and broadly (2) dispersed populations as a function of laser fluence. The data were taken from [117].

1.4.2. Fabrication of Ultrapure Colloids in Chemical Solutions

Historically, studies of laser ablation in chemical solutions started by works of chemists. Although some applications such as the use of novel metal nanoparticles for Surface-Enhanced Raman Scattering (SERS) tasks were mentioned, these works were mostly fundamental and focused on a synthesis of new nanomaterials in different organic and aqueous solutions, as well as on a stabilization of produced solutions.

Fojtik and Henglein [84,85] were first to produce colloidal nanomaterials (Au, Ni, C) by ruby laser ablation of thin 35–50 nm films (Au and Ni) and suspended powders (C) in isopropanol, cyclohexane, and water. The size of produced nanoparticles varied from few nanometers to few tens of nanometers (for Au and Ni). The authors also managed to detect the decrease of the nanoparticle mean size as the laser fluence increased. It should be noted that this tendency is opposite to experimental results obtained in many other studies. Probably, such a contradiction is explained by the fact that Fojtik and Henglein ablated thin films, while other groups dealt with solid targets. For gold, the produced colloidal solutions exhibited the absorption peak around 520 nm, associated with the generation of localized plasmons in individual nanoparticles. Finally, particles of C-60, C-70, and C-(80–90) were obtained during the ablation of carbon microparticles, suspended in toluene. Although a mechanism of material ablation was not clarified, the authors proposed that the liquid environment plays an important role in cooling the ablated material to protect them from the disintegration.

In 1993, Neddersen, Chumanov, and Cotton [86] used laser ablation technique to prepare solutions of different colloidal nanoparticles (Ag, Au, Pt, Pd, Cu). They evaporated solid targets by a Nd:YAG laser (1064 nm) in water, methanol, and acetone. The aim of this research was the production of metallic nanomaterials with a clean surface for their subsequent application as substrates for SERS. This technique consists of the employment of plasmon-related properties of metal nanoparticles to enhance Raman spectroscopy signals (by several orders of magnitude) from different absorbed molecules. The method requires an ultrapure metal surface, since the presence of residual ions or other impurities can strongly influence the final signal. The authors showed that laser ablation can be used to rapidly produce a variety of ultrapure materials for SERS.

Next studies were focused on the fabrication and stabilization of colloidal nanoparticles from a variety of materials (see, e.g. [87–101,119–122]). In particular, Yang *et al.* managed to synthesize diamond nanoparticles during laser ablation of carbon in water and acetone [90,91], while similar ablation in 25% ammoniac lead to the formation of C_3N_4 nanoparticles [91]. The same group also managed to produce colloidal nanoparticles of Ni–Ag alloy by the ablation of Ni in a solution of silver nitride [119]. Shafeev *et al.* were first to synthesize a variety of semiconductor-based nanoparticles (Si, ZnSe, CdS) [115]. Yeh *et. al.* [92] managed to produce pure Cu nanoparticles during the ablation of CuO powders in isopropanol. Stepanek *et al.* [87,89] showed that different salts can be used to control the growth of nanoparticles and stabilize colloidal solutions. In particular, an efficient size control was obtained by laser ablation in solutions of NaCl (in concentration less than 70 mM, but more than 0.07 mM) and phtalazine (0.01 mM).

Recently, Mafuné *et al.* achieved significant progress in the development of chemical methods to control the nanofabrication during laser ablation in liquids [97–100]. It was shown that the nanoparticle size can be drastically reduced by the use of aqueous solutions of surfactants, which cover the particles just after their ablation and thus prevent them from a further agglomeration. Sodium dodecyl sulfate (SDS) was found to be the most efficient among the surfactants to reduce the mean size of nanoparticles down to 12 and 5 nm during nanosecond laser ablation of silver [98] and gold [99], respectively. An example of Ag nanoparticle size reduction by SDS is shown in Fig. 15. Such a result could be explained in terms of the dynamic formation mechanism suggested by Mafuné and co-workers [97,98]. In brief, a dense cloud of gold atoms (plume) was accumulated in the laser spot of the gold target during the course of ablation. This core was made of a number of small gold atoms that were agglomerated accidentally due to the density fluctuation to form embryonic nanoparticles. Even when the ablation process had been terminated, the agglomeration continued at a significantly slower growth rate until all atoms in the vicinity (~40 nm) of the embryonic nanoparticles were depleted. As both ablated atoms and embryonic nanoparticles diffuse through the solution towards each other to form larger clusters, this consecutive nanoparticle growth was slow, random, and could not be controlled. Yeh *et al.* [110] later showed an efficient size reduction by the use of other surfactant cetyltrimethylammonium bromide (CTAB).

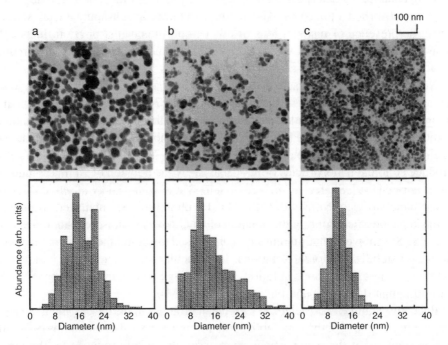

Figure 15: Electron micrographs and size distributions of the silver nanoparticles produced by laser ablation in SDS aqueous solutions. The concentrations of the solution in panels a–c are 0.003, 0.01, and 0.05 M, respectively. The average size decreases with an increase in the SDS concentration. The data were taken from [98].

As it was pointed out in the introduction of this section, the interest to laser ablation in liquids is now more and more reoriented toward the fabrication of nanofunctional materials, which could be used in biosensing and bioimaging applications. This requires the use of biologically compatible reducers in aqueous solutions to control the size of produced nanoparticles and stabilize colloids. In principle, SDS or other surfactants could be rather efficient reducers for these applications, providing the nanoparticle size in the range of 5–15 nm. However, gold nanoparticles covered with surfactants do not have any chemical groups available on the surface, which strongly complicates a subsequent biomolecule immobilization step. Some materials such as SDS are also known to denature biomolecules or, at least, impair their biological activity.

Up to now, only first attempts on the use of biocompatible materials in combination with laser ablation are known. Kabashin *et al.* [123] and Sylvestre *et al.* [124–125] tested a variety of biologically compatible materials for the control of nanoparticle growth during laser ablation in aqueous solutions. In particular, very impressive results were obtained using cyclodextrins (α-cyclodextrin (CD), β-CD, and γ-CD). CDs are torus-like macrocycles built up from D-(+)-glucopyranose units, which are linked by α-1-4-linkages (the most common ones are α, β, and γ-CDs, consisting of six, seven, and eight units, respectively). Their interior cavity, as well as the primary face, is hydrophobic, while the secondary face and exterior surface are hydrophilic. In spite of the absence of any obvious chemical interactions between CDs and gold colloids, a drastic size reduction was observed when the concentration of CDs increased, with the smallest size and narrowest dispersion for β-CD, followed by γ-CD and α-CD. In particular, the ablation in 10 mM β-CD produced particles with the mean size of 2.1–2.3 nm with dispersion less than 1 nm FWHM, as shown in Fig. 16. This distribution corresponded to a deep red color of the solution. The colloidal solutions were extremely stable, while no traces of glucose, a major degradable products of CDs were present in the solution after the ablation process, suggesting that CD molecules remained intact during the course of experiment.

In [124,125], a model of chemical interactions during the ablation in CDs was developed. It was postulated that the reduction of nanoparticle size during laser ablation in aqueous solutions of CDs is a result of two simultaneous effects: 1) the hydrophobic interaction between the primary face of the CD molecule and the unoxidized gold nanoparticle surface, and 2) the hydrogen bonding of the –OH groups present on the same face of the CDs and the –O– at the gold surface. The CD molecules cover gold nanoclusters just after ablation and act like 'bumpers', limiting the contact between particles and preventing their coalescence (when the particles are still 'hot') and aggregation (when the particles are 'cold'). It is important to mention that gold nanoparticles prepared with CDs make possible an easy biofunctionalization step. Indeed, the interior hydrophobic cavity of CDs can still be

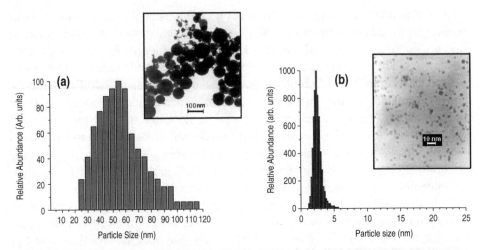

Figure 16: TEM micrograph images and corresponding size distributions of gold particles prepared by the femtosecond laser ablation in deionized water (a) and in 0.01 M β-CD (b). The data were taken from [123].

used for immobilization of different biological species such as some hydrophobic drugs. This gives a promise for potential applications of nanoparticle-CD complexes in biosensing tasks.

Later, tests with the use of other biocompatible solutions such as dextranes, some amine compounds etc. [125,126] revealed some additional regularities of the nanoparticle growth. First, it was found that an efficiency of size reduction strongly depends on the reactivity of ablation environment. In fact, stronger is a reaction between the ablated species and the environment, more efficient is the reduction at fixed laser fluence. Second, it was shown that physical mechanisms could also be used to improve size properties of ablated species. In fact, the best size control by physical mechanisms was achieved by the use of femtosecond radiation (Fig. 14). Such a property of femtosecond ablation helps to produce small, low-dispersed nanoparticles, even if there is no strong chemical interaction between the nanoparticles and the environment.

1.5. Conclusions

Developed during last 10 years, laser ablation methods have become one of top-down approaches for the nanofabrication in the gaseous and liquid environment. These methods take advantage of a natural property of laser ablation process to eject the material in the form of nanoclusters and have at least three following advantages over conventional chemical fabrication methods:

1. Low selectivity to ablated material. The laser ablation can perform effective nanostructuring for a variety of materials.
2. Possibility of nanostructure synthesis in clean, well controlled environment.
3. Possibility of a chemical modification of nanostructures during the ablation process.

Materials synthesized by laser ablation were found to exhibit unique properties and characteristics, which make them very important for photonics and biological sensing/imaging applications.

Acknowledgments

The authors would like to thank collaborators, Edward Sacher and D.-Q. Yang from École Polytechnique de Montréal, John Luong from the Biotechnology Research Institute, and students, J-P. Sylvestre, S. Patskovsky, and S. Besner from the Laser Processing Laboratory of École Polytechnique for their important contributions and useful discussions. This work was supported by Canada Research Chair on Laser micro/nanoengineering of materials and the NSERC (Natural Science and Engineering Research Council) of Canada.

References

[1] Kittel, C., *Introduction to Solid State Physics*, Wiley, New York, **2005**.
[2] Steigerwald, M.L. and Brus, L.E., *Annu. Rev. Mater. Sci.*, **1989**, *19*, 471.
[3] Goldstein, A.N., Echer, C.M. and Alivisatos, A.P., *Science,* **1992**, *256*, 1425.
[4] Canham, L.T., *Appl. Phys. Lett.*, **1990**, *57*, 1046.
[5] Takagi, H., Ogawa, H., Yamzaki, Y., Ishizaki, A., Nakagiri, T., *Appl. Phys. Lett.*, **1990**, *56*, 2379.
[6] Kerker, M., *The Scattering of Light and other Electromagnetic Phenomena,* Academic Press, New York, **1969**.
[7] Kreibig, U. and Vollmer, M., *Optical Properties of Metal Clusters*, Springer-Verlag, Berlin, **1996**.
[8] Agranovich, V.M. and Mills, D.L. (Eds.), *Surface Polaritons – Electromagnetic Waves at Surfaces and Interfaces*, Amsterdam, North-Holland, **1982**.
[9] Her, T.-H., Finlay, R.J., Wu, C., Deliwala, S. and Mazur, E., *Appl. Phys. Lett.*, **1998**, *73*, 1673.
[10] Wu, C., Crouch, C.H., Zhao, L. and Mazur, E., *Appl. Phys. Lett.*, **2002**, *81*, 1999.
[11] Crouch, C.H., Carey, J.E., Warrender, J.M., Aziz, M.J., Mazur, E. and Genin, F.Y., *Appl. Phys. Lett.*, **2004**, *84*, 1850.

[12] Pedraza, A.J., Fowlkes, J.D. and Lowndes, D.H., *Appl. Phys. Lett.*, **1999**, *74*, 2322.

[13] Pedraza, A.J., Fowlkes, J.D. and Lowndes, D.H., *Appl. Phys. Lett.*, **2000**, *77*, 1629.

[14] Pedraza, A.J., Fowlkes, J.D. and Lowndes, D.H., *Appl. Phys. Lett.*, **2000**, *77*, 3018.

[15] Fowlkes, J.D., Pedraza, A.J., Blom, D.A. and Meyer III, H.M., *Appl. Phys. Lett.*, **2002**, *80*, 3799.

[16] Costache, F., Henyk, M. and Reif, J., *Appl. Surf. Sci.*, **2002**, *186*, 352.

[17] Reif, J., Costache, F., Henyk, M. and Pandelov, S.V., *Appl. Surf. Sci.*, **2002**, *197–198*, 891.

[18] Daminelli, G., Krüger, J. and Kautek, W., *Thin Solid Films*, **2004**, *467*, 334.

[19] Kabashin, A.V. and Meunier, M., *Appl. Surf. Sci.*, **2002**, *186*, 578–582.

[20] Kabashin, A.V. and Meunier, M., *Appl. Phys. Lett.*, **2003**, *82*, 1619–1621.

[21] Kabashin, A.V. and Meunier, M., *Mat. Sci. Eng. B*, **2003**, *101*, 60–64.

[22] Yang, D.-Q., Kabashin, A.V., Pilon-Marien, V.-G., Sacher, E. and Meunier, M., *J. Appl. Phys.*, **2004**, *95*, 5722.

[23] Chiu, L.A., Seraphin, A.A. and Kolenbrander, K.D., *J. Electron. Mater.*, **1994**, *23*, 347.

[24] Movtchan, I.A., Marine, W., Dreyfus, R.W., Le, H.C., Sentis, M. and Autric, M., *Appl. Surf. Sci.*, **1996**, *96–98*, 251.

[25] Muramoto, J., Nakata, Y., Okada, T. and Maeda, M., *Jpn. J. Appl. Phys. Part 2*, **1997**, *36*, L563.

[26] Geohegan, D.B., Puretzky, A.A., Duscher, G. and Pennycook, S.J., *Appl. Phys. Lett.*, **1998**, *73*, 438.

[27] Geohegan, D.B., Puretzky, A.A., Duscher, G. and Pennycook, S.J., *Appl. Phys. Lett.*, **1998**, *72*, 2987.

[28] Makimura, T., Mizuta, T. and Murakami, K., *Appl. Phys. A*, **1999**, *69*, S213.

[29] Perrière, J., Millon, E., Seiler, W., Boulmer-Leborgne, C., Craciun, V., Albert, O., Loulergue, J.C. and Etchepare, J., *J. Appl. Phys.*, **2002**, *91*, 690.

[30] Foltyn, S.R., in *Pulsed Laser Deposition of Thin Films*, Chrisey, D.B. and Hubler, G.K. (Eds.), Wiley, New York, **1994**, Chap. 4, and references therein.

[31] Krajnovich, D.J. and Vazquez, J.E., *J. Appl. Phys.*, **1993**, *73*, 3001.

[32] Klimentov, S.M., Kononenko, T.V., Pivovarov, P.A., Garnov, S.V., Konov, V.I., Prokhorov, A.M., Breitling, D. and Dausinger, F., *Quantum Electron.*, **2001**, *31*, 378.

[33] Vogel, A., Noack, J., Nahen, K., Theisen, D., Busch, S., Parlitz, U., Hammer, D.X., Noojin, G.D., Rockwell, B.A. and Birngruber, R., *Appl. Phys. B*, **1999**, *68*, 271.

[34] Raizer, Yu.P., *Laser-Induced Discharge Phenomena*, Consultants Bureau, New York, **1977**.

[35] Abraham, F.F., *Homogeneous Nucleation Theory: The Pretransition Theory of Vapor Condensation*, Academic Press, New York, **1974**.

[36] Kashchiev, D., *Nucleation: Basic Theory with Applications*, Butterworth-Heinemann, Oxford, **2000**.

[37] Landau, L.D. and Lifshitz, E.M., *Statistical Physics*, Butterworth-Heinemann Ltd, Oxford, **1996**.

[38] Friedlander, S.K. and Wu, M.K., *Phys. Rev. B*, **1994**, *49*, 3622.

[39] Kanemitsu, Y., Ogawa, T., Shiraishi, K. and Takeda, K., *Phys. Rev. B*, **1993**, *48*, 4883.

[40] Edelberg, E., Bergh, S., Naone, R., Hall, M. and Aydil, B.S., *Appl. Phys. Lett.*, **1996**, *68*, 1415.

[41] Min, K.S., Shcheglov, K.V., Yang, C.M., Atwater, H.A., Brongersma, M.L. and Polman, A., *Appl. Phys. Lett.*, **1996**, *69*, 2033.

[42] Prokes, S.M., *Appl. Phys. Lett.*, **1993**, *62*, 3244.

[43] Iijima, S., Yudasaka, M., Yamada, R., Bandow, S., Suenaga, K., Kokai, F. and Takahashi, K., *Chem. Phys. Lett.*, **1999**, *309*, 165.

[44] Nisha, J.A., Yudasaka, M., Bandow, S., Kokai, F., Takahashi, K. and Iijima, S., *Chem. Phys. Lett.*, **2000**, *328*, 381.

[45] Puretzky, A.A., Geohegan, D.B., Fan, X. and Pennycook, S.J., *Appl. Phys. A*, **2000**, *70*, 153.

[46] Puretzky, A.A., Geohegan, D.B., Fan, X. and Pennycook, S.J., *Appl. Phys. Lett.*, **2000**, *76*, 182.

[47] Werwa, E., Seraphin, A.A., Chiu, L.A., Zhou, C. and Kolenbrander, K.D., *Appl. Phys. Lett.*, **1994**, *64*, 1821.

[48] Movtchan, I.A., Dreyfus, R.W., Marine, W., Sentis, M., Autric, M., Le Lay, G. and Merk, N., *Thin Solid Films*, **1995**, *255*, 286.

[49] Yamada, Y., Orii, T., Umezu, I., Takeyama, Sh. and Yoshida, T., *Jpn. J. Appl. Phys. Part 1*, **1996**, *35*, 1361.

[50] Makimura, T., Kunii, Y. and Murakami, K., *Jpn. J. Appl. Phys. Part 1*, **1996**, *35*, 4780.

[51] Makimura, T., Kunii, Y., Ono, N. and Murakami, K., *Appl. Surf. Sci.*, **1998**, *127–129*, 388.

[52] Umezu, I., Shibata, K., Yamaguchi, S., Sugimura, A., Yamada, Y. and Yoshida, T., *J. Appl. Phys.*, *84*, **1998**, 6448.

[53] Patrone, L., Nelson, D., Safarov, V.I., Sentis, M., Marine, W. and Giorgio, S., *J. Appl. Phys.*, **2000**, *87*, 3829.

[54] Marine, W., Patrone, L., Luk'yanchuk, B. and Sentis, M., *Appl Surf Sci.*, **2000**, *154–155*, 345.

[55] Suzuki, N., Makino, T., Yamada, Y., Yoshida, T. and Onari, S., *Appl. Phys. Lett.*, **2000**, *76*, 1389.

[56] Kabashin, A.V., Charbonneau-Lefort, M., Meunier, M. and Leonelli, R., *Appl. Surf. Sci.*, **2000**, *168*, 328.

[57] Kabashin, A.V., Meunier, M. and Leonelli, R., *J. Vacuum Sci. and Tech. B*, **2001**, *19*, 2217.

[58] Kabashin, A.V., Sylvestre, J-Ph., Patskovsky, S. and Meunier, M., *J. Appl. Phys.*, **2002**, *91*, 3248–3254.

[59] Inada, M., Nakagawa, H., Umezu, I. and Sugimura, A., *Appl. Surf. Sci.*, **2002**, *197–198*, 666.

[60] Chen, X.Y., Lu, Y.F., Wu, Y.H., Cho, B.J., Liu, M.H., Dai, D.Y. and Song, W.D., *J. Appl. Phys.*, **2003**, *93*, 6311.

[61] Kim, J.H, Jeon, K.A., Kim, G.H., Lee, S.Y., *Opt. Mater.*, **2005**, *27*, 991.

[62] Li, S., Germanenko, I.N. and El-Shall, S., *J. Phys. Chem. B*, **1998**, *102*, 7319.

[63] Seto, T., Orii, T., Hirasawa, M. and Aya, N., *Thin Solid Films*, **2003**, *437*, 230.

[64] Orii, T., Hirasawa, M. and Seto, T., *Appl. Phys. Lett.,* **2003**, *83*, 3395.

[65] Ready, J.F. and Farson, D.F., (Eds.), *LIA Handbook of Laser Materials Processing,* Springer-Verlag and Heidelberg GmbH & Co., Berlin, **2001**.

[66] Bulgakov, A.V., Ozerov, I. and Marine, W., *Thin Solid Films,* **2004**, *453–454*, 557.

[67] Amoruso, S., Ausanio, G., Bruzzese, R., Vitiello, M. and Wang, X., *Phys. Rev. B,* **2005**, *71*, 033406.

[68] Parratt, L.T., *Phys. Rev.,* **1954**, *95*, 359–369.

[69] Cullity, B.D., *Elements of X-ray Diffraction,* Addison-Wesley, Reading, MA, **1978**.

[70] Yoshida, Y., Yamada, Y. and Orii, T., *J. Appl. Phys.,* **1998**, *83*, 5427.

[71] Chen, Y.Q., Zhang, K., Miao, B., Wang, B. and Hou, J.G., *Chem. Phys. Lett.,* **2002**, *358*, 396–400.

[72] Whaley, S.R. *et al.,* *Nature,* **2000**, *405*, 665.

[73] Bruchez, M. *et al.,* *Science,* **1998**, *281*, 2013.

[74] Mitchell, G.P. *et al.,* *J. Am. Chem. Soc.,* **1999**, *121*, 8122.

[75] Elghanian, R. *et al.,* *Science,* **1997**, *277*, 1078.

[76] Chan, W.C.W. and Nie, S.M., *Science,* **1998**, *281*, 2016.

[77] Hyatt, M.A. (Ed.), *Colloidal Gold: Principles, Methods, and Applications,* Academic Press, New York, Vol. 3., **1989**.

[78] Fojtik, A. *et al.,* *Bunsenges. Phys. Chem.,* **1984**, *88*, 969.

[79] Dameron, C.T. *et al.,* *Nature,* **1989**, *338*, 596.

[80] Murray, C.B. *et al.,* *J. Am. Chem. Soc.,* **1987**, *115*, 8706.

[81] Thompson, D.W. and Collins, I.R., *J. Colloid Interface Sci.,* **1992**, *152*, 197.

[82] Derfus, A.M., *Nano Letters,* **2004**, *4*, 11.

[83] Michalet, X. *et al.,* *Single Mol.,* **2001**, *2*, 261.

[84] Fojtik, A. and Henglein, A., *Ber. Bunsen-Ges. Phys. Chem.,* **1993**, *97*, 252.

[85] Henglein, A., *J. Phys. Chem.,* **1993**, *97*, 5457.

[86] Nedderson, J., Chumanov, G. and Cotton, T.M., *Appl. Spectrosc.,* **1993**, *47*, 1959.

[87] Prochazka, M., Stepanek, J., Vlckova, B., Srnova, I. and Maly, P., *J. Mol. Struct.,* **1997**, *410–411*, 213.

[88] Jeon, J.-S. and Yeh, C.-S., *Journal of the Chinese Chemical Society,* **1998**, *45*, 721–726.

[89] Srnova, I., Prochazka, M., Vlckova, B., Stepanek, J. and Maly, P., *Langmuir,* **1998**, *14*, 4666.

[90] Yang, G.-W., Wang, J.-B. and Liu, Q.-X., *J. Phys. Condens. Matter,* **1998**, *10*, 7923.

[91] Yang, G.W. and Wang, J.B., *Appl. Phys. A,* **2000**, *71*, 343.

[92] Yeh, M.-S., Yang, Y.-S., Lee, Y.-P., Lee, H.-F., Yeh, Y.-H. and Yeh, C.-S., *J. Phys. Chem. B,* **1999**, *103*, 6851–6857.

[93] Lee, Y.-P., Liu, Y.-H. and Yeh, C.-S., *Phys. Chem. Chem. Phys.,* **1999**, *1*, 4681–4686.

[94] Kao, H.-M., Wu, R.-R., Chen, T.-T., Chen, Y.-H. and Yeh, C.-S., *J. Mater. Chem.,* **1999**, *10*, 2802–2804.

[95] Hwang, C.-B., Fu, Y.-S., Lu, Y.-L., Jang, S.-W., Chou, P.-T., Wang, C.R.C. and Yu, S.J., *Journal of Catalysis,* **2000**, *195*, 336–341.

[96] Chen, C.-D., Yeh, Y.-T. and Wang, C.R.C., *Journal of Physics and Chemistry of Solids*, **2001**, *62*, 1587–1597.

[97] Mafuné, F., Kohno, J.-Y., Takeda, Y., Kondow, T. and Sawabe, H., *J. Phys. Chem. B*, **2000**, *104*, 8333–8337.

[98] Mafuné, F., Kohno, J.-Y., Takeda, Y., Kondow, T. and Sawabe, H., *J. Phys. Chem. B*, **2000**, *104*, 9111–9117.

[99] Mafuné, F., Kohno, J.-Y., Takeda, Y. and Kondow, T., *J. Phys. Chem. B*, **2001**, *105*, 5114–5120.

[100] Mafuné, F., Kohno, J.-Y., Takeda, Y. and Kondow, T., *J. Phys. Chem. B*, **2001**, *105*, 9050–9056.

[101] Liu, C.H., Peng, W. and Sheng, L.M., *Carbon*, **2001**, *39*, 137–158.

[102] Tsuji, T., Tsuboi, Y., Kitamura, N. and Tsuji, M., *Appl. Surf. Sci.*, **2004**, *229*, 365.

[103] Takami, A., Yamada, H., Nakano, K. and Koda, S., *Jpn. J. Appl. Phys.*, **1996**, *35*, L781.

[104] Tsuji, T., Iryo, K., Ohta, H. and Nishimura, Y., *Jpn. J. Appl. Phys.*, **2000**, *39*, L931–L983.

[105] Tsuji, T., Iryo, K., Watanabe, N. and Tsuji, M., *Appl. Surf. Sci.*, **2002**, *202*, 80–85.

[106] Mafuné, F., Kohno, J.-Y., Takeda, Y. and Kondow, T., *J. Phys. Chem. B*, **2002**, *106*, 7575–7577.

[107] Mafuné, F., Kohno, J.-Y., Takeda, Y. and Kondow, T., *J. Phys. Chem. B*, **2002**, *106*, 8555–8561.

[108] Mafuné, F., Kohno, J.-Y., Takeda, Y. and Kondow, T., *J. Am. Chem. Soc.*, **2003**, *125*, 1686–1687.

[109] Mafuné, F., Kohno, J.-Y., Takeda, Y. and Kondow, T., *J. Phys. Chem. B*, **2003**, *107*, 12589–12596.

[110] Chen, Y.-H. and Yeh, C.S., *Colloids and Surfaces A*, **2002**, *197*, 133.

[111] Liu, W., Kosareva, O., Golubtsov, I.S., Iwasaki, A., Becker, A., Kandidov, V.V. and Chin, S., *Appl. Phys. B*, **2003**, *76*, 215–229.

[112] Daminelli, G., Krüger, J. and Kautek, W., *Thin Solid Films*, **2004**, *467*, 334.

[113] Simakin, A.V., Voronov, V.V., Shafeev, G.A., Brayner, R. and Bozon-Verduraz, F., *Chem. Phys. Lett.*, **2001**, *348*, 182–186.

[114] Dolgaev, S.I., Simakin, A.V., Voronov, V.V., Shafeev, G.A. and Bozon-Verduraz, F., *Appl. Surf. Sci.*, **2002**, *186*, 546–551.

[115] Anikin, K.V., Melnik, N.N., Simakin, A.V., Shafeev, G.A., Voronov, V.V. and Vitukhnovsky, A.G., *Chem. Phys. Lett.*, **2002**, *366*, 357.

[116] Tsuji, T., Kakita, T. and Tsuji, M., *Appl. Surf. Sci.*, **2003**, *206*, 314.

[117] Kabashin, A.V. and Meunier, M., *J. Appl. Phys.*, **2003**, *94*, 7941.

[118] Sylvestre, J.-P., Kabashin, A.V., Sacher, E. and Meunier, M., *Appl. Phys. A*, **2005**, *80*, 753–758.

[119] Liu, Q.X., Wang, C.X., Zhang, W. and Yang, G.W., *Chem. Phys. Lett.*, **2003**, *382*, 1.

[120] Bae, C.H., Nam, S.H. and Park, S.M., *Appl. Surf. Sci.*, **2002**, *197–198*, 628–634.

[121] Compagnini, G., Scalisi, A.A. and Puglisi, O., *Phys. Chem. Chem. Phys.*, **2002**, *4*, 2787–2791.

[122] Compagnini, G., Scalisi, A.A. and Puglisi, O., *J. Appl. Phys.*, **2003**, *94*, 7874–7877.
[123] Kabashin, A.V., Meunier, M., Kingston, C. and Luong, J.H.T., *J. Phys. Chem. B*, **2003**, *107*, 4527.
[124] Sylvestre, J.-P., Kabashin, A.V., Sacher, E., Meunier, M. and Luong, J.H.T., *J. Am. Chem. Soc.*, **2004**, *126*, 7176.
[125] Sylvestre, J.-P., Poulin, S., Kabashin, A.V., Sacher, E., Meunier, M. and Luong, J.H.T., *J. Phys. Chem. B*, **2004**, *108*, 16864.
[126] Besner, S., Kabashin, A.V. and Meunier, M., *J. Phys. Chem.*, **2005**, in press.

Chapter 2

Metal–Dielectric Nanocomposites Produced by Pulsed Laser Deposition: A Route for New Functional Materials

C.N. Afonso, J. Gonzalo, R. Serna and J. Solís

Laser Processing Group, Instituto de Óptica, CSIC, Serrano 121, E-28006 Madrid, Spain

Abstract

This chapter aims at providing a comprehensive review on the production by pulsed laser deposition of nanocomposite films formed by metal nanoparticles embedded in a dielectric host. A general introduction to the deposition technique, including its main features and advantages and a description of the parameters that have a direct impact on the quality and properties of the nanocomposite films is first made. From the material structure point of view, it is desirable to control both the morphology of the nanoparticles (size or shape) and their distribution within the host. From the process point of view, the control of the kinetic energy, the flux and ionization degree of the species arriving to the substrate is essential either to achieve further control on the morphology of the nanoparticles or to avoid undesirable processes such as metal implantation in the host or self-sputtering of the arriving species. The final section of the chapter provides a description of the properties of metal based nanocomposites produced by pulsed laser deposition, including optical, magnetic and thermal properties.

Keywords: Metal–dielectric nanocomposites; Magnetic properties; Nanocomposite films; Nanoparticles; Optical properties; Pulsed Laser Deposition (PLD); Thermal properties; Thin film technologies.

2.1. Introduction

Physical and chemical properties of nanocomposite materials formed by metallic nanoparticles (NPs) embedded in a dielectric host are strongly dependent on their

Recent Advances in Laser Processing of Materials
J. Perrière, E. Millon and E. Fogarassy (Editors)

morphological features such as size, shape or size dispersion as well as on their distribution within the matrix. Thus, the control of these features during the production has been a challenge for long time and, in fact, the lack of production methods with the required degree of control has been one of the major drawbacks for the development of practical applications based on this type of materials. A strong experimental effort has been made in recent years to develop methods with the required level of control, the most widely used ones being ion implantation and thin film technologies. While the former introduces the metal into a bulk dielectric, the second is based on the production of both the host and the NPs using the same deposition technique. Sputtering is among the more promising methods within the second approach, and more recently, the inherent characteristics of pulsed laser deposition (PLD) have turned it in an excellent alternative for producing metal–dielectric nanocomposites in thin film configuration. This chapter intends to provide a comprehensive overview of the production of nanocomposite films prepared by PLD by describing the main parameters influencing their production as well as the properties of the nanocomposite films.

2.2. Production of Metal–Dielectric Nanocomposites

A wide variety of wavelengths and pulse durations have been used for thin film deposition by PLD. The reports published in the literature on material ablation and deposition include laser pulses ranging from infrared to ultraviolet and from continuous to femtosecond pulses. Nevertheless, the conditions that have demonstrated up to date to be more successful for material deposition and for which more results are available are what we would refer to as *classical PLD*, in which a laser with wavelength in the ultraviolet range ($\approx$193–351 nm) and laser pulses in the nanosecond regime is used as ablation source. Thus this chapter will focus on nanocomposites and metal NPs obtained using these conditions.

2.2.1. Features, Advantages and Limitations of Pulsed Laser Deposition

The success of PLD is indeed related to a unique combination of properties that make the technique suitable for the production of a wide range of materials and/or for structuring the film configuration in the nanometer scale. The more widely recognized properties of PLD are its *quasi-stoichiometric character*, that allows to reproduce the composition of complex targets provided that appropriate experimental parameters are used, and the *compatibility of the ablation process with any environment*, either ultra-high vacuum ($<10^{-10}$ mbar) or high gas pressures (>1 mbar) that allows producing a broad range of materials, from pure metals to complex oxides, in thin film configuration. These two properties are responsible for the superior ability of PLD to produce complex oxide films when compared to other physical vapor deposition methods, i.e. sputtering.

There are three additional intrinsic characteristics that confer PLD with unique properties, namely the *high kinetic energy* (10's of eV) of the species generated during the ablation process, the possibility of having very *low average deposition rates* (less than a monolayer deposited per pulse) and very *high instantaneous deposition rates*. Table 1 compares the typical values of these parameters in the case of PLD with those of two other broadly used physical vapor deposition methods, namely sputtering and thermal evaporation.

The kinetic energy of the species involved in the PLD process is at least one order of magnitude higher than in the case of sputtering and two orders of magnitude higher than in the case of thermal evaporation. The values of kinetic energies included in Table 1 have to be understood as typical values under standard conditions, since the range of values achievable by PLD through an appropriate control of the deposition conditions can be very broad (1–500 eV). This high kinetic energy has a positive impact effect on the properties of the films since the density of the films and their adherence to the substrate is enhanced [1,2]. This feature is also beneficial for the production of metal NPs on surfaces since the mobility of species at the substrate surface is greatly enhanced [3]. However, this high kinetic energy has a negative impact in sensitive materials such as semiconductors since it is responsible for the production of defects at the substrate level by ion bombardment that can be detrimental for the properties of the films [4].

The pulsed nature of the PLD process has two important consequences. The first one is the very low average deposition rate, as shown in Table 1, that it is comparable to that achieved by thermal evaporation. This feature is essential to achieve a control of the deposition process at the monolayer level and allows PLD to be

Table 1: Comparison of the kinetic energy of the species and typical deposition rates in two standard thin film deposition techniques (thermal evaporation and sputtering) to PLD. The deposition rate in PLD is given as the average thickness deposited per second or per pulse duration (in brackets)

	Kinetic energy (eV)	Deposition rate (nm·s^{-1})
Thermal evaporation	<1	0.01–0.1
Sputtering	5–10	0.1–1.0
Pulsed laser deposition	10–100	0.1/($>10^3$)

seen as a *kind of pulsed* molecular beam epitaxy for oxides. This layer-by-layer growth has successfully been applied to produce complex multilayer and epitaxial oxide heterostructures [5,6]. The second consequence is the very high instantaneous deposition rate included in Table 1, in brackets, that represents a conservative minimum value. This high instantaneous rate or high flux of species arriving at the substrate leads to high quenching rates at the substrate that favor the production of metastable materials or phases [7,8].

The major limitation of PLD addressed in the literature relates to the presence in the films of large particulates (having typically diameters of ~µm) [9] due to the ejection of molten droplets from the target. In the particular case of metals, the formation of particulates is related to the thermal nature of the ablation process as the melting of the target surface favors the ejection of liquid droplets in most cases. Their formation depends very much on the type of metal ablated. In particular, it has been found to be negligible for the case of noble metals (Ag, Au) and some transition metals (Cu, Pt), whereas it has been reported to be quite significant for the case of Al [10]. However, for such cases as Al, different experimental approaches have been proposed in the literature to reduce, and even suppress, the presence of particulates in the deposited films [9].

2.2.2. Control of the Deposition Sequence: Alternate PLD

Two main approaches have been reported in the literature for producing NPs by PLD, the main difference being the place where the NPs are produced. The first approach, that has mainly been applied to produce semiconductor NPs, is based on the ablation of the target in a high pressure (10's of mbar) of an inert gas, typically Ar. The NPs are condensed in the gas phase and subsequently deposited on the substrate [11–14]. The second approach takes advantage of the Vollmer–Weber nucleation mechanism (i.e. island growth) [3,15,16] that is favored in the PLD process by the high kinetic energy of the species arriving at the substrate that favors their mobility. This approach has widely been applied to produce metal NPs on dielectric surfaces by physical deposition techniques, the process being controlled in order to deposit an amount of metal at the substrate that is below the percolation threshold. In the case of PLD, ablation of the metal target is performed in vacuum. The main advantage of any of these two approaches over other production methods such as ion implantation is its single-step character that requires no post-deposition treatment.

In what follows, we will restrict to the second approach, i.e. nucleation at a surface, to produce metal nanocomposite films. To achieve an independent control of the deposition of the host and the NPs, a modification of the *classical* PLD technique is usually considered. *Alternate*-PLD (a-PLD) that was first used to produce CdTe NPs embedded in SiO$_2$ [17], basically consists of the sequential ablation of the host and metal targets to produce a nanocomposite film that is composed by alternate layers of the host material and the metal NPs. This sequence is repeated several times until the desired number of NPs layers is produced. Typically, the nanocomposite film is covered with a final layer of the host that acts as a protecting or capping layer. Figure 1 shows a scheme illustrating the '*layered*' character of the nanocomposite films produced by a-PLD and how the structure of the film can be controlled through the number of pulses on each target. The thickness of the host spacing layers can be controlled at the nanometer scale through the number of laser pulses on the host target, thus providing control of the in-depth separation between consecutive NP layers. An increase of the number of laser pulses on the metal target instead allows increasing the amount of deposited metal and thus enables the control of the in-plane dimensions and concentration of NPs.

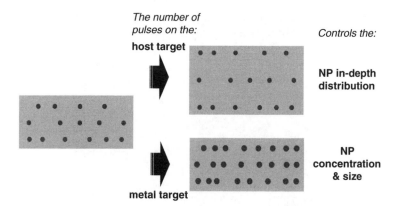

Figure 1: Schematic of the layered configuration of nanocomposite films produced by a-PLD and the effect that the number of laser pulses on host and metal targets have on the in-depth distribution, concentration, and size of metal NPs.

2.2.3. The Role of PLD Parameters on Morphological and Structural Properties of Nanocomposite Films

2.2.3.1. Influence of the Number of Laser Pulses
Figure 2 shows transmission electron microscopy (TEM) plan view images of nanocomposite films in which a single layer of Cu NPs is embedded between two 10 nm thick Al$_2$O$_3$ layers ('*sandwich*' films). The images show dark areas that

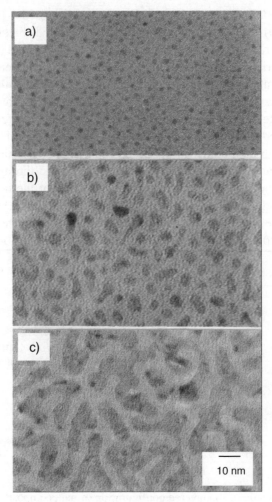

Figure 2: Plan view TEM images of Cu:Al$_2$O$_3$ nanocomposite films having different Cu areal densities per layer: (a) 2.2×10^{15} at. cm^{-2}, (b) 10.7×10^{15} at. cm^{-2} and (c) 17.9×10^{15} at. cm^{-2}. (Adapted from Afonso *et al.* [22].)

correspond to the metal NPs surrounded by a homogeneous background that corresponds to the host. The corresponding high resolution transmission electron microscopy (HREM) images show that whereas the dark areas exhibit lattice fringes from crystalline Cu, the background is structureless which is consistent with the amorphous character of the host (a-Al$_2$O$_3$). The different images correspond to specimens having different Cu contents. The images show that the average dimensions of the NPs increase whereas their shape evolves from a circular to an elongated one as the metal content increases. Similar evolution in the NPs features with the metal content has been reported for other metals like Au, Ag, Bi and Fe [18–21].

The images in Fig. 2 thus show that well-defined Cu NPs are produced by PLD with their average in-plane dimensions being controlled through the number of pulses in the metal target. In general it is observed that the metal content (metal areal density) in the films follows a linear dependence with the number of laser pulses on the metal target [22]. Figure 3 shows the evolution of the NPs morphological features with the Cu areal density up to values close to the percolation limit (12×10^{15} at. cm^{-2} or 500 pulses, for the conditions used in Ref. [22]). From this figure

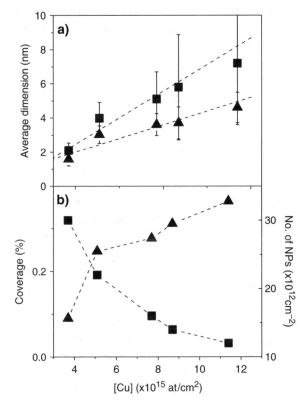

Figure 3: (a) Average in-plane NP dimensions ((▲) breadth and (■) length) of NPs as a function of Cu areal density per layer in Cu:Al_2O_3 nanocomposite films. The error bars relate to the spread in NPs dimensions. (b) Number of NPs per unit area (■) and coverage (▲) as a function of Cu areal density. (Adapted from Serna *et al.* [16].)

it can easily be deduced that 1.8×10^{13} at. cm^{-2} per pulse have been deposited on the a-Al$_2$O$_3$ surface, thus evidencing that a control of the amount of metal deposited on the substrate down to the sub-monolayer level is indeed achieved.

The evolution of the average NPs length (longest dimension, L) and breadth (shortest dimension perpendicular to the length, B) is shown in Fig. 3a as a function of the Cu areal density for Cu:Al$_2$O$_3$ nanocomposite films. Similarly to what has already been deduced from TEM images in Fig. 2, Fig. 3 evidences the evolution of the shape of the NPs from circular (aspect ratio $L/B = 1.0$), with an average diameter of 2.0 ± 0.4 nm, to elongated shapes with aspect ratios up to 1.6. The error bars included in the figure correspond to the spread of dimensions rather than to errors in the measurements. It becomes then clear that the dispersion on the NP dimensions achieved by PLD for small NPs (typically <3 nm in diameter) is small but it increases as the average dimensions increase. The coverage of the surface by the metal NPs is seen to increase in Fig. 3b as the metal content increases while the areal density of NPs decreases. The analysis of the data presented in Fig. 3 clearly evidences that the production of NPs occurs in two main stages. In the first stage, nucleation is the main process occurring upon arrival of Cu species to the a-Al$_2$O$_3$ surface. This process is most likely dominated by the low adsorption energy of metals on oxides and insulating substrates. As an example, the sticking coefficient of Cu on SiO$_2$ has been reported to be as low as 0.35 [23]. Most of the metal arriving to the substrate is reflected back and small NPs (2 nm in diameter) are only observed once a relatively high amount of Cu has been deposited (3.7×10^{15} at. cm^{-2} according to Fig. 3b). If the metal is assumed to be distributed in 2D islands, this means that the metal coverage should be close to 25%. From Fig. 3b it becomes clear that the actual coverage is much smaller, therefore 3D islands have been formed and the process is consistent with a growth mode of the Vollmer–Weber type. The NPs formed at these initial stages are quasi-spherical since this helps to reduce the interface energy of the Cu in contact with a-Al$_2$O$_3$. This is confirmed by the cross section image of a '*sandwich*' specimen containing a single layer of Cu NPs shown in Fig. 4a. The comparison of the in-plane dimensions from the images in Fig. 2 to the cross section dimensions in Fig. 4a, leads to the conclusion that the small NPs are indeed 3D structures that can be considered as slightly oblate spheroids since they are approximately round in the film plane while their transversal dimension is slightly smaller.

The second stage is dominated by the growth of the NPs by coalescence and coarsening, consistently with the fact that the increase of the average dimensions of the NPs (Fig. 3a) occurs along with a decrease in the number of NPs per unit area (Fig. 3b). This coalescence process becomes very evident in Fig. 2b where several elongated NPs formed by coalescence of some NPs are observed. At the same time that coalescence occurs, the surface coverage increases rapidly

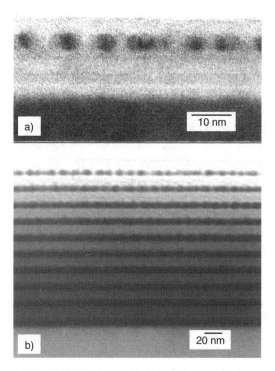

Figure 4: Cross section TEM images of Cu:Al$_2$O$_3$ nanocomposite films having a Cu areal density per layer of $13.0 \pm 0.5 \times 10^{15}$ at. cm^{-2} and containing (a) a single layer and (b) 10 layers of Cu NPs.

(Fig. 3b), whereas the transversal dimension has been reported to show only a small increase [24]. This behavior provides thus clear evidence for the growth of NPs occurring preferentially along the film plane and being dominated by coalescence.

The cross section image in Fig. 4a also shows that the Al$_2$O$_3$ host indeed fills in the space between the NPs and that the surface of the nanocomposite film layer is flat in the nanometer scale. This result is extremely important for practical devices based on these nanocomposite films such as those based on optical waveguides. Figure 4b includes a cross section TEM image of a film containing 10 NPs layers grown under similar conditions than the '*sandwich*' film shown in Fig. 4a. The layers are equally spaced evidencing the excellent control that the PLD technique offers in the nanometer scale. The fact that the NPs layers seem thicker when approaching the substrate is an artifact related to the wedge shape of the cross section specimen caused by the preparation method [25].

2.2.3.2. Influence of Laser Fluence

The nucleation process of the NPs depends not only on the number of pulses on the metal target but also on the density of nucleation centers at the host layer acting as substrate and the surface mobility of the species that reach its surface. These two factors are strongly related to the laser fluence used to ablate the targets that determines the ablation rate and thus, the *flux*, the *kinetic energy* and the *ionization degree* of species arriving to the substrate surface. Figure 5 shows as an example that the ablation rate of Co increases linearly with laser fluence for fluences up to ≈6 J cm^{-2} and tends to saturate for fluences above this value [26]. This is a general behavior observed in most metals, although the value for which saturation starts depends on the laser wavelength and the metal itself. The tendency to saturation has been related to the onset of partial absorption and reflection of the incident laser beam by the ejected material during the initial stages of the expansion of the ablated material [27].

The pulsed character of the PLD process must be considered when analyzing the nucleation rate of NPs at the substrate surface. There is a short lapse of time within which a high flux of species arrive to the substrate followed by a much longer lapse of time (at least 2 orders of magnitude longer) in which no further species arrive [15]. As it was deduced from the results presented in Fig. 3, the amount of metal deposited per pulse is much smaller than that required to generate stable NPs and therefore the amount of metal deposited during the first pulse will lead to *non-stable* metal nucleation centers. During the lapse of time between

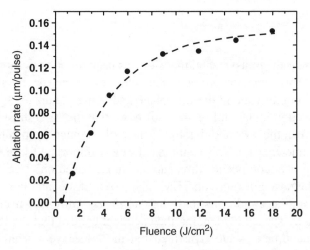

Figure 5: Ablation rate as a function of the laser fluence used for ablating a metallic Co target at λ = 308 nm. (From Tiejun Li *et al.* [26], with permission.)

consecutive pulses, these *nuclei* tend to dissociate into mobile atoms that help to stabilize some of them. Upon arrival of next 'pulse' of species, a fraction of the incoming atoms adds to the already *stable* nucleation centers while the rest of atoms re-start the process [15]. After a few cycles, a stationary regime is reached in which most of the incoming material adds to the stable growing centers that eventually become NPs. According to this picture, the effect of increasing the arrival flux of species should be to increase the density of metal *nuclei* that reach the critical size to become stable. A complete description of the cluster nucleation process during PLD is out of the scope of the present chapter and further details can be found in Refs [5,16].

The kinetic energy and degree of ionization of the ablated species in the case of metals are important parameters that are strongly correlated. Difficulties in estimating the ionization fraction in a laser generated plasma are evidenced by the wide range of values that can be found in the literature: from 10 to 100% [28]. Low ionization fractions are typically observed at low fluences, while ionization strongly increases with laser fluences above the plasma formation threshold. Recent works on UV ablation of metals have reported ionization fractions larger than 50% for several metals: Ti ($>50\%$), Fe (≈50–90%), Al ($\approx60\%$), Ag ($\approx60\%$) [28,29]. In addition, ions are significantly accelerated in the plasma and kinetic energies in the range from 0.1 eV to more than 100 eV, with a strong dependence on the laser fluence, have been reported in the case of metals [30]. As an illustrative example, the dependence of the kinetic energy of Fe species on the laser fluence reported elsewhere [31] is shown in Fig. 6 where it is clearly seen that kinetic energies as high as 100 eV are reached for laser fluences of just ≈8 J cm^{-2}.

Two clear regimes can be distinguished when analyzing the impact of the kinetic energy of the impinging species on the nucleation and growth of the NPs. The first one is the low kinetic energy regime in which the kinetic energies are lower than the displacement threshold energy [10] or the bond strength [32] (≤30–50 eV). Under this regime, the species have no significant impact on the surface quality of the growing film since the excess of kinetic energy is used to increase surface mobility and reactivity, thus promoting coarsening of the metal NPs. The second regime corresponds to kinetic energies above these values that lead to phenomena such as implantation or surface sputtering that might have a clear impact on the properties of nanocomposite films. Implantation phenomena in PLD have unambiguously been evidenced for the case of high mass species such as Bi [33] or Au [18]. Figure 7 shows a cross section TEM image of a nanocomposite film containing three layers (a, b, c) of Bi NPs embedded in a-Al$_2$O$_3$, the growth direction being from bottom to top and each NPs layer being grown at increasing laser fluences in the growth direction. A continuous and very thin layer of dark contrast

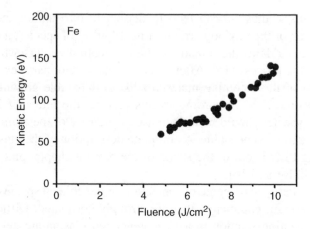

Figure 6: Average kinetic energy of Fe ions as a function of the laser energy used for ablating a Fe target at $\lambda = 248$ nm. (From Fahler *et al.* [31], with permission.)

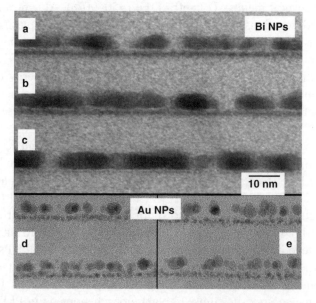

Figure 7: Cross section TEM images of nanocomposite films containing (a, b, c) Bi NPs and (d, e) Au NPs produced using different laser energy densities: (a) 5 J cm^{-2}, (b) 2 J cm^{-2} and (c) 0.4 J cm^{-2} in the case of Bi NPs and (d) 2.7 J cm^{-2} and (e) 8.9 J cm^{-2} in the case of Au NPs. The metal areal densities of the different layers are: (a, b, c) [Bi] = 1.9×10^{15} at. cm^{-2}, (d) [Au] = 4.9×10^{15} at. cm^{-2} and (e) [Au] = 4.2×10^{15} at. cm^{-2}. In all cases, the growth direction is from bottom to top of the images.

can be seen below the NP layers (a and b) produced at the two higher laser fluences (5 J cm^{-2} and 2 J cm^{-2}). These layers are located at a depth from their corresponding NP layers that increases with the laser fluence used for deposition (1.6 nm at 5 J cm^{-2} and 1.1 nm at 2 J cm^{-2}). No such a layer is observed for the Bi NPs layer (c) deposited at the lowest fluence (0.4 J cm^{-2}). Since the plan view TEM images do not show evidence of this rich Bi layer, it has been suggested that it is formed by Bi implanted in the a-Al$_2$O$_3$ matrix. These results are in contrast with those observed when producing Au NPs by PLD in which the metal implanted actually forms NPs with typical sizes close to ≈2 nm in diameter (layers d and e) [18]. The fact that implantation effects are very noticeable in Bi and Au is not surprising since both metals have similar mass and thus, under similar laser fluences with respect to their ablation threshold, the kinetic energy of the species are expected to be similar. This is consistent with the implantation effects becoming less noticeable as the mass decreases (i.e. for Ag NPs) [19]. However, the reasons why Bi gets dissolved in the host while Au forms implanted NPs are not yet clear.

Self-sputtering of a fraction of the metal already deposited at film surface induced by the bombardment of incoming species has been reported in the literature [28,34–36]. This phenomena is caused by the fact that the mean kinetic energy of ions present in the laser generated plasma exceeds the threshold for sputtering of the metal that is in the range 10–25 eV for most metals [28]. Figure 8 shows self-sputtering yields reported when growing different metal layers by PLD as a function of the corresponding cohesive energy of the metal. In all cases, the yields

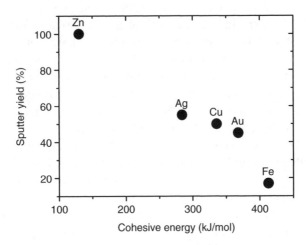

Figure 8: Self-sputtering yields measured for Ag, Fe, Cu and Zn and calculated for Au as a function of the cohesive energy of each element. (Data taken from Hidalgo *et al.* [34] and references therein.)

have been determined for a laser fluence close to 4–5 J cm^{-2} [28,34–36]. The self-sputtering yield is found to increase as the cohesive energy decreases and it can be as high as 1 (100% sputtering and thus no effective deposition) in the case of Zn. Sputtering phenomena have been observed to compete with growth in the case of Au NPs embedded in a-Al$_2$O$_3$ and it becomes the dominant process at high laser fluences [18]. Using classical models for self-sputtering [37], a yield of 0.72 has been calculated for Au when the target is ablated with ≈9 J cm^{-2}, this value becoming much smaller (0.37) at low fluences (≈3 J cm^{-2}).

2.2.3.3. Influence of a Gas Environment

A possible means to control the kinetic energy of the species arriving to the substrate as well as to overcome the undesired effects related to its high values (implantation and self-sputtering) is the use of low laser fluences. However, for fluences close to the ablation threshold, PLD loses part of its unique properties and approaches a thermal deposition process. An alternative route to reduce the kinetic energy while keeping acceptable deposition rates (both peak and average values), is to introduce a gas background in the vacuum chamber during the ablation process. If the gas pressure is kept below 1 mbar, the main effect of the gas background is to modify the expansion dynamics of the ejected material [38]. This leads to broader angular distributions than in vacuum and to a slow down of the expansion process that eventually decreases the kinetic energy of the species reaching the substrate.

Figure 9 shows an example of the effect that a moderate *inert* gas pressure has on the morphology of the NPs. The images in Fig. 9a and b correspond to plan view images of *sandwich* films having approximately the same Cu content (≈7.0 ± 0.3 × 10^{15} at. cm^{-2}) and grown either in 1 × 10^{-4} mbar (Fig. 9a) or in 7 × 10^{-2} mbar of Ar (Fig. 9b). Most of the NPs grown at the lowest pressure have a quasi-spherical shape with only a small fraction having elongated shapes as a consequence of coalescence. Instead, most of the NPs grown in 7 × 10^{-2} mbar of Ar have elongated shapes, looking like bend strings lying on the surface. These features are quantified in Fig. 9c where the breadth (B), length (L) and height (H), the latter defined as the dimension in the direction perpendicular to the substrate, are plotted as a function of the gas pressure used during deposition [22]. It is clearly seen that for this metal content, whereas the NPs produced in vacuum are approximately spherical ($B ≈ L ≈ H$), the NPs produced in gas become elongated in the substrate plane while preserving an approximately constant cross section. This behavior has been related to the reduction of the mobility of the species that reach the substrate due to their reduced kinetic energy, as this decreases coarsening and favors the aggregation of neighboring NPs once they get close enough [22].

If the gas environment is reactive, additional effects related to chemical reactions between the ablated or the already deposited species and the gas environment can

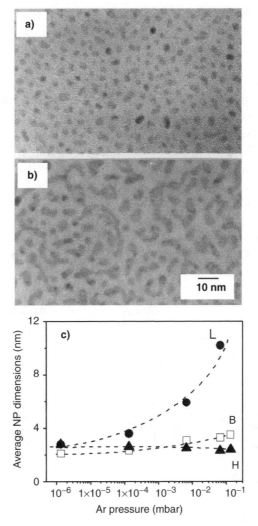

Figure 9: (a, b) Plan view TEM images of Cu:Al$_2$O$_3$ nanocomposite films having approximately the same Cu areal density ($\approx 7.0 \pm 0.3 \times 10^{15}$ at. cm^{-2}) and grown at (a) 1×10^{-4} mbar and (b) 7×10^{-2} mbar of Ar. (c) Average NP dimensions: ($\square$) breadth B, ($\bullet$) length L and ($\blacktriangle$) height H as a function of the Ar pressure used during growth. (Adapted from Afonso *et al.* [22].)

take place [39]. NPs are known to have an enhanced reactivity with respect to bulk metals as evidenced for instance by the fact that Ag or Au NPs oxidize much faster than the corresponding thin films [40]. Therefore the use of a reactive gas such as O_2 provides a route for producing complex NPs, i.e. metal-oxide or metal core-oxide shell NPs [41]. The degree of oxidation can be controlled through the gas pressure either during or after ablation of the metal. As an example, Fig. 10 shows the X-ray absorption near edge spectroscopy (XANES) spectra taken from $Cu:Al_2O_3$ nanocomposite films prepared in vacuum and in oxygen atmospheres (10^{-6}–10^{-1} mbar). The spectra of bulk Cu and CuO reference films are included for comparison. Qualitatively the XANES spectrum of the film grown at the highest O_2 pressure is similar to that of CuO and therefore it can be concluded that Cu is in oxidized state. In contrast, the XANES spectra for films deposited in vacuum, at 2.6×10^{-4} mbar of O_2 or exposed to oxygen after metal deposition, resemble that of bulk Cu. The quantitative analysis of these results have provided evidence for the fraction of Cu-oxidized phases in the films being higher the higher the oxygen pressure [41].

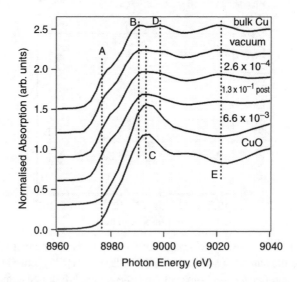

Figure 10: XANES spectra of $Cu:Al_2O_3$ nanocomposite films having a Cu areal density per layer of $10.0 \pm 0.5 \times 10^{15}$ at. cm^{-2} and grown at different gas environments. Spectra obtained for metallic (bulk Cu) and Cu oxide (CuO) references are included for comparison. The letters A, B and D mark the position of characteristic peaks of metallic Cu, C marks the position of the white line of CuO and E marks when metal and oxide signals are in phase opposition. The numbers on the curves indicate the oxygen pressure at which the Cu was deposited (in mbar) or was exposed to after deposition (curve marked post). (From Serna *et al.* [41]. Copyright 2002 by the American Physical Society.)

2.3. Properties and Applications of Metal–Dielectric Nanocomposites

2.3.1. Optical Properties

Nanocomposite materials formed by metal nanoparticles (NPs) embedded in a transparent dielectric have two important characteristics that determine their properties in the optical frequency range [42]. First, if NPs dimensions are much smaller than the wavelength of an external applied electromagnetic field, the electric field of the electromagnetic wave polarizes the free charges of the NPs, leading to a greatly enhanced effective field that modifies the effective dielectric constant of the nanocomposite. This effect is called dielectric or *classical confinement*, and is responsible for the well known coloring of many decorative glasses having embedded metal NPs. Second, the electrons in the NPs are confined to regions much smaller than their mean free path, which is of the order of a few micrometers for the case of bulk noble metals. This effect, widely referred to as *quantum confinement*, is responsible for the appearance of optical non-linearities in dielectric nanocomposites. As a consequence of these confinement effects, the nanocomposites show linear and non-linear optical properties very different than those corresponding to their individual components. This leads to a wide range of interesting applications based either on their linear optical properties (filters, polarizers or selective absorbers) or on their non-linear optical properties (optical limiters or all-optical switches) [43–45].

2.3.1.1. Linear Optical Properties

The most widely known optical feature of nanocomposite materials is the appearance of optical resonances or enhanced absorption at the wavelength of the surface plasmon resonance (SPR) due to *classical confinement* effects. These resonances appear whenever the condition $\varepsilon_m(\lambda) + 2\varepsilon_d(\lambda) = 0$ is satisfied, where λ is the wavelength, ε_m is the real part of the dielectric constant of the metal and ε_d is the real part of the linear refractive index of the dielectric host. Since the imaginary part of the refractive index of the host is negligible, the previous condition can be also expressed by substituting ε_d by n_d, where n_d is the refractive index of the dielectric.

Figure 11 shows the real (n, Fig. 11a) and imaginary (k, Fig. 11b) parts of the refractive index of Cu:Al$_2$O$_3$ and Ag:Al$_2$O$_3$ nanocomposite films. The data have been obtained by spectroscopic ellipsometry. The n values of a reference a-Al$_2$O$_3$ film with no NPs are also shown for comparison, while no corresponding curve for k is included since it is negligible. It is clearly seen that both n and k values of the nanocomposite differ significantly from those of the host. The clearest effect of having NPs embedded in a dielectric is that k (and thus the absorption) is

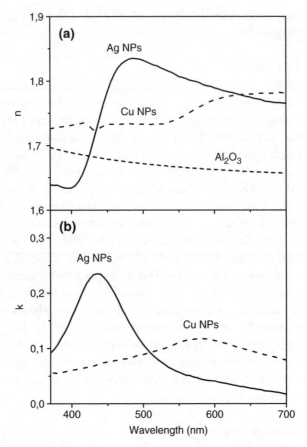

Figure 11: (a) Real (n) and (b) imaginary (k) parts of the refractive index of Cu:Al$_2$O$_3$ and Ag:Al$_2$O$_3$ nanocomposite films as a function of wavelength. The real part of the refractive index of an a-Al$_2$O$_3$ reference film is included for comparison.

no longer negligible and shows a band peaking at the SPR wavelength. Figure 11b shows that the SPR bands for Cu and Ag NPs embedded in a-Al$_2$O$_3$ peak at ≈590 nm and at ≈430 nm, respectively. The n values show instead anomalous dispersion behavior at the SPR wavelength, that makes the n of the nanocomposite to be generally higher than that of the pure host, the higher enhancement occurring at wavelengths longer than that of the SPR. It is clearly seen from Fig. 11 that neither the real nor the imaginary part of the refractive index of the nanocomposite can be approximated by those of the host [46]. Nevertheless, as the real part of the refractive index has seldom been measured in nanocomposite systems, it has been a common practice to make this approximation, even if this can affect the accuracy of the non-linear refractive index values experimentally obtained [47].

Effective medium theories, such as Maxwell-Garnett or Bruggeman, have widely been used to describe qualitatively the spectral behavior of nanocomposites around the SPR as a function of the metal volume fraction (p), especially for the case of low metal volume fractions ($p \ll 0.01$) [42,48,49]. Nevertheless, these theories fail to predict the detailed dependence of the absorption spectra on the NPs dimensions and shape, particularly for high metal volume fractions. Additional difficulties arise from the dispersion of both the dimensions and shapes of the NPs that depend very much on the preparation method and on the occurrence of coalescence. This latter effect is known to enhance the effects related to electromagnetic interactions among NPs that are usually neglected in these theories [48,50].

The main influence of the NPs size on the optical properties of the nanocomposite is to shift the SPR to longer wavelengths as the NPs become bigger. This effect is clearly seen in Fig. 12a, where the spectral position of the SPR as a function of the average NPs diameter is shown for the case of Ag:Al_2O_3 [50] and Cu:Al_2O_3 [51] nanocomposite films. The modification of the NP dimensions thus provide a method to tune the optical absorption over a range that can be of 100 nm at the most, since the range of accessible diameters in a deposition process is limited by the occurrence of percolation. However, this wavelength shift is accompanied by a significant increase (higher than a factor of 3) of the absorption of the nanocomposite as shown in Fig. 12b for the case of Ag:Al_2O_3. This increased absorption is detrimental for some applications such as those involving waveguide propagation, and thus alternative routes to tune the SPR in a broader range while keeping the absorption values sufficiently low are desirable.

One possible route is the production of mixed metal NPs by combining two metals with well defined and clearly separated SPR resonances. The key idea within this approach is to shift the position of the SPR between the positions corresponding to each single metal while keeping the NPs dimensions small enough in order to achieve enough low absorption. Such NPs formed by a mixture of Ag and Cu have successfully been produced by PLD by alternating the ablation of the host and of two metal targets [52]. Figure 13 shows the position of the SPR of the resulting Ag-Cu:Al_2O_3 nanocomposite films as a function of Ag content in percentage, the NPs dimensions being in all cases in the range of 3–4 nm. The results achieved in samples containing pure metal NPs of similar dimensions are also shown for 0 and 100% Ag contents. The SPR wavelength varies from that of pure Cu NPs to that of pure Ag NPs. Similar results have been reported for the case of Au-Ag NPs embedded in Al_2O_3 prepared by laser ablation in a high gas pressure that condensate the NPs in the gas phase [53].

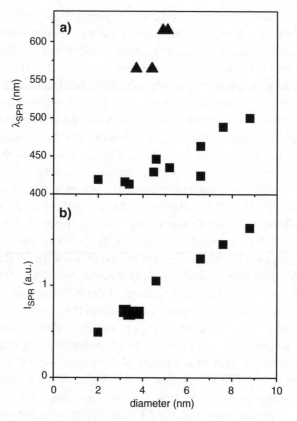

Figure 12: (a) Spectral position and (b) intensity of the SPR as a function of the in-plane average diameter of metal NPs for the case of (■) Ag:Al$_2$O$_3$ or (▲) Cu:Al$_2$O$_3$ nanocomposite films.

In addition to the size, shape and composition of the NPs, the oxidation state of the metal forming the NPs plays a significant role on the absorption of nanocomposite films. This is illustrated in Fig. 14 where the optical absorption spectra of Cu:Al$_2$O$_3$ nanocomposite films in which the metal NPs were produced in different gas environments (vacuum, 2.6×10^{-4} mbar and 6.6×10^{-3} mbar of oxygen). The absorption spectrum for the case in which the NPs were grown in vacuum and subsequently exposed to an oxygen pressure of 1.3×10^{-1} mbar for 5 minutes before being covered by the host is also included (post in Fig. 14) [41]. The comparison of the spectra of the films in which the metal NPs were produced in vacuum with or without post growth exposure to oxygen suggests that in the former case a metallic core remains in the NPs while the surface becomes oxidized, the reduced absorption being a combined result of the smaller dimensions of the '*effective*' metal NPs

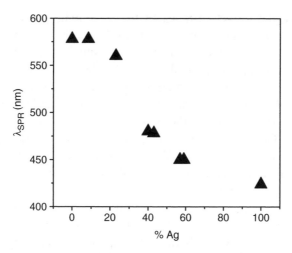

Figure 13: Spectral position of the SPR of Ag-Cu:Al$_2$O$_3$ as a function of the Ag content in the mixed Ag-Cu NPs. (Adapted from Gonzalo *et al.* [52].)

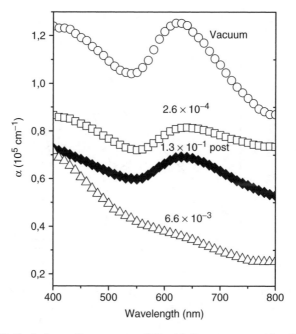

Figure 14: Optical absorption spectra of Cu:Al$_2$O$_3$ nanocomposite films having a Cu areal density per layer of $10.0 \pm 0.5 \times 10^{15}$ at. cm^{-2} and being grown at different gas environments. The numbers on the curves refer to the oxygen pressure at which the Cu was deposited (in mbar), or was exposed to after deposition (curve marked post). (Adapted from Serna *et al.* [41].)

and the fact that the refractive index of the surrounding host (i.e. the oxidized shell) is different than that of the embedding host. This interpretation is further confirmed by the results achieved in nanocomposite films in which the NPs were produced in an oxygen pressure. In this case, the absorption at the SPR wavelength decreases until it disappears for a pressure of 6.6×10^{-3} mbar. From the comparison of these results with those presented in Fig. 10, it can be concluded that the metallic core Cu still dominates the optical response at low O_2 pressures, while the formation of Cu oxides at high oxygen pressures induces a strong decrease of the optical absorption and eventually the disappearance of the SPR band.

The control of the overall absorption of the nanocomposite film in the neighborhood of the SPR wavelength is in itself an important issue. A possible way to tailor its value to the particular application envisaged is the optimization of the in-depth distribution of the NPs within the host [54]. The method uses photonic units formed by a few layers of NPs as building blocks for the nanocomposite film. Figure 15 illustrates, as an example, the transmission of nanocomposites formed by two of these photonic units. The metal forming the NPs is Cu, the host is a-Al_2O_3 and

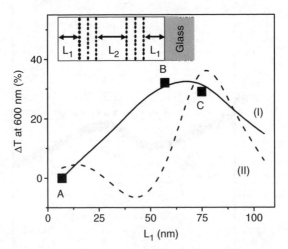

Figure 15: Calculated change of the transmission (ΔT) at $\lambda = 600$ nm of Cu:Al_2O_3 nanocomposite films having the NPs layers distributed in depth as indicated in the inset, as a function of L_1 and having $L_2 = 7$ nm (—, type I) or $L_2 = L_1/2$ (- - -, type II). The dark dots and the background in the inset represent the Cu NPs and the a-Al_2O_3 host, respectively, each group of three layers of Cu NPs forming the so called '*photonic unit*'. Experimental values (■) obtained for samples (A, B, C) grown with the different configurations indicated in the text are included. (Reprinted with permission from Suarez-García *et al.* [54].) Copyright 2003 American Institute of Physics.)

each photonic unit is formed by 3 NPs layers spaced 7 nm each from the other. The change of transmission calculated for such structured systems with respect to a reference specimen is plotted as a function of the parameter L_1, where L_1 is the separation between each photonic unit and the interfaces of the nanocomposite film (the surface and the substrate). The reference specimen is a nanocomposite film having 6 NPs layers equally spaced (7 nm), located at the center of the nanocomposite film and separated 7 nm from both surface and substrate interfaces (film A).

The parameter L_2 that represents the separation between the two photonic units is set equal to 7 nm for type I nanocomposite films and $L_1/2$ for type II nanocomposite films. Optical simulations show that at a wavelength of 600 nm, an enhancement of the transmission higher than 30% with respect to the reference specimen is expected for certain values of L_1 (67 nm for type I (film B) and 77 nm for type II (film C)). Nanocomposite films having these L_1 values have been produced and their resulting transmission agrees very well with the calculated one, as also shown in Fig. 15.

2.3.1.2. Non-linear Optical Properties

The existence of both dielectric and quantum confinement also determines the non-linear optical properties of metal–dielectric nanocomposites. Such effects are particularly relevant at wavelengths close to the surface plasmon resonance (SPR) of the nanocomposite due to the strong enhancement of the local field in the vicinity of the metal NPs associated to the collective oscillation of their conduction band electrons. The centro-symmetric nature of the nanocomposites makes the local field resonant enhancement to be mirrored in the non-linear properties of these materials through the third order susceptibility term ($\chi^{(3)}$), that can reach values many orders of magnitude higher than those of the bulk dielectric host and the metal forming the NPs [55].

Metal–dielectric nanocomposites thus show intense third order optical non-linearities which make them particularly attractive for the development of non-linear photonic devices, like all-optical switches [56]. In general, devices based on highly efficient non-linear optical materials are expected to become important in future high-capacity communication networks in which ultrafast switching, signal regeneration and high speed demultiplexing will be performed in an all-optical way through the use of third order optical non-linearities [57]. In this context, PLD has been proven as an excellent tool for producing extremely efficient non-linear metal–dielectric nanocomposites in thin film configuration [51]. This is illustrated in Fig. 16 that shows the evolution of the modulus of the $\chi^{(3)}_{\text{eff}}$ component of the effective third order susceptibility of a $Cu:Al_2O_3$ nanocomposite film as a function of the metal volume fraction (p). The effective third order susceptibility has been determined by degenerate four wave mixing experiments using 10 ps laser pulses at

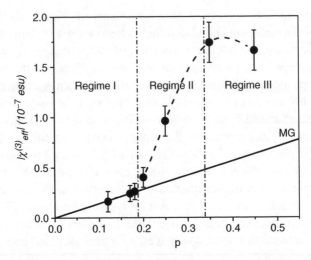

Figure 16: Modulus of the $\chi_{\text{eff}}^{(3)}$ component of the effective third order susceptibility at $\lambda = 585$ nm as a function of the volume fraction (p) of Cu NPs. The continuous line is a linear fit of the data with $p < 0.20$ to the Maxwell-Garnett (MG) model. The dashed line is a guide to the eye. (Reprinted with permission from del Coso *et al.* [51]. Copyright 2004 American Institute of Physics.)

a wavelength centered at the SPR of the nanocomposites and is seen to reach values above 10^{-7} esu [51].

The intrinsic third order susceptibility of metal NPs ($\chi_{\text{m}}^{(3)}$) is associated to the dynamic changes in their dielectric function caused by interband and intraband transitions as well as by the so-called *hot-electron* contribution in which the conduction band electrons of the NPs absorb energy and get heated loosing their energy ultimately by electron–phonon scattering. The contribution showing larger quantum confinement effects is that associated to intraband transitions while the one expected to be dominant at wavelengths close to the SPR is the one associated to hot-electrons. Their expected contributions to the $\chi_{\text{m}}^{(3)}$ value have been analyzed in detail elsewhere [58]. Although there are large uncertainties in the parameters used for providing the corresponding estimates, it is generally accepted that $\chi_{\text{m}}^{(3)}$ can reach values above 10^{-10} esu for group IB metals (Ag, Au, Cu). The combination of such strong intrinsic NP susceptibilities with the presence of intense field enhancement leads to the possibility of producing artificial materials with effective susceptibilities $\chi_{\text{eff}}^{(3)}$ that are six orders of magnitude higher than that of SiO_2 [51,59,60].

Similarly to the linear case, a first approach to describe the non-linear optical properties of these nanocomposites can be made in the frame of effective medium theories in the low-dilution regime (i.e. when $p \ll 1$) [48]. These theories allow predicting the value of $\chi_{\text{eff}}^{(3)}$ for the nanocomposite directly from the $\chi_{\text{m}}^{(3)}$ value of the isolated NPs [58,61]. This leads to expressions in which $\chi_{\text{eff}}^{(3)}$ is proportional to $\chi_{\text{m}}^{(3)}$ and p, and scales with the fourth power of the field enhancement factor (f) [62], the latter being related to the dielectric functions of the host and the metal. Although these theories predict strong third order optical non-linearities in metal dielectric nanocomposites in the low dilution regime, only few theoretical models have attempted a rigorous description of the optical behavior of nanocomposites with high metal volume fraction ($p \gg 0.01$) [63]. This regime is instead most interesting, since electromagnetic interactions among NPs as well as local giant field electromagnetic fluctuations that can potentially lead to even larger enhancements of different non-linear optical processes, like Raman scattering, Kerr refraction or Four Wave Mixing are expected to occur close to the percolation limit [64].

Three distinct regimes for the evolution of $\chi_{\text{eff}}^{(3)}$ with p can be clearly observed in Fig. 16. For low p values (Regime I), the surface-to-surface separation among NPs is >3 nm and their average dimensions are smaller than 4.4 nm. Within this regime, $\chi_{\text{eff}}^{(3)}$ increases linearly with p, consistently with what is predicted by the Maxwell-Garnet effective medium model [58]. When the metal volume fraction is further increased (Regime II), $\chi_{\text{eff}}^{(3)}$ increases sharply to a maximum value of 1.8×10^{-7} esu that is achieved for $p = 0.35$. This is among the higher values ever observed in metal–dielectric nanocomposite materials [51]. Since Regime II starts once coalescence becomes significant, the observed enhancement of $\chi_{\text{eff}}^{(3)}$ is most likely related to electromagnetic coupling within coalesced NPs that behave as two or more electromagnetically coupled entities. Finally, for $p > 0.35$ (Regime III), $\chi_{\text{eff}}^{(3)}$ starts to decrease. This decrease is expected since $\chi_{\text{eff}}^{(3)}$ should approach the value for bulk Cu that is orders of magnitude smaller, once the metal forms a continuous layer (i.e. p approaches 1). Around the percolation threshold, the presence of multiple particle scattering leads to an enhancement of the local electric field [65,66] and, accordingly, of $\chi_{\text{eff}}^{(3)}$. This effect is overlapped to local giant field enhancement effects caused by the presence of a network of metallic filaments that concentrate the electric field in the point-like edges of the metallic filaments [64,67]. Both effects, multiple particle scattering and giant local enhancement, compensate up to some extent the decrease in $\chi_{\text{m}}^{(3)}$ that occurs within the nearly percolated metal structure, giving rise to the wide maximum of $\chi_{\text{eff}}^{(3)}$ observed in Regime III.

Globally, the behavior shown in Fig. 16 indicates that beyond the low dilution regime, the metal volume fraction itself is not the only factor conditioning the non-linear response of nanocomposites. The particular meso- and nanoscopic morphology of the nanocomposite is at the end what determines the presence of multiple particle electromagnetic interactions and giant local resonance effects which can substantially enhance the non-linear response of the material beyond the predictions of effective medium models. These effects, although very important in terms of the strength of the non-linear response, seem to have little impact on the non-linear optical time response of the material. This can be seen in Fig. 17 where the time evolution of the conjugated signals in the four wave mixing experiments associated to Fig. 16 are plotted. The time evolution of the signal for all samples with $p < 0.35$ is the same (only the one corresponding to $p = 0.25$ is plotted), with fast build-up and decay times (<2 ps), indicating that the mechanisms associated to the time evolution of $\chi_{eff}^{(3)}$ are not affected by the presence of multiple particle interactions. For specimens with $p \geq 0.35$, an additional slow decay component of hundreds of picoseconds is observed that is related to the cooling of the metal NPs lattice by thermal diffusion to the host matrix. This interpretation is consistent with the fact that this process is more efficient for the smaller NPs due to their higher surface to volume ratio and thus only becomes significant for samples close to the percolation threshold.

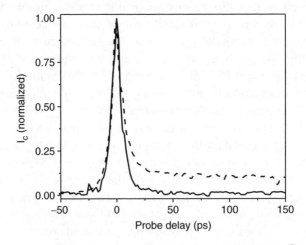

Figure 17: Time evolution of the conjugated signal associated to $\chi_{eff}^{(3)}$ in Fig. 16 for Cu:Al$_2$O$_3$ nanocomposite films for films having metal volume fractions of $p = 0.25$ (——) and $p = 0.45$ (- - -). (Reprinted with permission from del Coso et al. [51]. Copyright 2004 American Institute of Physics.)

2.3.2. Magnetic Properties

Magnetic NPs are promising materials for applications in magneto-optics and ultra-high density recording. In addition, magnetic NPs are used in other fields such as medicine [68–70], the main applications being drug delivery and clinical diagnostic techniques. However, they are used in colloidal suspensions and are thus out of the scope of the present chapter that deals with nanocomposite films.

The special magnetic properties of nanocomposites containing ferromagnetic NPs are mainly related to the magnetic anisotropy of the NPs and to magnetic interactions among the NPs [71,72]. Whereas the former are related to both the structure and morphology (size and shape) of the NPs, the latter depend on both the NPs separation and the magnetic properties of the embedding host. Most of the envisaged magnetic applications are based on the use of single-domain non-interacting NPs in order to maintain the magnetic order with time. These requirements impose thus restrictions on the structural and morphological properties that magnetic NPs must satisfy. To date, different types of magnetic NPs have been produced by a-PLD namely Fe, Ni and Co [21,73–75]. In all cases, the importance of NPs dimensions on the magnetic response of the nanocomposites has been evidenced. This is illustrated in Fig. 18 where it is shown the magnetization of Fe:Al$_2$O$_3$ nanocomposite films produced by a-PLD as a function of the magnetic field applied (M-H loops). The measurements were performed at room temperature in films containing 5 layers of Fe NPs and as a function of the NPs dimensions [73]. Both the coercitivity (Hc) and the saturation magnetization decrease significantly when decreasing the average dimension of the NPs. These results are a clear indication for the superparamagnetic (Hc = 0) behavior of nanocomposite films containing NPs with average dimensions ≤5 nm. Films containing NPs with dimensions ≥7 nm are instead ferromagnetic (FM).

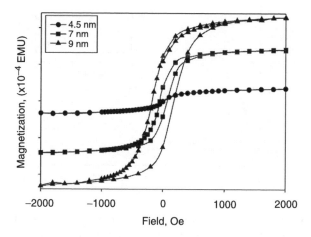

Figure 18: M-H loops at 300 K recorded in Fe:Al$_2$O$_3$ nanocomposite films having NPs with the different average dimensions (in nm) indicated. (From D. Kumar *et al.* [73], with permission.)

Superparamagnetism is an undesired effect since NPs are unable to maintain the magnetic order. The appearance of superparamagnetism is related to the reduction of the magnetic anisotropy energy per NP that is responsible for holding the magnetic moments aligned along certain directions. When this energy becomes comparable to the thermal energy ($k_B T$), the thermal fluctuations in the orientation of the magnetic moments average out the magnetization and the magnetic order is lost [21,71,76]. The temperature at which the ferromagnetic–superparamagnetic transition occurs depends on the morphology of the NPs itself. Figure 19 shows the temperature dependence of the magnetization for two nanocomposite films containing either isolated quasi-spherical ($L = B = 2$–3 nm) or ellipsoidal ($B = 3$ and $L = 5$ nm) NPs, where L is the length and B the breadth of the NPs in the film plane [21]. Data collected when cooling the nanocomposite films in a magnetic field of 5 mT (FC) and in zero magnetic field (ZFC) are both included.

The ZFC magnetization curves increase with increasing temperature to reach a maximum at the so called blocking temperature, T_B. This temperature is 25 and 55 K for samples containing quasi-spherical and ellipsoidal Fe NPs, respectively. The dependence of T_B on the magnetic field for the former type of nanocomposite films is plotted in the inset of Fig. 19 where it is seen to decrease when the applied

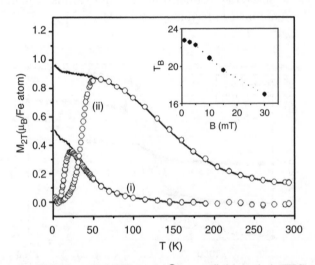

Figure 19:   (—) 5 mT field cooled (ZF) and (○) zero field cooled (ZFC) magnetization of Fe:Al$_2$O$_3$ nanocomposite films containing isolated spherical (i) and ellipsoidal (ii) Fe NPs as a function of temperature. The inset shows the blocking temperature as a function of the magnetic field for the specimen containing spherical NPs. (Reprinted with permission from Dempsey *et al.* [21]. Copyright 2001 American Institute of Physics.)

magnetic field is increased. The ZFC measured T_B corresponds to the ferromagnetic–superparamagnetic transition temperature. As the dimensions of NPs are increased, T_B can be above room temperature as shown in Fig. 18 for the case of Fe NPs with dimensions ≥ 9 nm [73]. However, the NPs are no longer single magnetic domain and the NPs separation is in general not large enough to ensure the condition of non-interacting NPs for these dimensions.

A possible way to overcome this limitation is to induce an enhancement of the coercivity of the NPs by embedding them in an antiferromagnetic (AFM) material since in such a case, the magnetic exchange coupling induced at the interface contributes to increase the magnetic anisotropy energy of the NPs [21,77]. This concept has proven to be valid in the case of Co core-CoO shell NPs embedded in a CoO matrix produced by sputtering [76]. In this case, the Co cores were found to remain ferromagnetic up to the Néel temperature of CoO (≈ 290 K) and therefore, the NPs were magnetically stable up to this temperature. Although Co core-CoO shell NPs embedded in Al_2O_3 produced by a-PLD have recently been reported, no similar magnetic behavior has been observed [78]. Instead, it has been found that there is a critical NP size below which exchange anisotropy is absent for hybrid ferromagnetic–antiferromagnetic NPs.

2.3.3. Thermal Properties

Metal NPs embedded in a solid matrix have also been used to study the thermo-dynamical properties of materials in the nanometric scale. It has generally been reported that the temperature for solid–liquid transition of small metal NPs is lower than that of the respective bulk material and decreases with the dimension of the NPs, a phenomenon known as *thermodynamic size effect* [79]. Qualitatively, this behavior can be understood by assuming that for a spherical NP containing N atoms, the number N_s of atoms located at the surface is inversely proportional to its radius r. This is for example the case for an fcc crystalline NP of lattice parameter a, for which $N_s/N \approx 3a/2r$. For a spherical NP having a diameter of 3 nm, about 50% of the atoms are on the surface. The bonds of the surface atoms differ from those of the bulk ones. The surface atoms constitute the interface between the NPs and the embedding host and they reflect the interaction with the media. The melting dynamics thus depends on surface effects as well as on shape effects.

Although the full theoretical understanding of the different phenomena that yield to a decrease of the melting temperature of NPs compared to the bulk metals is far from being complete, there are phenomenological theories that have been able to give an insight on the thermodynamic size effects. Whatever the starting hypothesis, they all lead to the relationship $1 - T_{mNp}/T_0 \propto r^{-1}$, where T_0 is the melting temperature of the bulk material and T_{mNp} is the melting temperature of the NP. For example, for the case of a 3 nm diameter Pb NP, the melting temperature is only 1/3 of that of the bulk material, that is indeed a very significant change [79].

Many different experimental techniques have been used to study the solid–liquid transition in NPs including X-ray diffraction [80], transmission electron microscopy [79], RBS-channeling [82] and optical measurements [81–83]. The choice of an experimental technique determines the physical parameter whose variation is used as a criterion for the transition, giving thus a particular insight of the phenomenon. Raman spectroscopy of Bi:Ge nanocomposite films is a powerful non-invasive technique that has proven to be very sensitive to the measurement of structural changes, and this makes it interesting to analyze the melting of metal NPs. It has been used in situ to follow the melting of Bi NPs embedded in a-Ge matrix produced by PLD [84]. This nanocomposite system is particularly well suited to study thermal properties since embedding NPs in a matrix present various advantages with respect to supported NPs such as avoiding any contamination and making the NPs independent of each other. As a consequence, the successive cycles (melting–solidification) exhibit no drift with time. Moreover, Bi has a relatively low melting point (574 K) and Bi and Ge form an eutectic system with low mutual solubility, thus facilitating the production of pure, isolated and unreacted Bi crystalline NPs. Figure 20 shows the Stokes Raman spectra of Bi:Ge nanocomposite films at three temperatures (440, 520, and 600 K). At low temperatures, the well-known modes E_g (70 cm^{-1}) and A_{1g} (97 cm^{-1}) of the Bi are observed [85]. As expected, when the temperature increases, the peaks shift towards lower wave numbers because of anharmonic effects. Once the melting point is reached, the peaks disappear and are replaced by broad bands characteristic of the liquid state. When the temperature decreases, the inverse process occurs and the peaks relate to crystalline NPs re-appear, thus completing a reversible cycle. However, this occurs at a significantly lower temperature than the temperature of melting.

Figure 21 shows the frequencies of the modes plotted as a function of temperature during the heating–cooling cycle. The optical phonons E_g (70 cm^{-1}) and A_{1g} (97 cm^{-1}) converge in one band (single frequency) as temperature increases (gray circles). One important feature that can be inferred from Fig. 21 is the hysteresis of the melting–solidification cycle [79,80]. This implies that there is a significant undercooling of the liquid phase upon cooling, the difference between the melting and solidification temperatures being at least of 50 K. Note that melting of Bi NPs according to Fig. 21 occurs at a lower temperature with respect to that of the bulk material in agreement to theoretical predictions [79]. Regarding the solidification transition it is most likely that liquid bismuth remains in a supercooled state which disappears once the first crystal seed is formed in the liquid, this seed being often

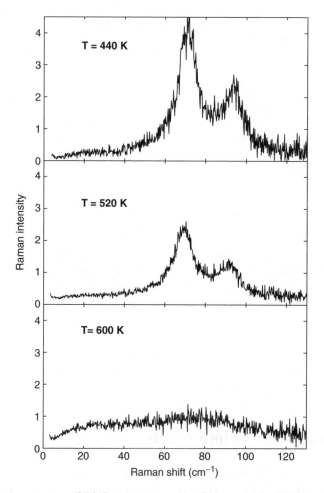

Figure 20: Raman spectra of Bi:Ge nanocomposite films at three different temperatures. (From Haro-Poniatowski *et al.* [84], with permission.)

an impurity. According to this interpretation, the probability of finding such an impurity increases with increasing the dimension of the NPs and thus bigger NPs should crystallize before small ones and lead to a broadening of the transition [86,87].

Although the study of the solid–liquid transition in systems of reduced dimensions is of intrinsic interest, it has also been pointed out that it has a potential interest for the development of thermally driven optical switches [81]. The idea for this application is based on the fact that significant reflectivity changes are observed

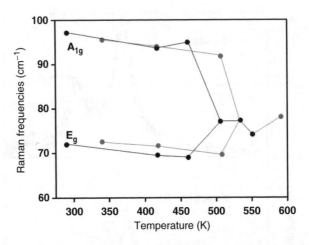

Figure 21: Raman frequencies of the A_{1g} and E_g peaks as a function of temperature. The gray lines correspond to data acquired during the heating process whereas the black ones correspond to data acquired during the cooling process. (From Haro-Poniatowski *et al.* [84], with permission.)

in Ga deposited on silica upon melting and related to the fact that the solid phase behaves as a semiconductor and the liquid phase as a metal. This same concept can be applied to NPs of other materials.

2.4. Summary and Future Trends

The different metal NPs produced so far by PLD in vacuum or low gas pressures are summarized in Table 2. There is a broad range of metal NPs successfully grown by PLD that extends from noble metals to transition metals and semi-metals. Most of the reports relate to NPs embedded in an a-Al_2O_3 host, optical and magnetic properties being the more studied ones. There are a few examples in which the NPs have been produced and studied on substrates (supported as opposed to embedded), most of them related to studies of the catalytic properties of the NPs.

In all cases, PLD has been shown as a versatile promising technique for producing metal NPs and nanocomposite films formed by metal NPs embedded in a dielectric host. When compared to its closest thin film deposition technique competitor, i.e. sputtering, PLD offers on the one hand, a superior performance for the production of dielectric oxide hosts. On the other hand, it allows the production of metal NPs in (ultra) high vacuum conditions as opposed to the high Ar pressure typically required in sputtering processes.

Table 2: Summary of the metal NPs produced by PLD by nucleation and growth at the substrate, either embedded in a host or supported on a substrate, together with the properties reported and the references

Metal	Host/Supported	Properties studied	References
Fe	Al_2O_3	Structural, magnetic, magneto-optic	[21,73]
Co	Al_2O_3	Structural, magnetic	[73,78]
Ni	Al_2O_3	Structural, magnetic	[73]
Cu	Al_2O_3	Structural, optical	[22,24,41,51,54]
Zn	Supported	Structural	[34]
Pd	Supported	Structural, catalytic	[88]
Ag	Al_2O_3, supported	Structural, optical	[19,50,89]
Pt	Supported	Structural, catalytic	[88,90]
Au	Al_2O_3, supported	Structural, optical, catalytic	[18,91]
Bi	Al_2O_3, Ge	Structural, optical, thermal	[20,33]
Ag-Cu	Al_2O_3	Structural, optical	[52]
Ag-Au	Al_2O_3	Structural, optical	[53]

The use of independent targets for the host and the metal is an important added value for the production of nancocomposite films by PLD since it allows controlling the nanocomposite structure in the nanometer scale. This is done through the organization of NPs in layers whose spacing can be varied through the control of the deposition of the host. This ability for nanostructuring the nanocomposite is essential for both fundamental studies and applications. Examples of the former are the possibility of analyzing the dependence of the nanocomposite properties on the separation of NPs to explore the role of NPs interactions and the limits at which these interactions become important. Although this dependence has been studied to some extent in the case of optical and magnetic properties, there are still many aspects and possibilities that remain unexplored. Examples of the importance of nanostructuring of nanocomposite films for applications are the possibility to design the distribution of NPs layers to achieve an improved response. This has been reported for the case of Cu NPs using 'photonic units' in order to achieve an enhanced optical transmission at a certain wavelength. This example evidences the high potential of nanostructuring for achieving high performance nanocomposites, at least for optical applications.

The results achieved so far in the production of metal–dielectric nanocomposite films by alternate PLD can be considered as very promising but preliminary. Future trends spread in several directions: exploiting the nanostructuring capabilities to progress in the knowledge of interactions among NPs and the role of such interactions on nanocomposite properties; exploiting further the capabilities of PLD to produce complex NPs such as NPs formed by new alloys or core-shell structures; exploring new and more complex hosts that can lead to enhanced responses through interactions among NPs and hosts; and finally, exploiting some of the excellent materials produced for real applications.

Acknowledgment

This work was partially sponsored by CICYT (Spain) under TIC1999-0866, TIC2002-03235 and MAT2003-01490 Projects and EU under BRPR-CT98-0616 and HPRN-CT-2002-00328 Projects. We wish to thank past and present members of the Laser Processing Group at Optics Institute who have contributed with their work to the progress and application of PLD. The enthusiastic collaboration and fruitful discussions with many other researchers at several foreign institutions is similarly recognized and appreciated, among which the ones with A. Petford-Long Group from University of Oxford (UK), P. D. Townsend Group from University of Sussex, I. Vickridge from University of Paris Universite de Paris 6 et 7 (France), and R. F. Haglund Jr. from Vanderbilt University (USA) have been particularly long lasting.

References

[1] de Sande, J.C.G., Afonso, C.N., Escudero, J.L., Serna, R., Catalina, F. and Bernabeu, E.E., *Appl. Optics.*, **1992**, *31*, 6133.
[2] Vega, F., de Sande, J.C.G., Afonso, C.N., Ortega, C. and Siejka, J., *Appl. Optics,* **1994**, *33*, 1203.
[3] Hubler, G.K., in *Pulsed Laser Deposition of Thin Films*, Chrisey, D.B. and Hubler, G.K. (Eds.), John Wiley and Sons Inc., New York, **1994**, 327.
[4] Morris, D., Benson, E. and Dubowski, J.J., *Surf. Sci.*, **1993**, *294*, 373.
[5] Rijnders, G.J.H.M., Koster, G., Blank, D.H.A. and Rogalla, H., *Appl. Phys. Lett.*, **1997**, *70*, 1888.
[6] Koinuma, H., Koida, T., Ohnishi, T., Komiyama, D., Lippmaa, M. and Kawasaki, M., *Appl. Phys. A,* **1999**, *69*, S29.
[7] Krebs, H.U., *Int. J. Non-eq. Proc.*, **1997**, *10*, 3.
[8] Gonzalo, J., Sanz, O., Perea, A., Fernández-Navarro, J.M., Afonso, C.N. and García López, J., *Appl. Phys. A,* **2003**, *76*, 943.
[9] Li-Chyong Chen, in *Pulsed Laser Deposition of Thin Films*, Chrisey, D.B. and Hubler, G.K. (Eds.), John Wiley and Sons Inc., New York, **1994**, 167.

[10] Kools, J.C.S., in *Pulsed Laser Deposition of Thin Films*, Chrisey, D.B. and Hubler, G.K. (Eds.), John Wiley and Sons Inc., New York, **1994**, 455.

[11] Yoshida, T., Takeyama, S., Yamada, Y. and Mutoh, K., *Appl. Phys. Lett.*, **1996**, *68*, 1772.

[12] Geohegan, D.B., Puretzky, A.A., Duscher, G. and Pennycook, S.J., *Appl. Phys. Lett.*, **1998**, *72*, 2987.

[13] Li, Q., Sasaki, T. and Koshizaki, N., *Appl. Phys. A.*, **1999**, *69*, 115.

[14] Marine, W., Patrone, L., Luk'yanchuk, B. and Sentis, M., *Appl. Surf. Sci.*, **2000**, *154–155*, 345 and references therein.

[15] Horwitz, J.S. and Sprague, J.A., in *Pulsed Laser Deposition of Thin Films*, Chrisey, D.B. and Hubler, G.K. (Eds.), John Wiley and Sons Inc., New York, **1994**, 229.

[16] Serna, R., Gonzalo, J., Suarez-Garcia, A., Afonso, C.N., Barnes, J., Petford-Long, A., Doole, R.C. and Hole, D., *J. Microscopy*, **2001**, *201*, Pt. 2, 250.

[17] Ohtsuka, S., Koyama, T., Tsunetomo, K., Nagata, H. and Tanaka, S., *Appl. Phys. Lett.*, **1992**, *61*, 2953.

[18] Gonzalo, J., Perea, A., Babonneau, D., Afonso, C.N., Beer, N., Barnes, J.P., Petford-Long, A.K., Hole, D.E. and Townsend, P.D., *Phys. Rev. B*, **2005**, *71*, 125420.

[19] Barnes, J.-P., Petford-Long, A.K., Doole, R.C., Serna, R., Gonzalo, J., Suarez-Garcia, A., Afonso, C.N. and Hole, D., *Nanotechnology*, **2002**, *13*, 465.

[20] Serna, R., de Sande, J.C.G., Ballesteros, J.M. and Afonso, C.N., *J. Appl. Phys.*, **1998**, *84*, 4509.

[21] Dempsey, N.M., Ranno, L., Givord, D., Gonzalo, J., Serna, R., Fei, G.T., Petford-Long, A.K., Doole, R.C. and Hole, D.E., *J. Appl. Phys.*, **2001**, *90*, 6268.

[22] Afonso, C.N., Gonzalo, J., Serna, R., de Sande, J.C.G., Ricolleau, C., Grigis, C., Gandais, M., Hole, D.E. and Townsend, P.D., *Appl. Phys. A*, **1999**, *69*, S201.

[23] Xu, X.P., and Goodman, D.W., *Appl. Phys. Lett.*, **1992**, *61*, 1799.

[24] Serna, R., Afonso, C.N., Ballesteros, J.M., Naudon, A., Babonneau, D. and Petford-Long, A.K., *Appl. Surf. Sci.*, **1999**, *138-139*, 1.

[25] Serna, R., Afonso, C.N., Ricolleau, C., Wang, Y., Zheng, Y., Gandais, M. and Vickridge, I., *Appl. Phys. A*, **2000**, *71*, 583.

[26] Tiejun Li, Qihong Lou, Jingxing Dong, Yunrong Wei and Jingru Liu, *Appl. Surf. Sci.*, **2001**, *172*, 356.

[27] Singh Rajiv, K. and Narayan, J., *Phys. Rev. B*, **1990**, *41*, 8843.

[28] Fahler, S., Sturm, K. and Krebs, H-U., *Appl. Phys. Lett.*, **1999**, *75*, 3766 and references therein.

[29] Thestrup, B., Toftmann, B., Schou, J., Doggett, B. and Lunney, J.G., *Appl. Surf. Sci.*, **2002**, *197-198*, 175 and Sasaki, K., Matsui, S., Ito, H. and Kadota, K., *J. Appl. Phys.*, **2002**, *92*, 6471.

[30] Elam, J.W. and Levy, D.H., *J. Appl. Phys.*, **1997**, *81*, 539, and Irissou, E., Le Drogoff, B., Chaker, M. and Guaya, D., *Appl. Phys. Lett.*, **2002**, *80*, 1716.

[31] Fahler, S. and Krebs, H-U., *Appl. Surf. Sci.*, **1996**, *96-98*, 61.

[32] Willmott, P.R. and Huber, J.R., *Rev. Mod. Phys.*, **2000**, *72*, 315.

[33] Barnes, J.P., Petford-Long, A.K., Suarez-Garcia, A. and Serna, R., *J. Appl. Phys.*, **2003**, *93*, 6396.

[34] Hidalgo, J.G., Serna, R., Haro-Poniatowsky, E. and Afonso, C.N., *Appl. Phys. A*, **2004**, *79*, 915 and references therein.

[35] Lunney, J., *Appl. Surf. Sci.*, **1995**, *86*, 79.

[36] Sturm, K. and Krebs, H-U., *J. Appl. Phys. Lett.*, **2001**, *90*, 1061.

[37] van de Riet, E., Kools, J.C.S. and Dieleman, J., *J. Appl. Phys.*, **1993**, *73*, 8290.

[38] Geohegan, D.B., in *Pulsed Laser Deposition of Thin Films*, Chrisey, D.B. and Hubler, G.K. (Eds.), John Wiley and Sons Inc., New York, **1994**, 115.

[39] Gupta, A., in *Pulsed Laser Deposition of Thin Films*, Chrisey, D.B. and Hubler, G.K. (Eds.), John Wiley and Sons Inc., New York, **1994**, 266.

[40] Bi, H.J., Cai, W.P., Kan, C.X., Zhang, L.D., Martin, D. and Trager, F., *J. Appl. Phys.*, **2002**, *92*, 7491.

[41] Serna, R., Babonneau, D., Suárez-García, A., Afonso, C.N., Fonda, E., Traverse, A., Naudon, A. and Hole, D.E., *Phys. Rev. B*, **2002**, *66*, 205402.

[42] Kreibig, U. and Vollmer, M. (Eds.), *Optical Properties of Metal Custers*. Springer Series in materials Science 25, Springer, Berlin, **1995**.

[43] Stegeman, G.I., Wright, E.M., Finlaysoin, N., Zanoni, R. and Seaton, C.T., IEEE *J. Lightwave Tech.* **1988**, *6*, 953.

[44] Haglund, R.F. Jr., in *Handbook of Optical Properties*, Vol. II, Hummel, R.E. and Wismann, P. Eds., CRC Press, Boca Raton, **1997**, 191.

[45] Hirao, K., Mitsuyu, T., Si, J. and Qiu, J. (Eds.), *Active Glass for Photonic Devices*, Springer-Verlag Berlin Heidelberg, **2001**, 208.

[46] Serna, R., Gonzalo, J., Afonso, C.N. and de Sande, J.C.G., *Appl. Phys. B*, **2001**, *73*, 339.

[47] del Coso, R. and Solis, J., *J. Opt. Soc. Am. B*, **2004**, *21*, 640.

[48] Choy, T.C., *Effective Medium Theory. Principles and Applications.* Oxford University Press, Oxford, **1999**.

[49] Dalacu, D. and Martinu, L., *J. Appl. Phys.*, **2000**, *87*, 228.

[50] Gonzalo, J., Serna, R., Solís, J., Babonneau, D. and Afonso, C.N., *J. Phys.: Cond. Matter*, **2003**, *15*, S3001.

[51] del Coso, R., Requejo-Isidro, J., Solis, J., Gonzalo, J. and Afonso, C.N., *J. Appl. Phys.*, **2004**, *95*, 2755.

[52] Gonzalo, J., Babonneau, D., Afonso, C.N. and Barnes, J.P., *J. Appl. Phys.*, **2004**, *96*, 5163.

[53] Gaudry, M., Lermé, J., Cottancin, E., Pellarin, M., Vialle, J-L., Broyer, M., Prével, B., Treilleux, M. and Mélinon, P., *Phys. Rev. B*, **2001**, *64*, 085407.

[54] Suárez-García, A., del Coso, R., Serna, R., Solís, J. and Afonso, C.N., *Appl. Phys. Lett.*, **2003**, *83*, 1842.

[55] Vogel, E.M., Weber, M.J. and Krol, D.M., *Phys. Chem. Glasses.*, **1991**, *32*, 231.

[56] Haglund, R.F. Jr., *Mat. Sci. Eng. A*, **1998**, *253*, 275.

[57] Cotter, D., Manning, R.J., Blow, K.J., Ellis, A.D., Kelly, A.E., Nesset, D., Phillips, I.D., Poustie, A.J. and Rogers, D.C., *Science*, **1999**, *286*, 1523.

[58] Hache, F., Ricard, D., Flytzanis, C. and Kreibig, U., *Appl. Phys. A*, **1988**, *47*, 347.

[59] Liao, H.B., Xiao, R.F., Wang, H., Wong, K.S. and Wong, G.K.L., *Appl. Phys. Lett.*, **1998**, *72*, 1817.

[60] Hamnaka, Y., Fukuta, K., Nakamura, A., Liz-Marzan, L.M. and Mulvaney, P., *Appl. Phys. Lett.*, **2004**, *84*, 4938.

[61] Sipe, E. and Boyd, R.W., *Phys. Rev. A*, **1992**, *46*, 1614.

[62] Stroud, D. and Hui, P.M., *Phys. Rev. B*, **1988**, *37*, 8719.

[63] Sarychev, A.K. and Shalaev, V.M., *Phys. Rep.*, **2000**, *335*, 275.

[64] Shalaev, V.M. and Sarychev, A.K., *Phys. Rev. B*, **1998**, *57*, 13265.

[65] Messinger, B.J., von Raben, K.U., Chang, R.K. and Barber, P.W., *Phys. Rev. B*, **1981**, *24*, 649.

[66] Xu, H., Aizpurua, J., Käll, M. and Apell, P., *Phys. Rev. E*, **2000**, *62*, 4318.

[67] Barber, P.W., Chang, R.K. and Massoudi, H., *Phys. Rev. B*, **1983**, *27*, 7251.

[68] Tartaj, P., Morales, M.P., Veintemillas-Verdaguer, S., González-Carreño, T. and Serna, C.J., *J. of Phys. D: Applied Physics*, **2003**, *36*, R182.

[69] Häfeli, U., Schütt, W., Teller, J., Zborowski, M. (Eds.), *Scientific and Clinical Applications of Magnetic Carriers*, Plenum Press, New York, **1997**.

[70]   Merbach, A.E. and Tóth, E., *The Chemistry of Contrast Agents in Medical Magnetic Resonance Imaging*, John Wiley and Sons, Chichester, **2001**.

[71] Chien, C.L., *J. Appl. Phys.*, **1991**, *69*, 5267.

[72] Bian, B. and Hirotsu, Y., *Jpn. J. Appl. Phys.*, **1997**, *36*, L1232.

[73] Kumar, D., Yarmolenko, S., Sankar, J., Narayan, J., Zhou, H. and Tiwari, A., *Composites: Part B,* **2004**, *35*, 149.

[74] Kumar, D., Pennycook, S.J., Lupini, A., Duscher, G., Tiwari, A. and Narayan, J., *Appl. Phys. Lett.*, **2002**, *81*, 4204.

[75] Dureuil, V., Ricolleau, C., Gandais, M., Grigis, C., Lacharme, J.P. and Naudon, A., *J. of Cryst. Growth*, **2001**, *233*, 737.

[76] Skumryev, V., Stoyanov, S., Zhang, Y., Hadjipanayis, G., Givord, D. and Nogués, J., *Nature*, **2003**, *423*, 850.

[77] Kodama, R.H. and Edelstein, A.S., *J. Appl. Phys.*, **1999**, *85*, 4316.

[78] Dobrynin, A.N., Ievlev, D.N., Temst, K., Lievens, P., Margueritat, J., Gonzalo, J., Afonso, C.N., Zhou, S.Q., Vantomme, A., Piscopiello, E. and Van Tendeloo, G., *Appl. Phys. Lett.,* **2005**, *87*, 012501.

[79] Kofman, R., Cheyssac, P., Aouaj, A., Lereah, Y., Deutscher, G., Ben-David, T., Penisson, J.M. and Bourret, A., *Surf. Sci.*, **1994**, *303*, 231.

[80] Henrik Andersen, H. and Jonson, E., *Nucl. Instrum. Method, B*, **1995**, *106*, 480.

[81] MacDonald, K.F., Fedotov, V.A., Zheludev, N.I., Zhdanov, V. and Knize, V., *Appl. Phys. Lett.*, **2001**, *79*, 2375.

[82] Stella, A., Migliori, V., Cheyssac, P. and Kofman, R., *Europhys. Lett.*, **1994**, *26*, 256.

[83] Garrigos, R., Kofman, R., Cheyssac, P. and Perrin, M.Y., *Europhys. Lett.*, **1986**, *1*, 355.

[84] Haro-Poniatowski, E., Serna, R., Afonso, C.N., Jouanne, M., Morhange, J.F., Bosch, P. and Lara, V.H., *Thin Solid Films*, **2004**, *453-54*, 467.

[85] Haro-Poniatowski, E. Jouanne, M., Morhange, J.F., Kanehisa, M., Serna, R. and Afonso, C.N., *Phys. Rev. B*, **1999**, *60*, 10080.

[86] Cheyssac, P., Kofman, R. and Garrigos, R., *Phys. Scr.*, **1988**, *38*, 164.

[87] Bosio, L., Defrain, A. and Epelboin, I., *J. Phys.*, **1966**, *27*, 61.

[88] Eppler, A.S., Rupprechter, G., Guczi, L. and Somorjai, G.A., *J. Phys. Chem. B*, **1997**, *101*, 9973.

[89] Warrender, J.M. and Aziz, M.J., *Appl. Phys. A*, **2004**, *79*, 713.

[90] Dolbec, R., Irissou, E., Chaker, M., Guay, D., Rosei, F. and El Khakani, M.A., *Phys. Rev. B*, **2004**, *70*, 201406.

[91] Guczi, L., Horvath, D., Paszti, Z., Toth, L., Horvath, Z.E., Karacs, A. and Peto, G., *J. Phys. Chem. B*, **2000**, *104*, 3183.

Chapter 3

Carbon-Based Materials by Pulsed Laser Deposition: From Thin Films to Nanostructures

T. Szörényi

Research Group on Laser Physics of the Hungarian Academy of Sciences, University of Szeged, P.O. Box 406, H 6701 Szeged, Hungary

Abstract

The peculiarities of pulsed laser deposition in determining the growth and properties of carbon-based materials is highlighted using film-based approaches rather than spectroscopy measurements. The benefits and limitations of ablation in different ambients and pressure domains, and the contribution of the target material, the relative position and distance of the target and the substrate, the laser parameters, including pulse duration and energy, spot size, energy distribution over the ablated area, effects of pulse trains and different laser sources in tailoring the properties of the pure carbon forms are analyzed. The general trend of changes in film microstructure with increasing pressure is exemplified by the case of carbon nitride. The exhaustive list of the most relevant papers cited should serve as a solid base for forthcoming studies.

Keywords: Carbon nitride; Chemical structure; DLC; Excimer lasers, Microstructures; Nanophase diamond; Nanotubes.

3.1. Introduction

Carbon is the sixth most abundant element in the universe and the only one building molecules necessary to sustain life. Its four valence electrons allow carbon to form a rich variety of compounds of very different properties. This gives the flavor of

Recent Advances in Laser Processing of Materials
J. Perrière, E. Millon and E. Fogarassy (Editors)

carbon research: though carbon has been known since the use of fire, we are still discovering new facets of this element. The pure carbon forms, diamond, graphite, amorphous carbon, diamond-like carbon (DLC), fullerenes and nanotubes differ in bond configuration and bond hybridization. Differences in their chemical structures materialize in the diversity of physical and chemical properties. This diversity in properties broadens even further when considering the millions of organic compounds and carbon-based materials in general.

Over the last decades, different forms of carbon, including inorganic carbon compounds, emerged as a new class of materials possessing remarkable properties and, accordingly, a wide range of potential technological applications. Pulsed Laser Deposition (PLD) proved to be one of the most successful preparation techniques, being able to fabricate all members of this class in thin film form.

Writing a thorough review on recent achievements of carbon research in general would clearly be an impossible task. Restricting our interest to scientific papers on pulsed laser deposition published in archival journals available in the database of ISI Web of Science as of March 2005 resulted in manageable numbers after all. Figure 1 shows the time evolution of the cumulated number of papers published on different forms of carbon and their compounds, prepared by pulsed laser deposition, including reactive PLD and combination with other techniques, since 1990.

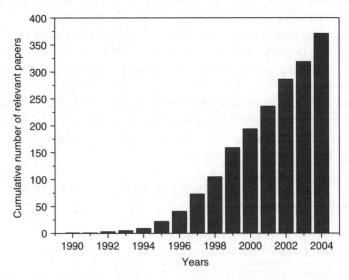

Figure 1: Cumulated number of papers published on PLD of carbon and carbon compounds between 1990 and 2004.

The bar chart, representing the initial part of a logistic function, convincingly illustrates that this field of research is still in its growing phase. On the one hand, the absolute number of papers prove that there are already enough information available, and on the other hand, the scientific interest has not declined yet. This warrants the timeliness of a critical review on PLD of carbonaceous materials.

The variety of properties of carbon allotropes and compounds, the diversity of their potential uses, together with the impressive arsenal of deposition techniques that have successfully been applied for fabrication of the different forms in general, including pulsed laser deposition in particular, have been discussed in a number of reviews [1–5]. However, the concept of this chapter differs from all earlier approaches. Our aim is to highlight the peculiarities of one technique, namely PLD in determining the growth and properties of carbon-based materials of a wide range of forms. We will discuss the benefits and limitations of ablation in different ambients and pressure domains, and the contribution of the laser parameters in tailoring the properties of the material produced.

Due to the multitude of process parameters, notably the nature and pressure of the atmosphere, the target material, the relative position and distance of the target and the substrate, the laser parameters (pulse duration and energy, spot size, energy distribution over the ablated area, effect of pulse trains, combination of laser sources), the existence of interplay between some of them, possible effects of special geometries, assistance of atom- or ion beams, etc., presentation of the subject in the most logical way with minimal overlap has not been an easy task at all. The problem is well illustrated with the case of the atmosphere: some effects of the nature and pressure of the ambient are independent of the material to be deposited, while others are material specific. These practical constrains are best fulfilled by presenting the results according to the effects of each principal process parameter for each materials group, separately.

According to this logic, a brief summary of the peculiarities of pulsed laser deposition (Section 3.2) will be followed by discussing the effect of the nature and the pressure of the ambient on the growth and properties of the pure carbon forms in Section 3.3. Finally, the general trend of changes in film microstructure with increasing pressure will be exemplified by the case of carbon nitride. Due to the lack of convincing evidence of the existence of well established correlation between the results derived from film-based and spectroscopic studies, the following discussion will focus on the 'ultimate goal', i.e. the grown layer, and will summarize literature results on film-based studies rather than spectroscopy measurements. As far as the cited works are concerned, choosing references is always somewhat arbitrary. Nevertheless I tried to collect an exhaustive list in which the most relevant papers are compiled, and which, therefore could serve as a solid base for forthcoming studies.

3.2. The Peculiarities of Pulsed Laser Deposition

Pulsed laser deposition (PLD) is an extremely versatile laboratory technique for the synthesis of material prototypes in thin film form. The concept is simple (Fig. 2): a target placed in a vacuum chamber is ablated by a train of high energy/power laser pulses focused onto the target surface. The ablated species form a plasma plume moving away from the target while expanding in three dimensions. The expansion velocities in the three directions are governed by the initial dimensions of the plasma at its birth, i.e. when it is still confined to the vicinity of the ablated area [6 and references therein]. The velocities are the highest in the direction of the smallest dimensions, and vice versa. In the usual case of not tightly focused beams, typically when ablating with excimer lasers, the lateral dimensions of the plasma cloud in the initial stage of expansion are much larger than its spread along the normal of the target (surface). Consequently in most cases the forward velocity is the highest, materializing in a strongly forward peaked material flux, cited frequently as one of the characteristics of PLD. The ablated species are collected on a substrate where they condense and build a layer.

Pumping the energy of the subsequent laser pulses (typically tens to hundreds of mJ/pulse) into very small volumes (of the order of $10^{-4}-10^{-7}$ mm^3) within nanoseconds or even shorter time results in congruent material removal irrespective of the composition of the target. The initial kinetic energy of the plasma species may range from tenths of eV to even hundreds of keV, depending essentially on the laser parameters. Landing of such energetic species on the substrate surface may

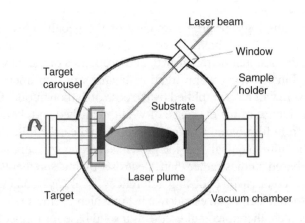

Figure 2: A schematic outline of PLD.

initiate crystal growth at temperatures much lower than attainable with competing techniques. Vacuum processing is not a prerequisite. Inert gases act as moderators: via the pressure of the atmosphere and the target-to-substrate distance the kinetic energy of the species at landing can be controlled. In reactive atmospheres, besides energy dissipation, collision of the ablated species with the ambient molecules may result in chemical interaction, as well. Freedom in choosing the material of the substrate to be coated together with heating or cooling broadens further the potential for controlling film growth.

All these characteristics render PLD an ideal tool for tailoring the chemical composition and/or chemical structure together with the micro- and nanostructure and thereby the physico-chemical properties of thin films of thicknesses ranging from monolayers up to several micrometers. Moderate acquisition and running costs, as compared to MBE or MOCVD, and high throughput are further benefits of the PLD technique. During recent years, PLD matured to a well-established laboratory technique of modern materials research and is on the way of becoming an economically viable alternative for the production of special classes of materials, e.g. oxides in general and next generation oxide semiconductors, in particular [7–10].

In production of carbon-based materials, in particular the allotropic forms of pure carbon, the kinetic energy of the plasma species plays a decisive role in determining the sp^3/sp^2 ratio and thereby the chemical structure of the carbon form to be synthesized. In PLD, both tuning the laser parameters and adjustment of the ambient pressure offer convenient, yet efficient control of the kinetic energy without interplay. The possibility of ablation in atmospheres promotes the formation of a particular form at the expense of another and gives a further degree of freedom in tailoring the properties of the product.

3.3. Pulsed Laser Deposition of Pure Carbon Forms

The term diamond-like carbon (DLC), used in the broadest sense interchangeably with amorphous hydrogenated carbon (a-C:H), denotes an amorphous carbon network with sp^3, sp^2 and even sp^1 hybridized chemical bonds with the possible presence of hydrogen. The structure and thereby the properties of these films depend essentially on the fraction of tetrahedrally coordinated (sp^3-hybridized) and three-fold coordinated (sp^2-hybridized) carbon atoms and the hydrogen content. More precisely, the key parameters are i) the sp^3 content, ii) the degree of clustering of the sp^2 phase, iii) the orientation of the sp^2 phase, iv) the nanostructure and v) the H content [4,11]. The great versatility of carbon materials validates labeling the different compositions on the Gibbs triangle of ternary phase diagram of the graphite–diamond–hydrogen system, as shown in Fig. 3.

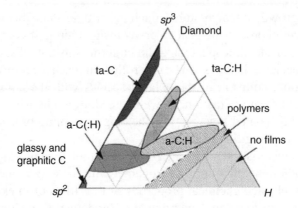

Figure 3: Ternary phase diagram of carbon and hydrogen, emphasizing the domains of various amorphous phases [4,11].

In a strict sense, DLC is an amorphous carbon network with a significant fraction of sp^3 bonds and low hydrogen content, in contrast to a-C:H, that consists mostly of sp^2 bonded C atoms. In the following, in harmony with the less strict terminology used in the laser community, all non-crystalline carbon forms will be denoted as DLC. Tetrahedral amorphous carbon (ta-C) is the form of DLC with the highest sp^3 (80–90%) and very low hydrogen content; its hydrogenated analogue is ta-C:H. When moving from graphite to diamond, along the diamond–graphite composition scale of the ternary phase diagram, we first reach the amorphous carbon domain and then the sp^3 dominated ta-C phase. While moving in this direction first the sp^2 groups become smaller, then topologically disordered, and finally change from ring to chain configurations [11].

3.3.1. PLD of DLC in Vacuum

In principle, ablation of a pure carbon target, in most cases graphite, in high vacuum results in practically hydrogen-free diamond-like carbon films with properties varying from a-C to 'nanophase diamond' [1,12, and references therein]. When performing PLD in high vacuum, typically at pressures between 10^{-2} and 10^{-6} Pa, the film properties are determined solely by the laser parameters. The key parameter determining the sp^2/sp^3 bond ratio is the kinetic energy of the ablated species, that can be controlled by the laser wavelength, pulse duration, and the energy/power density on the target surface [1,13–17]. The influence of the pulse repetition rate on the characteristics of the films produced has also been noted recently [18–20].

In the following, the effect of each laser parameter on film properties will be discussed separately. To facilitate comparison between the results quoted, the ablation parameters will be given in terms of both energy- and power density in all cases whenever the data available allow conversion.

3.3.1.1. The Effects of Laser Wavelength

There is a consensus in the literature that the size of ionic carbon clusters, C_n^+ produced by ablating a carbon target in vacuum decreases with decreasing laser wavelength. Nd:YAG laser pulses of 1064 nm produce C_n^+ ions with $5 < n < 15$. Clusters with $n = 3, 5, 7$ and $n = 1, 3$ dominate the spectra recorded at 532 and 248 nm, respectively [14]. Ablation with an ArF excimer laser ($\lambda = 193$ nm), however, produces mainly C atoms and C^+ ions of high kinetic energy [21,22].

In terms of film properties, arrival of energetic carbon ions, atoms or small clusters at a low temperature substrate initiates the formation of films with high sp^3 fraction while landing of larger clusters favors the stabilization of the amorphous carbon network via mostly sp^2 bonded C atoms [23]. Correspondingly, the sp^3 content of the films increases with decreasing laser wavelength. Variation in the sp^3 content of DLC films as a function of the wavelength is shown in Fig. 4 for nanosecond (ns) pulse duration. Since variation of the energy/power density at any fixed wavelength, as expounded in the next Section (3.3.1.2.), causes significant changes in the sp^3 content, only data points measured on films deposited within the ~3–11 Jcm^{-2} fluence window have been included in Fig. 4. Even with this

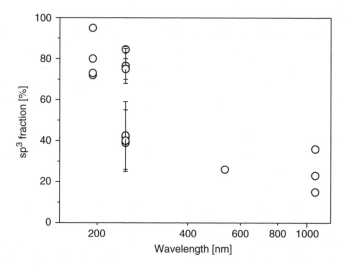

Figure 4: sp^3 content of DLC films as a function of wavelength [1,16,23–28].

limitation, the range of sp^3/sp^2 ratios attainable at a given wavelength is rather broad. Nevertheless, the trend is clear: the sp^3 content decreases with increasing wavelength.

The dominance of neutral and ionized monoatomic C species in the plasma produced by UV lasers renders the determination of the kinetic energy of the ions from time of flight measurements straightforward. By following changes in the bonding and optical properties of the growing films as a function of the kinetic energy of the ablated carbon ions within the 0–225 eV range, Geohegan and coworkers determined that the most diamond-like character (i.e. films of maximum sp^3 fraction, highest density and band gap values) was obtained at approx. 90 eV C^+ ion kinetic energy when ablating at 193 nm wavelength [24]. This result defines the process window for fabricating tetrahedral amorphous carbon, ta-C films with UV lasers of nanosecond pulse duration.

3.3.1.2. The Effects of Energy/Power Density

Since the kinetic energy of the ablated carbon species depends not only on the laser wavelength but also on the laser fluence, it is evident that the diamond-like character of the films can also be controlled by changing the energy/power density of the laser pulses impinging on the target surface. The linear dependence of ion energy on both the laser energy density (at constant spot diameter) and spot size (at constant energy density) offers a convenient way for tuning film properties [13,14,16]. In harmony with the wavelength dependence, detailed in the previous section (3.3.1.1.), the minimum power/energy density needed to form DLC in general and highly diamond-like ta-C films in particular increases with increasing laser wavelength [1].

Threshold intensities in the order of 10^{10} Wcm^{-2} (~10^2 Jcm^{-2}) on the target surface are necessary when ablating graphite with Nd:YAG lasers emitting at the fundamental frequency in the IR [25,29]. In the experiments performed at University of Texas at Dallas in the early nineties, intensities approaching 5×10^{11} Wcm^{-2} (10^4 Jcm^{-2}(!)) have been used [15,30]. Such extreme intensities produce ions up to keV energies which grow films of unique properties, namely chemically bonded to a wide variety of substrates via interfacial layers. The latter are formed when highly energetic ions actually penetrate into the surface layer of the substrate. The hydrogen-free material produced consists of densely packed nanocrystalline diamond-like nodules of approx. 100 nm diameter, interlinked by mainly non-sp^3 bonds, and therefore was termed as nanophase diamond [31]. The unique combination of hardness (in excess of 78 GPa, in the best case), apparently greater than that of natural diamond, low coefficient of friction around 0.1 and firm bonding to substrates render nanophase diamond films superior to natural diamond in some mechanical applications [15,30,31].

These results reveal that well above the upper limit of the kinetic energy range investigated in [24], i.e. 225 eV, there exists at least one more energy domain in which a material of perfect diamond character/structure can be produced. It is therefore sensible to mention that the authors of [24] also noted, that the 90 eV C^+ ion kinetic energy represented (only) a local maximum in the diamond-like properties, and was expected to increase above ≈ 200 eV.

In the experiments performed at 532 nm power densities between 6×10^9 and 10×10^9 Wcm^{-2} have been used [1]. Entering the UV domain, the lower limit of the power densities producing DLC shifts down to $1-3 \times 10^8$ Wcm^{-2} for 308 nm [1]. When ablating with KrF pulses, power densities of $\sim 10^8$ Wcm^{-2} (two or three Jcm^{-2}) are sufficient to produce high quality hydrogen-free DLC films [1,12,16,32].

From the fact that the lower limit of the power/energy density range examined is the same for both KrF and ArF lasers (e.g. Fig. 8 and references in [1] and Fig. 2 in [33]), one might conclude that the step from 248 down to 193 nm does not result in appreciable difference in the properties of the films fabricated at comparable power/energy densities. Nevertheless, the comparison of the C^+ ion energies produced by both lasers convincingly demonstrates that the ArF laser has a clear advantage in producing considerably higher kinetic energies than the KrF at any given power/energy density [33]. Therefore the correct conclusion is that films possessing very similar properties can be fabricated at both wavelengths, indeed; at 193 nm, however, at somewhat lower fluences. Since intensities of the order of 10^8 Wcm^{-2} can easily be produced by standard excimer lasers, fabrication of practically hydrogen-free DLC coatings with mainly sp^3 bonds characterized by high density ($2-3$ gcm^{-3}), hardness ($60-70$ GPa) and elastic modulus (600 GPa), together with low coefficient of friction (< 0.1) became a routine.

Parallel experiments with a Lambda Physik Compex 301i and a Questek 2960 ArF excimer laser revealed that differences in the temporal and spatial profiles of pulses of identical energy density (8 Jcm^{-2}, in terms of power density: 3.1×10^8 Wcm^{-2} and 4.2×10^8 Wcm^{-2}, respectively, in the two cases) might result in strikingly different C^+ kinetic energies and consequently different film properties [33]. This result of the ORNL group emphasizes the necessity of precise definition of all process parameters and extreme precaution in comparing results.

The broadness of the dynamic range of energy/power density control over sp^3 bond fraction is well documented for the case of ablation with KrF laser. Yamamoto *et al.* reported an increase in sp^3 fraction from ~ 26 to 59% when raising the fluence from 0.2 to 7.9 Jcm^{-2} (from $\sim 10^{-2}$ to 0.4 $GWcm^{-2}$) [26]. Approaching from the high intensity end, Minami *et al.* revealed, by means of evaluating EELS spectra, an increase in the sp^3 fraction of their DLC films from 27 ± 15 to $40 \pm 15\%$ when decreasing the intensity (fluence) of the ablating pulses from 10 $GWcm^{-2}$ (~ 150 Jcm^{-2}) to 0.7 $GWcm^{-2}$ (~ 10.5 Jcm^{-2}) [27]. Comparison of these results suggests that maximum sp^3 content can be attained at relatively low fluences, i.e. within the 5–10 Jcm^{-2} window. On the other hand, Bonelli and coworkers concluded from Raman and FTIR spectroscopic, EELS and specular X-ray reflectivity measurements, as well as from changes in the apparent growth rate as a function of fluence, which is actually an appealingly simple approach, that films, produced by ablation of HOPG with 0.5–37 Jcm^{-2} pulses of a KrF laser in moderate vacuum (10^{-2} Pa), underwent a sharp transition from mainly disordered graphitic to ta-C at around 5–7 Jcm^{-2} (0.25–0.35 $GWcm^{-2}$) [12,28]. The sp^3 content of the films deposited at 0.5 and 20 Jcm^{-2} (25 $MWcm^{-2}$ and 1 $GWcm^{-2}$) has directly been measured by EELS to be 42 and 80%, respectively. Based on the comparison of the Raman spectra, the authors estimate that all graphitic and ta-C films contain ~40 and 70–80% sp^3 fraction, respectively.

No doubt, due to uncertainties in measuring pulse energy, spot dimensions and defining pulse width (pulse duration versus FWHM) and possible effects of the other process parameters, perfect agreement between results coming from different laboratories can hardly be expected and disagreement between fluence values of different groups is acceptable. What complicates the picture here is the stark contrast between the step-like [12,28] and the gradual [26] change. Moreover, literature data can also be used to demonstrate that identical sp^3 content cannot guarantee identical film properties either [28] versus [4,11,32].

3.3.1.3. The Effects of Pulse Duration

With sub-ps lasers we enter a domain of orders of magnitude higher intensities. Commercial Ti:sapphire systems routinely produce 10^{14}–10^{15} Wcm^{-2}. Ablation with pulses of such high intensity opens another window for studying the effect of carbon ion kinetic energy on the properties of the films produced. When recalling that in the case of deposition with nanosecond pulses, relatively high kinetic energies proved to be favorable to produce high sp^3 fractions (cf. the preceding sections), and combine it with the fact that ablation with ultrashort pulses produces highly ionized plasmas [34], one naturally expects that DLC films deposited by femtosecond (fs) lasers should be superior to that grown with nanosecond ones. However, recognizing the non-monotonous nature of the sp^3/sp^2 versus power density function in the nanosecond case, it is not so much shocking that – though the number of reports on DLC preparation using ultrashort-pulse lasers is far less yet and therefore, coming to any final conclusion seems to be too early – the results presented below suggest that the orders of magnitude higher kinetic energies do not necessarily yield films with a perfect diamond character.

In a comprehensive joint paper, the groups of R.K. Singh and P.P. Pronko correlated the kinetic energy of the plasma species with the surface topography, optical constants, bonding structure and sp^3 content of the DLC films formed, as a function of power density of 100 fs Ti:sapphire laser pulses [35]. An increase in the power density from 3×10^{14} to 6×10^{15} Wcm^{-2} resulted in i) an increase in the number density of micron-sized particulates, covering the entire surface area at high intensities, ii) a decrease in optical transparency and the Tauc band gap from 1.5 to 0.8 eV and iii) a decrease in the sp^3 fraction from ~60 to 50%. Parallel analysis of the ion TOF spectra revealed the existence of two ionic components of the expanding carbon plasma: the kinetic energy of the faster 'suprathermal' ions was measured to increase from 3 to over 20 keV following $I^{0.33}$ dependence, while that of the slower one increased from ~60 eV to 2 keV according to $I^{0.55}$ [35]. Weakening of the diamond-like characteristics of the films with increasing power density suggests that the highly energetic ions in the femtosecond plasma may have a detrimental impact on DLC film formation.

The relatively low, 41% sp^3 fraction of the films deposited by ablating frozen acetone with tightly focused 10 mJ, 130 fs pulses reaching 4×10^{15} Wcm^{-2} power density on the target surface leads to a similar conclusion [36]. The results of Garrelie *et al.* are also in line with this notion [37]. The experimental conditions here were quite similar: their mode-locked Ti:sapphire laser emitted 1.5 mJ pulses of 150 fs duration at a repetition rate of 1 kHz. The power density range stretched from 6.7×10^{12} to 4×10^{13} Wcm^{-2} (1–6 Jcm^{-2}). They also report the existence of micron-sized particulates covering the entire surface area, while noting that their average size and number were lower than those obtained when using nanosecond lasers. The sp^3 content of the films deposited at 1.35, 2.82 and 5.18 Jcm^{-2} (9×10^{12}, 1.9×10^{13} and 3.5×10^{13} Wcm^{-2}) were 71, 73 and 70%, respectively, i.e. identical within experimental error as determined by XANES. Interestingly, the authors obtained a surprisingly good correlation between the sp^3 contents and the hardness and Young's modulus of the films. These data suggest that deposition with pulses of power densities within the 10^{12}–10^{13} Wcm^{-2} range could be more advantageous to produce DLC films in terms of both sp^3 content and mechanical properties, indeed. Nevertheless, this trend is not without an exception either. The 40–50% sp^3 content reported by Banks and coworkers for thick DLC films grown within the 2.5×10^{12}–2.5×10^{14} Wcm^{-2} window [38] is apparently not in harmony with the above statement.

Due to the somewhat longer pulse length and, more importantly, larger spot size applied, Köster and Mann mapped an even lower power density domain, ranging from ~5×10^{10} to ~2×10^{13} Wcm^{-2} (~2.5×10^{-2} to 10 Jcm^{-2}) [14]. From the analysis of time-of-flight measurements, these authors concluded that 500 fs pulses at 248 nm resulted in kinetic energies increasing linearly with increasing power density up to 10 keV at about 9 Jcm^{-2}, as compared to the 50 eV at similar energy densities in the nanosecond regime. Unfortunately, results of relevant film-based studies have not been reported. Since the 10^{10}–10^{11} Wcm^{-2} range is attainable with traditional

KrF lasers, as well, comparison of the properties of films deposited at the same wavelength, 248 nm and the same (or comparable) power densities could be a challenging yet feasible approach to examine similarities and differences of nanosecond versus femtosecond PLD of DLC.

Nevertheless, from the data available up to now one may conclude that ablation with femtosecond pulses apparently does not lead to significant improvement in film characteristics at all. Therefore, while from the basic science point of view, application of femtosecond lasers for DLC growth may remain a challenge, from the point of view of practical applications, the future of ultrashort pulse PLD does not look appealing at all [39]. In any case, instead of striving to reach maximum possible power densities by tight focusing of tens of mJ pulses [35,36], ablation with pulses of lower power densities conveniently attainable with commercial femtosecond lasers seems to be a more practical approach.

The overview presented above is certainly not complete. The results cited, however, unambiguously demonstrate that in high vacuum, the properties of films are solely determined by the laser parameters. By varying the principal parameters, i.e. the laser wavelength, the pulse duration, and the energy/power density on the target surface, the properties of the films can be tuned from soft, graphitic a-C to ta-C with properties approaching that of diamond. Therefore, instead of quoting further publications describing particular cases, we refer to an excellent recent analysis of both theoretical and experimental aspects of the effect of laser intensity, wavelength and pulse duration in determining optimal scenarios for ablation of graphite with Nd:YAG, KrF and ArF lasers [40]. Finally, PLD with high-repetition-rate lasers is mentioned, as a special case, which deserves particular attention from the viewpoint of both basic science and practical applications. This approach should seriously be considered in the future because of the superior quality of the films produced [19,20].

3.3.2. Ablation in Inert Gas Atmospheres: He, Ar, Xe

The presence of ambient gas has a profound effect on the nature and energy of the plasma species and thereby on the properties of the films formed. The study of the effect of background gas is important not only to get a better understanding of the mechanisms involved but also to be able to control and optimize the PLD process.

In principle, in inert atmospheres, the trajectory and the absolute value of the velocity (the kinetic energy) of the plasma species are controlled by simple collision kinetics. While the increase in pulse energy/power or spot dimensions produce more energetic plasma components, the effect of increasing ambient pressure is just the opposite. With increasing number of collisions with the background gas, the ablated species continuously lose kinetic energy until complete thermalization. The main consequence of this is a concomitant change in the microstructure of the deposits. As it will be shown below, while following this trend in general, ablation of carbon in inert atmospheres shows again peculiarities: the apparently simple mechanism results in a wide diversity of material properties and forms depending on the initial velocity distribution of the carbon species and the characteristics of the actual gas ambient. A few examples are given below.

The most comprehensive description on the evolution of the microstructure and morphology of carbon deposits with helium pressure increasing from tenths of pascals to kilopascals has been published recently by Ossi *et al.* [41]. Based on the analysis of SEM pictures of deposits grown by KrF excimer laser ablation of a HOPG target with 20 ns pulses of power densities ranging from 50 to 1900 MWcm^{-2} (1–38 Jcm^{-2}), the authors identify three morphologies: columnar, nodule-like and highly porous dendritic, that evolve in this order with increasing pressure. The films fabricated at the lowest He pressure investigated, 0.6 Pa, are dense, flat and laterally homogeneous. Practically, they are DLC. In the 30–70 Pa partial pressure range spherically capped, partly agglomerated nodules start to grow out of the high density columnar structure. The number density of the nodules increases while their cap radius decreases with rising pressure. Further increase in pressure results in loosening up the structure: at 250 Pa low density columnar growth coexists with the development of micrometer-size, ball-shaped nodules. The deposits become laterally inhomogeneous due to the large differences in the height of adjoining structural elements. Variation in the power density does not have any measurable influence either on the morphology or the poor adhesion at this pressure. At 1 kPa landing of completely thermalized clusters on the substrate surface results in a highly porous and extremely irregular deposit with peak heights of tens of micrometers, while leaving some areas almost uncovered. The adhesion, if it is appropriate to call it that way, between the substrate and the aggregate of clusters formed in the domain of kilopascals is very poor, if any, and such terms as film thickness (and even film) have no real meaning any more.

Though using different terminology, Thareja and coworkers describe a very similar sequence of structural changes with pressure [42]. Their paper is one of the very few ones correlating film characteristics with plume properties. SEM images of the films deposited in He ambient pressures between 1.33 and 1.33 × 10^4 Pa using 4.68 Jcm^{-2} (5.9 × 10^8 Wcm^{-2}) pulses at 532 nm revealed gradual increase in the number density of spherical features with increasing He pressure. Based on parallel analysis of UV–VIS–IR spectra of the carbon soot collected from the deposition chamber at various He pressures, the authors identify these features, termed as nodules in [41], as spherical aggregates of carbon clusters, dominantly C_{60} and C_{70}.

Carbon clusters, like C_{60} and higher fullerenes, are well known to be formed as a product of laser ablation of carbon in helium ambient anyway [43]. The authors explain the difference in the effect of He versus Ar atmospheres by correlating the evolution of film properties with the changes in the vibrational temperature of C_2 species in both atmospheres as a function of pressure. Above ≈100 Pa, the vibrational temperature becomes larger in Ar ambient because the heat removal from the confined plasma by the heavier Ar atoms is less effective. The He atmosphere acts as a heat sink. Increasing pressure results in more and more effective cooling and thereby clustering of the species, promoting the formation of larger carbon clusters, i.e. fullerenes [42,44].

The papers focusing on narrower domains of the pressure range serve with further details, sometimes, however, also with results apparently not fitting into the main trend sketched above.

According to Ebihara et al. [45], films deposited in 40 Pa He at 8 Jcm^{-2} consist of well defined, fairly uniform thin square grains of ~80 × 80 nm^2 producing a very smooth surface characterized by a RMS surface roughness of 2 nm. Though Ossi and coworkers do not quantify the surface quality of their films in terms of RMS, from the description of the evolution of the microstructure and in particular from the SEM pictures presented, it follows that the surface of the films grown at around 40 Pa should already be deteriorated by outgrowth of nodules, leading to higher RMS figures. One possible solution to this apparent contradiction could be that, in order to collect and give information on the nanostructure of the surface, the Japanese authors reported best RMS values derived by scanning minute (500 × 500 nm^2) areas free from nodules. If so, their figures cannot be considered as characteristic of the surface morphology in general. The optical gap values, E_{opt} of the films, determined from Tauc plots increase from 0.8 to 1.1 eV with increasing He pressure from 13.3 to 40 Pa. Though the optical band gap of DLC depends not exclusively on the sp^3/sp^2 ratio, and therefore the assumption of a simple linear relationship between E_{opt} and sp^3 content may well be false, the measured increase in E_{opt} suggests slight strengthening of the diamond-like character with increasing He pressure in the investigated range [45].

In Ar atmosphere, the same transition from atomic to cluster growth results in very similar changes in film microstructure. Comparison of the changes in the apparent growth rate, defined as film thickness per number of laser pulses, and carbon deposition rate, defined as the number of carbon atoms deposited onto unit film area per pulse, backed by parallel mapping of the changes in surface morphology with increasing argon pressure by AFM reveals that up to approx. 5 Pa atomic species of gradually decreasing kinetic energy build dense DLC films. At around 5 Pa formation and arrival of carbon clusters start to modify the growth process.

Above ≈5 Pa the pressure controls the actual abundance of fast carbon species and slow clusters, thereby controlling the contribution of atomic versus cluster growth to film formation, and resulting in films of entirely different (surface) structure and properties [46]. In terms used by Ossi and coworkers, this is the domain of nodular growth [41]. In line with the notion that less energetic larger clusters build films with smaller sp^3/sp^2 ratio, the sp^3 content of the films deposited keeping all parameters but the ambient pressure fixed decreases from 0.4 at 0.13 Pa below 0.1 at 13 Pa [47]. Exceeding 100–200 Pa, we just enter a pressure domain where the formation and interaction of clusters and particles, rather than atoms and ions determine both gas phase and condensation processes [46,48]. Within this domain the individual nanoparticles become the species of interest, and with increasing pressure, the standard terminology of film growth comes to be more and more meaningless [49].

Experiments approaching and performed at atmospheric pressures mark the upper end of the pressure domain examined. Since at these pressures, the mean free path of the ablated species reduces to micrometers, the terminology of PLD can hardly be used anymore. As a sign, target-to-substrate distances are no more specified in the reports [50,51]. The principal difference between this type of experiments and nanoparticle production by gas condensation in general is in the way of collecting and characterizing the product. The 'PLD-type' experiments are performed in static atmospheres and the products are collected on the spot. In the majority of gas condensation set-ups, however, nanoparticles are let to condense from the ablation plume in the gas volume around the target, and are carried away by the gas flow. In both cases, the characteristics of the products are determined by the target, the laser parameters, the nature and pressure of the atmosphere and, most importantly, the ambient temperature.

Shadowgraphic and emission imaging of the temporal evolution of carbon plasma generated by ablating a graphite target with 12 Jcm^{-2} pulses of a Nd:YAG laser (1064 nm, 8 ns) in static Ar ambient of 75,500 Pa revealed that the carbon species remained confined in the vicinity of the target within a <1 mm region, indeed. Frequent collisions of the carbon species with energies having dissipated into the Ar ambient within this restricted volume resulted in the formation of clusters and condensational growth of particles [50].

By ablating crystalline graphite and amorphous carbon targets at and near to atmospheric pressure in He, Ar and Xe with ≈10^6 Wcm^{-2} pulses of a rarely used laser source, a free-running Nd:YAG laser delivering millisecond pulse trains of 5–30 J energy Gnedovets and coworkers produced fine amorphous carbon particles with dimensions increasing with increasing pressure and atomic weight of the ambient gases. Increase in pressure resulted in outgrowth of graphite whiskers and spherulites on the surface of the large spherical particles. Unfortunately, the authors do not help the reader with dimensions of the amorphous particles. From the diameters of the whiskers (~100–500 nm) and spherulites (up to ~500 nm), one may guess diameters in excess of micrometers [51].

Similarly to PLD, in nanoparticle generation by laser ablation the increase of pulse energy and/or repetition rate leads to an increase in the aggregate size, primary particle size and particle number concentration. An increase in flow rate results in bigger number concentration of smaller primary particles and aggregates. As an example: ablation of a graphite target with 100–120 mJ pulses of an XeCl excimer laser of about 28 ns pulse duration focused onto ~3 mm² areas (3.3 Jcm⁻²) and running at 2 Hz in a flow system flushed with high purity N_2 at 1 liter per minute flow rate produces aggregates of ~86 nm diameter, built up of ~4.9 nm diameter primary particles [52].

A scarcely mentioned consequence of the loosening of the structure is that the a priori hydrogen-free films change into severely hydrogenated when leaving the deposition apparatus, and their *ex situ* oxygen content will also increase to a remarkable level. The measurement of both the hydrogen and oxygen content of the(se) unintentionally (*ex situ*) hydrogenated/oxygenated films can be used as a sensitive indicator, sign-posting changes in film microstructure [53,54]. H/C ratios at around 0.05 characterize the dense films fabricated at Ar partial pressures below approx. 5 Pa, while the microstructure of the films deposited at 200 Pa materialize in H/C ratios approaching 0.5, sign-posting increasing film porosity [54]. The H/O atomic ratio of ≈ 2 strongly supports the notion that both H and O contamination can be accounted for by *ex situ* water adsorption onto the increased surface area of the more structured films and contribution of other molecules like CO_2 or CO is negligible.

The possibility of easy control of the film microstructure by the ambient pressure could be a convincing argument for the application of PLD in the production of special functional layers. Due to their high adsorptivity, porous carbon films may act as ideal coating on surface acoustic wave and integrated optical chemical sensors for a wide range of analyte gases. Fine tuning of the porosity of carbon films during KrF excimer laser ablation of a graphite target in controlled Ar ambient, at RT is an elegant application of non-reactive PLD as demonstrated recently by Siegal *et al.* [55]. The ability to produce films of densities ranging from less than 0.1 to 2.0 gcm⁻³ with negligible residual stress and excellent reproducibility clearly indicates the potential of the technique.

High-repetition-rate laser ablation creates a quasi-continuous flow of carbon atoms and ions. Interaction of carbon vapor produced by ablating glassy carbon with 120 ns pulses from a Nd:YAG laser running at 10 kHz with Ar atmosphere of $13.3–1.33 \times 10^4$ Pa results in the formation of a network of interconnected carbon clusters of 6 nm average diameter. The foam-like material with a significant fraction of sp^3-bonding, extremely low bulk density (2–10 × 10^{-3} gcm^{-3}) and large specific surface area (300–400 m^{2}g^{-1}) displays semiconductor characteristics: band gap values between 0.1 and 0.5 eV and resistivities in the 10^9–10^{12} Ωcm range, depending on the Ar pressure [18,56]. This diamond-like cluster-assembled low density carbon foam represents another example of the possible nanostructured material forms attainable in the intriguing and only partly explored territory of high pressures.

Note, that the terminology used for the description of different carbon forms and their properties in general is far not as simple and standardized as the common terminology of all laboratories for the description of the changes in the (physical) microstructure. In the majority of the papers, even the term microstructure refers actually to the chemical structure of the material, as determined usually by spectroscopies, e.g. [1]. Nevertheless, independently of the sometimes confusing terminology, the papers of Thareja and Dwivedi for example [42,44,57,58] well exemplify that even drastic changes in the microstructure are not necessarily coupled to parallel changes in the chemical structure of the material, and vice versa.

All results discussed above refer to experiments performed at room temperature substrates. Papers dealing with the effect of substrate temperature on film properties are scarce [59]. One reason for this could be that materials engineers prefer low temperature processing, a priori. The other, more specific to the material is that deposition in vacuum and inert atmospheres at higher temperatures does not lead to improvement in material properties: higher temperatures promote graphitization and clustering at both the macroscopic and the microscopic level. In the majority of applications, this is not a favorable effect.

3.3.3. Ablation in Hydrogen and Oxygen Ambient

Carbon ablation in H$_2$ and O$_2$ ambients represents a particular case. Both atomic hydrogen and oxygen are known to preferentially etch sp^2 carbon, thereby to promote sp^3 stabilization and growth of mostly tetrahedrally coordinated DLC films. Several attempts have been made to exploit this effect for production of ta-C and diamond.

In Fig. 5, the optical gap values, E_{opt} of DLC films deposited in hydrogen and oxygen atmospheres at room temperature are shown as a function of the respective partial pressures. In H$_2$ ambient, the E_{opt} values determined from Tauc plots are constant below ~0.1 Pa. Above this threshold E_{opt} monotonously increases with increasing H$_2$ pressure. The highest E_{opt} value, 2.5 eV was reported for a film deposited in 133 Pa at 8 Jcm^{-2} [45,60]. Since the variation in E_{opt} is considered to be a reliable indicator of the degree of diamond-like character and consistent with

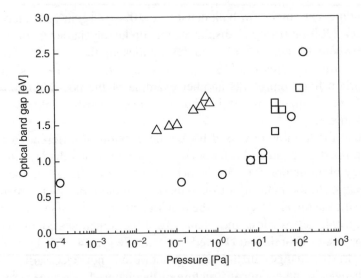

Figure 5: Optical band gap of DLC films grown by ablating graphite targets with a KrF excimer laser in H_2 (□) [45], and with an ArF excimer laser in H_2 (○) and O_2 (Δ) ambients [60], at room temperature.

the variation in sp^3 content, this result means that relatively high H_2 pressures provide optimum condition for depositing DLC films of approx. 70% sp^3 content. AFM images of such films revealed the existence of cubic structures of ~150×150 nm^2 with RMS surface roughness of 13 nm [45]. While the net result of introducing oxygen or hydrogen during film deposition at RT is the same, the mechanisms are rather different. Molecular hydrogen does not react directly with carbon atoms. In PLD collision with the ablated energetic carbon species dissociate hydrogen molecules into atoms, which then readily react with carbon atoms even at RT. Preferential etching of sp^2 bond fractions becomes therefore effective as soon as the mean free path decreases close to the actual target-to-substrate distance. For 25 mm this threshold lies at ≈1 Pa, exactly at the onset of the increase in the optical gap of the films deposited by Yoshitake and coworkers [60]. Oxygen molecules, to the contrary, have a high reaction rate with carbon atoms. In O_2 atmosphere, therefore, there is no need for collisions to initiate sp^2 etching, the reaction is effective from very low pressures and temperatures on. Just on the contrary, at higher pressures, collisions generate extreme reactive species which may etch away all carbon. This effect materializes in stopping of film growth (even at room temperature) when exceeding ≈1 Pa [60].

While at room temperature the effect of the H_2/O_2 pressure manifests itself only in the control of the sp^3/sp^2 ratio of the DLC films, at higher temperatures a big variety of different carbon forms can be formed depending on the actual process parameters.

Nucleation and growth of diamond crystals have been demonstrated within extremely narrow oxygen pressure windows at elevated temperatures. Yoshimoto *et al.* initiated heteroepitaxial growth on single crystal sapphire substrates in pure oxygen environment of 13–26 Pa at around 600°C. Similarly to the RT case, at lower oxygen partial pressures the graphitic phase could not be completely etched away, while above 26 Pa growth was no more possible [61]. At 550°C the pressure window narrowed even further: In a paper published three years later, Yoshimoto reported nucleation and growth of μm-size diamond crystals to occur at approximately 550°C under an oxygen pressure of 20 Pa on ultrasmooth sapphire substrates [62]. Though Chen *et al.* claim diamond growth in oxygen pressures ranging from 15 to 20 Pa, they present and discuss the characteristics of a single sample deposited by ablating a pyrolytic graphite target with 10^8 Wcm^{-2} pulses of a KrF excimer laser at 20 Pa [63]. Here, non-uniform and discontinuous nucleation resulted in small hexagonal and cubic diamond crystals with an average size of approx. 30 nm. Experiments performed at a fixed oxygen pressure of 6.7 Pa have verified that the optimum temperatures for PLD of diamond lie within the 600 ± 50°C window, indeed. Deposition at 400°C resulted in DLC films. Nucleation started at around 450°C. With increasing substrate temperature, the size of the crystallites rose continuously until at temperatures between 550 and 600°C the square crystallites of 1–2 μm diameter epitaxially grown on the diamond (100) substrate formed a continuous film. Further increase in substrate temperature led to the re-appearance of the amorphous carbon phase connecting the individual crystallites [64]. Taking into account that the uncertainty of the pressure values quoted above may well be even 20–50% when using standard gauges and not high accuracy capacitive manometers, the agreement between the above reports is fair and one may conclude that by (excimer laser) ablation of carbon targets in 15 ± 10 Pa oxygen nucleation and epitaxial growth of diamond films can be initiated on substrates heated up to 600 ± 50°C.

Following on from their successful experiment to grow single phase diamond films on diamond (100) in oxygen ambient [64], Yoshitake and coworkers made another attempt in hydrogen atmosphere [65]. Noting that at room temperature, oxygen and hydrogen are effective in different pressure ranges, this series of experiments has been executed between 133 and 800 Pa H$_2$, i.e. at pressures much higher than that of the oxygen atmosphere. Except the pressure, very similar process parameters (the same target, pulses of somewhat higher power densities from the same ArF excimer laser, substrates held at the optimum temperature of the previous experiments, 550°C), resulted in films of rather different characteristics. The films grown at the optimum H$_2$ pressure of 533 Pa consisted of diamond nanocrystals with maximum diameters of 20 nm in an amorphous carbon matrix, revealing that the existing pool of information is hardly enough to predict the result of the combined effect of the atmosphere and substrate temperature on the properties of the resulting carbon structures.

The results achieved up to now on PLD of pure carbon materials, sampled here, allow more or less consistent description and, in the majority of the cases, also understanding of the effects of the laser parameters and the ambient on the growth and properties of the different carbon forms obtained. Nevertheless, due to the exciting but sometimes confusing multitude of the carbon forms combined with the great number of process parameters, extrapolation of results of experiments to be performed in parameter domains not explored until now (especially at higher temperatures) is hardly possible. Recent production of non-identified, micrometer sized carbon particles with flat surfaces as a result of ultrashort pulse ablation in 100 Pa hydrogen at 350°C [36] is only one example suggesting that there is ample room for further systematic research yet.

The growth of carbon nanotubes by laser ablation provides another example. Carbon nanotubes are traditionally produced at (near to) atmospheric pressures and ambient temperatures around 1000°C. Nevertheless, some examples of the next section will demonstrate that all possibilities of carbon nanotube growth have far not been explored, either.

3.3.4. Production of Nanotubes

Synthesis of carbon nanotubes (CNTs) by pulsed laser ablation (PLA) is worth mentioning not only because it was the first method used to produce fullerene nanostructures in the gas phase [66] but also because this technique may produce nanotubes of fairly uniform diameter, self-organized into bundles consisting of hundreds of single-wall carbon nanotubes with remarkably high yield [67]. Growth of single-wall carbon nanotubes (SWNT) by the dual pulsed laser vaporization technique introduced by the group of R.E. Smalley at Rice University gives a good example for the exploitation of the flexibility offered by the absence of the substrate [68–70]. In the 4′-internal diameter flow-tube system, the laser beams propagate along the symmetry axis of the flow tube in the same direction as the Ar flow and ablate the coaxially placed carbon target containing Co and Ni, (1 at.% each). With 750 sccm Ar flow rate at 66,500 Pa ablation with two Nd:YAG lasers delivering 930 mJ pulses at 1064 nm, coincident on a 7.1 mm diameter spot with 4 ns delay between pulses, 20 g of 40–50 vol.% SWNT material can be produced in 48 h of continuous operation at 1100°C. Let us mention for its curiosity value that high quality SWNTs could also be grown by KrF excimer laser at process temperatures even lower than those required for the most commonly used Nd:YAG lasers [71].

Understandably, the approaches to produce carbon nanotubes in PLD configuration have been attempted in relatively low pressure atmospheres. Yap and coworkers [72] report on formation of aligned carbon nanotubes by RF plasma assisted PLD in hydrogen plasma of 50 Pa, presumably at 800°C. The authors succeeded in identifying one set of plasma parameters that allowed growth of multiwalled carbon nanotubes of 250–500 nm length and 40–80 nm diameter onto Fe particles of the same diameter pre-treated in the H_2 plasma at 800°C as catalysts. The successful attempt of Nakajima and coworkers [73] to grow carbon nanofibers in oxygen ambient suggests that, similarly to diamond growth, the hydrogen ambient could be substituted by lower pressure oxygen. The nanofibers of 10–50 nm diameter were grown at 550°C, in 13 Pa O_2 on a sapphire substrate 2 cm away from the surface of the graphite target containing 0.7 mol% NiO and ablated by 3×10^8 Wcm^{-2} pulses of a KrF excimer laser.

Though the authors do not specify the distance of the substrate from the carbon target, from the description of their 'all-laser' process for lateral growth of SWNTs on substrates prepatterned with CoNi catalyst nanoparticles, one may conclude that in this case the carbon nanotubes have also been produced by PLD. Here the pure carbon target was ablated in Ar atmosphere of 665 Pa with 100 pulses of 3×10^8 Wcm^{-2} power density from a KrF excimer laser. 1000°C was identified as the optimum temperature for the growth of the highest density of SWNT bundles connecting the CoNi nanoparticles [74].

Clearly, PLD cannot compete with PLA in mass production. Nevertheless, as directly demonstrated by the latter example, when localized growth onto patterned substrates is required, PLD of carbon nanotubes could be an alternative.

3.4. Pulsed Laser Deposition of Carbon Compounds – A Case Study of Carbon Nitride

The prediction of Liu and Cohen of a C_3N_4 phase 'harder than diamond' [75] initiated tremendous efforts to synthesize this magic compound. While the hypothetical phase, C_3N_4 has never been made [2,76,77], it has been realized that (amorphous) carbon nitrides, a-CN$_x$ with nitrogen content even well below the magic 57 at.% possess remarkable properties and promise a wide range of potential applications. The fabrication of a-CN$_x$ materials is a big challenge, indeed, since a high degree in tuning the structural, optical and electronic characteristics is possible, depending on the preparation conditions [2,78 and references therein]. Being triggered by the race for the synthesis of the hypothetical C_3N_4 phase, the overwhelming majority of the papers reports on the effect of process parameters: in particular nitrogen pressure and laser fluence on the chemical composition, most notably the nitrogen content, and the chemical structure of the films, paying much less attention to the concomitant changes in microstructure and the mechanism of film growth [2,53,54,78 and references therein]. In stark contrast to this general trend, here we will concentrate solely to the effect of changing N_2 pressure in influencing the growth process and determining the properties of the films produced. We will demonstrate that the trend of changes in the microstructure of pure carbon deposits with increasing pressure described in detail in Section 3.3.2. is general, and holds in the case of PLD of carbon nitride as well.

Comparison of the apparent growth (or deposition) rates, the deposition rates of the constituting elements and the mass densities of DLC and carbon nitride films grown by ablating identical graphite targets in the same optical and geometrical configuration in Ar and N_2 atmospheres, respectively, is a straightforward approach to separate the mechanical and chemical effects of collisions, follow possible changes in film microstructure, and thereby to give a consistent description of film formation in general and its peculiarities in the critical pressure range of 1–100 Pa, in particular [79].

Film growth is routinely characterized by the *apparent growth (or deposition) rate*, defined as the (maximum) film thickness per number of pulses. The pressure dependence of the apparent growth rates of DLC and carbon nitride films fabricated by ablating a graphite target with ArF excimer laser pulses of energy densities identical within experimental error (± 20% at best) in Ar and N_2 atmospheres, respectively, is shown in Fig. 6. Ablation in Ar atmosphere results in constant growth rates at around 0.046 nm/pulse for pressures below 0.5 Pa. In line with the expectations, between 0.5 and 5 Pa the growth rate drops down to 0.022 nm/pulse. The most salient feature of the pressure dependence recorded in Ar ambient is the unexpected increase between 5 and 100 Pa. The increase in N_2 pressure has apparently no effect on the growth rate up to approx. 50 Pa. The values, obtained on samples fabricated in separate deposition experiments scatter between 0.043 and 0.055 nm/pulse, without any clear trend. In N_2 ambient the onset of decrease is shifted to more than one order of magnitude higher pressures, as compared to the Ar case: the growth rate drops below 0.035 nm/pulse between 50 and 100 Pa. From the comparison of both growth rates the conclusion is that within the 10^{-5}–0.5 Pa pressure window, the nature of the atmosphere has apparently no effect on the growth rate. Above 0.5 Pa, however, ablation in Ar and N_2 atmospheres results in substantial differences in the pressure dependence of the apparent growth rates. Due to the variety and diversity of surface structures, constituting the meanwhile rather soft films, above 100 Pa reliable thickness measurements can hardly be achieved [46]. Therefore calculation of the apparent growth rates at pressures exceeding 100 Pa is not feasible.

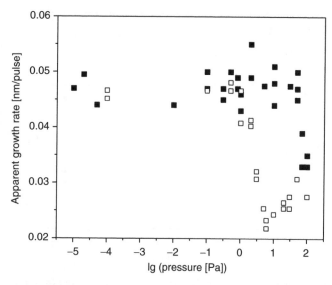

Figure 6: The apparent growth rate of DLC (□) and carbon nitride (■) films fabricated by ablating identical graphite targets with ArF excimer laser pulses in Ar and N_2 atmospheres, respectively, as a function of pressure. The data refer to 10 Jcm^{-2} fluence.

While being an apparently simple basic parameter, the interpretation of the dependence of the apparent growth rate on process parameters in general and ambient pressure in particular is far not trivial. One must always remember that in the changes in the apparent growth rate variations in both the absolute number of the constituting atoms and the way they fill the space (i.e. the microstructure) manifest themselves. Recording the change in the number of carbon and nitrogen atoms deposited over unit film area per pulse as a function of the ambient pressure by nuclear techniques (RBS, NRA, ERDA) is a straightforward approach to separate the two contributions [79].

Below approx. 0.1–1 Pa, the *carbon deposition rates* remain constant at around 6×10^{14} atoms cm^{-2} pulse^{-1} in both N_2 and Ar atmospheres (Fig. 7). The carbon deposition rate starts decreasing exactly in the pressure domain where the mean free path decreases below the target-to-substrate distance, $d = 30$ mm. A close-up of the change in carbon deposition rate in Ar atmosphere reveals that the decrease in the apparent growth rate between 0.5 and 5 Pa, shown in Fig. 6, is a direct consequence of the decrease in the number of carbon atoms reaching the substrate due to collisions with background gas atoms. However, the more than threefold decrease in the number of film building atoms results only in an approximately twofold decrease in the apparent growth rate, suggesting contribution of another effect: a concomitant decrease in the compactness of the films. In the 5–50 Pa Ar

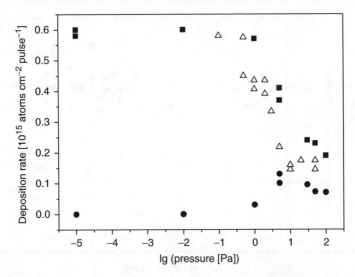

Figure 7: The deposition rate of carbon (■ and Δ) and nitrogen (●) atoms in DLC (Δ) and carbon nitride (■ and ●) films, as a function of pressure.

pressure domain the carbon deposition rate remains virtually unchanged at around 0.16×10^{15} atoms cm^{-2} $pulse^{-1}$ while the apparent growth rate increases, revealing that the decrease in film density continues. Keeping in mind that with increasing pressure the thickness distribution broadens, leading to a decrease in the maximum thickness, this effect becomes even more striking.

In N_2 atmosphere incorporation of nitrogen at least partly compensates for the initial decrease in carbon arrival rate at around 1 Pa. Between 1 and 5 Pa practically all species formed due to collisions between carbon atoms and N_2 molecules [80–83] may reach the substrate and contribute to the growth of films with nitrogen contents monotonously increasing up to 30–35 at.% [84,85]. Further increase in N_2 pressure results in the decrease in the number of both carbon and nitrogen atoms incorporated into the films. While the dependence of the carbon and nitrogen deposition rates on N_2 pressure accounts for the practically constant N/C ~ 0.35 ratios measured within the 5–50 Pa pressure domain [84,85], it cannot account for the unchanged apparent growth rate (cf. Fig. 6).

The dependence of the macroscopic mass density, calculated by dividing the sum of the atomic masses of all C, N, O and H atoms over unit area by the measured thickness, shown in Fig. 8, adds the missing piece of information by sheding light on the decisive role of the film micro- (and chemical-) structure in determining the variations in the apparent growth rates. The mass density versus ambient pressure plots display similar dependences for both DLC and carbon nitride, and indicate

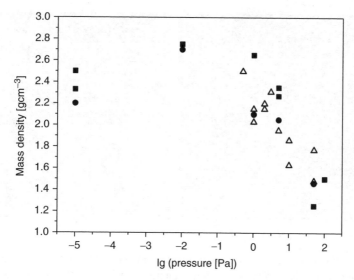

Figure 8: The density of DLC (Δ) and carbon nitride (■ and ●) films as a function of pressure with data calculated from NRA + ERDA (■) and RBS + ERDA (Δ and ●) measurements.

profound structural changes, indeed. Though the single, relatively low thickness value measured for the film deposited in 10^{-2} Pa N_2 (Fig. 6) may also contribute to the local maximum in mass density (Fig. 8), the results are in accord with the notion that the macroscopic density of carbon nitride films first increases with increasing N content and then starts to decrease when exceeding N/C ≈ 0.2 [86–89]. The sudden decrease at around 1 Pa explains i) the difference obtained in the pressure dependence of the apparent growth rate (Fig. 6) and the carbon deposition rate (Fig. 7) of the DLC films above 0.5 Pa and ii) the apparently constant growth rate of the carbon nitride films in the 5–50 Pa domain (Fig. 6) in spite of the continuous decrease in both C and N arrival rates (Fig. 7).

The fact that increasing pressure results in porous films – practically independently of both the material and the technique applied – is well documented (e.g. Refs 90–92 and references therein). As exemplified in Fig. 9, in the particular case of pulsed laser deposited DLC and carbon nitride films the critical pressure is at around 1–5 Pa. Below this threshold smooth dense films can be grown (Fig. 9a), while at higher pressures formation and arrival of clusters start to modify the growth process, resulting in profound changes in film microstructure and properties. Above ≈50 Pa the interaction of clusters rather than atoms and ions determines film growth. The film forming species undergo so many gas-phase collisions during

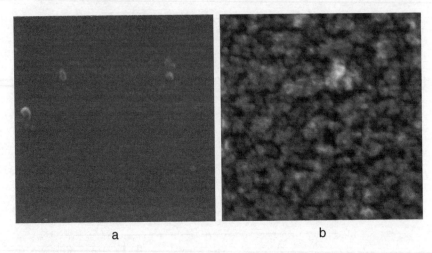

a   b

Figure 9: SEM micrographs of carbon nitride films grown by ablating a graphite target with ArF excimer laser pulses of 1 Jcm^{-2} energy density in 1 Pa N$_2$ (a) and 7.5 Jcm^{-2} energy density in 100 Pa N$_2$ (b). Each picture is 3×3 µm^2 in size.

their transport from the target to the substrate, that they become thermalized and, therefore, unable to build a dense, compact network when reaching the room temperature substrate (Fig. 9b).

Acknowledgments

I am thankful to Frederic Antoni, Eric Fogarassy and Zsolt Geretovszky for fruitful discussions and constant support.

References

[1] Voevodin, A.A. and Donley, M.S., *Surf. Coat. Technol.*, **1996**, *82*, 199.

[2] Muhl, S. and Méndez, J.M., *Diamond Relat. Mater.*, **1999**, *8*, 1809.

[3] Malkow, T., *Mat. Sci. Eng.*, **2001**, *A302*, 311.

[4] Robertson, J., *Mater. Sci. Eng.*, **2002**, *R 37*, 129.

[5] Perrone, A., *Jpn. J. Appl. Phys.*, **2002**, *41*, 2163.

[6] Giardini Guidoni, A., Kelly, R., Mele, A. and Miotello, A., *Plasma Sources Sci. Technol.*, **1997**, *6*, 260.

[7] Chrisey, D.B. and Hubler, G.K. (Eds.), Pulsed laser deposition of thin films, John Wiley & Sons, Inc., New York, **1994**.

[8] Gammino, S., Mezzasalma, A.M., Neri, F. and Torrisi, L. (Eds.), Plasma production by laser ablation, World Scientific Publishing Co., Singapore, **2004**.

[9] Boyd, I. and Chrisey, D.B., Pulsed laser deposition of thin films, in *Handbook of Laser Technology and Applications*, Webb, C.E. and Jones, J.D.C. (Eds.), IOP Publishing Limited, Bristol, **2004**.

[10] Blank, D.H.A., Doeswijk, L.M., Karakaya, K., Koster, G. and Rijnders, G., Pulsed laser deposition of dielectrics, in *High-κ Gate Dielectrics*, Houssa, M. (Ed.), IOP Publishing Limited, Bristol, **2004**.

[11] Ferrari, A.C., *Surf. Coat. Technol.*, **2004**, *180–181*, 190.

[12] Bonelli, M., Ferrari, A.C., Fioravanti, A., Li Bassi, A., Miotello, A. and Ossi, P.M., *Eur. Phys. J.*, **2002**, *B 25,* 269.

[13] Müller, F. and Mann, K., *Diamond Rel. Mater.*, **1993**, 2, 233.

[14] Köster, H. and Mann, K., *Appl. Surf. Sci.*, **1997**, *109–110*, 428.

[15] Davanloo, F., Lee, T.J., Jander, D.R., Park, H., You, J.H. and Collins, C.B., *J. Appl. Phys.*, **1992**, *71,* 1446.

[16] Pappas, D.L., Saenger, K.L., Bruley, J., Krakow, W., Cuomo, J.J., Gu, T. and Collins, R.W., *J. Appl. Phys.*, **1992**, *71*, 5675.

[17] Yoshitake, T., Nishiyama, T., Aoki, H., Suizu, K., Takahashi, K. and Nagayama, K., *Appl. Surf. Sci.*, **1999**, *141*, 129.

[18] Rode, A.V., Gamaly, E.G. and Luther-Davies, B., *Appl. Phys.*, **2000**, *A 70*, 135.

[19] Gamaly, E.G., Rode, A.V. and Luther-Davies, B., *J. Appl. Phys.*, **1999**, *85*(8), 4213.

[20] Rode, A.V., Luther-Davies, B. and Gamaly, E.G., *J. Appl. Phys.*, **1999**, *85*(8), 4222.

[21] Lade, R.J., Claeyssens, F., Rosser, K.N. and Ashfold, M.N.R., *Appl. Phys.*, **1999**, *A 69*, S935.

[22] Lade, R.J. and Ashfold, M.N.R., *Surf. Coat. Technol.*, **1999**, *120–121*, 313.

[23] Yamamoto, K., Koga, Y., Fujiwara, S., Kokai, F. and Heimann, R.B., *Appl. Phys.*, **1998**, *A 66*, 115.

[24] Merkulov, V.I., Lowndes, D.H., Jellison, G.E. Jr., Puretzky, A.A. and Geohegan, D.B., *Appl. Phys. Lett.*, **1998**, *73*, 2591.

[25] Lackner, J.M., Stotter, C., Waldhauser, W., Ebner, R., Lenz, W. and Beutl, M., *Surf. Coat. Technol.*, **2003**, *174–175*, 402.

[26] Yamamoto, K., Koga, Y., Fujiwara, S. and Kokai, F., *Jpn. J. Appl. Phys.*, **1997**, *36*, L1333.

[27] Minami, H., Manage, D., Tsui, Y.Y., Fedosejevs, R., Malac, M. and Egerton, R., *Appl. Phys.*, **2001**, *A 73*, 531.

[28] Bonelli, M., Miotello, A., Mosaner, P., Casiraghi, C. and Ossi, P.M., *J. Appl. Phys.*, **2003**, *93*, 859.

[29] Marquardt, C.L., Williams, R.T. and Nagel, J., *Mater. Res. Soc. Symp. Proc.*, **1985**, *38*, 325.

[30] Collins, C.B. and Davanloo, F., Noncrystalline carbon films with the bonding and properties of diamond, in *Pulsed Laser Deposition of Thin Films*, Chrisey, D.B. and Hubler, G.K. (Eds.), Wiley, New York, **1994**, Chapter 17.

[31] Collins, C.B., Davanloo, F., Jander, D.R., Lee, T.J., Park, H. and You, J.H., *J. Appl. Phys.*, **1991**, *69*, 7862.

[32] Voevodin, A.A. and Zabinski, J.S., *Diamond Relat. Mater.*, **1998**, *7*, 463.

[33] Lowndes, D.H., Merkulov, V.I., Puretzky, A.A., Geohegan, D.B., Jellison, G.E.Jr., Rouleau, C.M. and Thundat, T., in *Advances in Laser Ablation of Materials, Mater. Res. Soc. Symp. Proc.*, MRS, Pittsburgh, **1998**, *526*, 325.

[34] Stuart, B.C., Feit, M.D., Herman, S., Rubenchik, A.M., Shore, B.W. and Perry, M.D., *Phys. Rev.*, **1996**, *B 53*, 1749.

[35] Qian, F., Craciun, V., Singh, R.K., Dutta, S.D. and Pronko, P.P., *J. Appl. Phys.*, **1999**, *86*, 2281.

[36] Okoshi, M., Higuchi, S. and Hanabusa, M., *Appl. Surf. Sci.*, **2000**, *154–155*, 376.

[37] Garrelie, F., Loir, A.S., Donnet, C., Rogemond, F., Le Harzic, R., Belin, M., Audouard, E. and Laporte, P., *Surf. Coat. Technol.*, **2003**, *163–164*, 306.

[38] Banks, P.S., Dinh, L., Stuart, B.C., Feit, M.D., Komashko, A.M., Rubenchik, A.M., Perry, M.D. and McLean, W., *Appl. Phys.*, **1999**, *A 69*, S347.

[39] Voevodin, A.A., Donley, M.S. and Zabinski, J.S., *Diamond Relat. Mater.*, **1998**, 7, 463.

[40] Gamaly, E.G., Rode, A.V. and Luther-Davies, B., *Appl. Phys.*, **1999**, *A 69*, S121.

[41] Ossi, P.M., Bottani, C.E. and Miotello, A., *Thin Solid Films*, **2005**, *482*, 2.

[42] Thareja, R.K., Dwivedi, R.K. and Abhilasha, *Phys. Rev.*, **1997**, *B 55*, 2600.

[43] Arepalli, S., Scott, S.D., Nikolaev, P. and Smalley, R.E., *Chem. Phys. Lett.*, **2000**, *320*, 26.

[44] Dwivedi, R.K. and Thareja, R.K., *Surf. Coat. Technol.*, **1995**, *73*, 170.

[45] Ebihara, K., Nakamiya, T., Ohshima, T., Ikegami, T. and Aoqui, S., *Diamond Related Mat.*, **2001**, *10*, 900.

[46] Geretovszky, Zs., Haraszti, T., Szörényi, T., Antoni, F. and Fogarassy, E., *Appl. Surf. Sci.*, **2003**, *208–209*, 566.

[47] Suda, Y., Ono, T., Akazawa, M., Sakai, Y., Tsujino, J. and Homma, N., *Thin Solid Films*, **2002**, *415*, 15.

[48] Suda, Y., Nishimura, T., Ono, T., Akazawa, M., Sakai, Y. and Homma, N., *Thin Solid Films*, **2000**, *374*, 287.

[49] Heszler, P., *Appl. Surf. Sci.*, **2002**, *186*, 538.

[50] Kokai, F., Takahashi, K., Shimizu, K., Yudasaka, M. and Iijima, S., *Appl. Phys.*, **1999**, *A 69, (suppl.)* 223.

[51] Gnedovets, A.G., Kul'batskii, E.B., Smurov, I. and Flamant, G., *Appl. Surf. Sci.*, **1996**, *96–98*, 272.

[52] Ullmann, M., Friedlander, S.K. and Schmidt-Ott, A., *J. Nanoparticle Research*, **2002**, *4*, 499.

[53] Szörényi, T., Tóth, A.L., Bertóti, I., Antoni, F. and Fogarassy, E., *Diamond Relat. Mater.*, **2002**, *11*, 1153.

[54] Szörényi, T., Stoquert, J-P., Perriere, J., Antoni, F. and Fogarassy, E., *Diamond Relat. Mater.*, **2001**, *10*, 2107.

[55] Siegal, M.P., Yelton, W.G., Overmyer, D.L. and Provencio, P.P., *Langmuir*, **2004**, *20*, 1194.

[56] Rode, A.V., Hyde, S.T., Gamaly, E.G., Elliman, R.G., McKenzie, D.R. and Bulcock, S., *Appl. Phys.*, **1999**, *A 69*, S755.

[57] Thareja, R.K. and Dwivedi, R.K., *Physics Letters,* **1996**, *A 222*, 199.

[58] Dwivedi, R.K. and Thareja, R.K., *Phys. Rev.*, **1995**, *B 51*, 7160.

[59] Capelli, E., Orlando, S., Mattei, G., Zoffoli, S. and Ascarelli, P., *Appl. Surf. Sci.*, **2002**, *197–198*, 452.

[60] Yoshitake, T., Nishiyama, T. and Nagayama, K., *Diamond Relat. Mater.*, **2000**, *9*, 689.

[61] Yoshimoto, M., Yoshida, K., Maruta, H., Hishitani, Y., Koinuma, H., Nishio, S., Kakihana, M. and Tachibana, T., *Nature*, **1999**, *399*, 340.

[62] Yoshimoto, M., Furusawa, M., Nakajima, K., Takakura, M. and Hishitani, Y., *Diam. Rel. Mat.*, **2001**, *10*, 295.

[63] Chen, Z.Y., Zhao, J.P., Yano, T., Ooie, T., Yoneda, M. and Sakakibara, J., *J. Crystal Growth*, **2001**, *226*, 62.

[64] Yoshitake, T., Nishiyama, T., Hara, T. and Nagayama, K., *Appl. Surf. Sci.*, **2002**, *197–198*, 352.

[65] Yoshitake, T., Hara, T., Fukugawa, T., Zhu, L., Itakura, M., Kuwano, N., Tomokiyo, Y. and Nagayama, K., *Japanese J. Appl. Phys.*, **2004**, *43*, L240.

[66] Kroto, H.W., Heath, J.R., O'Brien, S.C., Curl, R.F. and Smalley, R.E., *Nature*, **1985**, *318*, 162.

[67] Kokai, F., Takahashi, K., Yudasaka, M. and Iijima, S., *J. Phys. Chem.*, **2000**, *B 104*, 6777.

[68] Thess, A., Lee, R., Nikolaev, P., Dai, H.J., Petit, P., Robert, J., Xu, C.H., Lee, Y.H., Kim, S.G., Rinzler, A.G., Colbert, D.T., Scuseria, G.E., Tomanek, D., Fischer, J.E. and Smalley, R.E., *Science*, **1996**, *273*, 483.

[69] Rinzler, A.G., Liu, J., Dai, H., Nikolaev, P., Huffman, C.B., Rodriguez-Macias, F.J., Boul, P.J., Lu, H., Heymann, D., Colbert, D.T., Lee, R.S., Fischer, J.E., Rao, A.M., Eklund, P.C. and Smalley, R.E., *Appl. Phys.*, **1998**, *A 67*, 29.

[70] Reed, B.W., Sarikaya, M., Dalton, L.R. and Bertsch, G.F., *Appl. Phys. Lett.*, **2001**, *78*, 3358.

[71] Braidy, N., El Khakani, M.A. and Botton, G.A., *Chem. Phys. Lett.*, **2002**, *354*, 88.

[72] Yap, Y.K., Yoshimura, M., Mori, Y., Sasaki, T. and Hanada, T., *Physica*, **2002**, *B 323*, 341.

[73] Nakajima, K., Furusawa, M., Yamamoto, T., Tashiro, J., Sasaki, A., Chikyow, T., Ahmet, P., Yamada, H. and Yoshimoto, M., *Diamond Relat. Mater.*, **2002**, *11*, 953.

[74] El Khakani, M.A. and Yi, J.H., *Nanotechnology*, **2004**, *15*, S534.

[75] Liu, A.Y. and Cohen, M.L., *Science*, **1989**, *245*, 841.

[76] Badzian, A., Badzian, T., Roy, R. and Drawl, W., *Thin Solid Films*, **1999**, *354*, 148.

[77] Rodil, S.E. and Muhl, S., *Diamond Relat. Mater.*, **2004**, *13*, 1521.

[78] Perrone, A., *Jpn. J. Appl. Phys.*, **2002**, *41*, 2163.

[79] Szörényi, T. and Fogarassy, E., *J. Appl. Phys.*, **2003**, *94*(3), 2097.

[80] Bulír, J., Novotny, M., Jelínek, M., Lancok, J., Zelinger, Z. and Trchová, M., *Diamond Relat. Mater.*, **2002**, *11*, 1223.

[81] Vivien, C., Dinescu, M., Meheust, P., Boulmer-Leborgne, C., Caricato, A.P. and Perriere, J., *Appl. Surf. Sci.*, **1998**, *127–129*, 668.

[82] Vivien, C., Hermann, J., Perrone, A., Boulmer-Leborgne, C. and Luches, A., *J. Phys. D: Appl. Phys.*, **1998**, *31*, 1263.

[83] Henck, R., Fuchs, C., Fogarassy, E., Hommet, J. and Le Normand, F., *Mat. Res. Soc. Symp. Proc.*, **1998**, *526*, 337.

[84] Szörényi, T., Antoni, F., Fogarassy, E. and Bertóti, I., *Appl. Surf. Sci.*, **2000**, *168*, 248.

[85] Cheng, Y.H., Sun, Z.H., Tay, B.K., Lau, S.P., Qiao, X.L., Chen, J.G., Wu, Y.P., Xie, C.S., Wang, Y.Q., Xu, D.S., Mo, S.B. and Sun, Y.B., *Appl. Surf. Sci.*, **2001**, *182*, 32.

[86] Hu, J., Yang, P. and Lieber, C.M., *Phys. Rev.*, **1998**, *B 57*, R3185.

[87] Alvarez, F., dos Santos, M.C. and Hammer, P., *Appl. Phys. Lett.*, **1998**, *73*, 3521.

[88] Walters, J.K., Kühn, M., Spaeth, C., Dooryhee, E. and Newport, R.J., *J. Appl. Phys.*, **1998**, *83*, 3529.

[89] Spaeth, C., Kühn, M., Richter, F., Falke, U., Hietschold, M., Kilper, R. and Kreissig, U., *Diamond Relat. Mater.*, **1998**, *7*, 1727.

[90] Bulír, J., Jelínek, M., Vorlícek, V., Zemek, J. and Perina, V., *Thin Solid Films*, **1997**, *292*, 318.

[91] Broitman, E., Zheng, W.T., Sjöström, H., Ivanov, I., Greene, J.E. and Sundgren, J.-E., *Appl. Phys. Lett.*, **1998**, *72*, 2532.

[92] Hellgren, N., Macák, K., Broitman, E., Johansson, M.P., Hultman, L. and Sundgren, J.-E., *J. Appl. Phys.*, **2000**, *88*, 524.

Chapter 4

Fabrication of Micro-optics in Polymers and in UV Transparent Materials

G. Kopitkovas, L. Urech and T. Lippert

Paul Scherrer Institut, CH–5232, Villigen-PSI, Switzerland

Abstract

Various laser-based approaches for the fabrication of micro-optical components in quartz and polymers have been studied. The fabrication of Fresnel micro-lens arrays in polymers is achieved by the combination of laser ablation and the projection of a *Diffractive Gray Tone Phase Mask* (DGTPM). Arrays of diffractive and refractive micro-lenses in quartz are fabricated by a laser assisted wet etching process and the projection of a DGTPM. An array of plano-convex micro-lenses was utilized as a beam homogenizer for high power Nd:YAG lasers.

Keywords: Diffractive Gray Tone Phase Mask (DGTPM); Excimer laser; Fresnel micro-lens; Lambert–Beer law; Laser ablation; Laser Induced Backside Wet Etching (LIBWE); Photothermal; Plano-convex micro-lenses.

4.1. Introduction

Recent developments in telecommunication and miniaturization of optoelectronic-mechanical devices stimulate the investigations of new techniques for fast micromachining of optical elements. Arrays of micro-lenses fabricated in polymers or in quartz are the key elements in modern optics and optoelectronics [1–8]. The unique properties of micro-lenses such as small dimensions and short focal length are utilized in high precision imaging systems, e.g. copiers, printers, and fax machines. The advantages of micro-optical components fabricated in polymers compared to large scale lenses are the flexibility and easy integration into complex micro-optical systems [4,7]. Arrays of micro-lenses are widely used in telecommunication and imaging systems [2,4,6,7,9–13].

Recent Advances in Laser Processing of Materials
J. Perrière, E. Millon and E. Fogarassy (Editors)

The most established techniques for the fabrication of micro-optical components in polymers and in plastics are hot embossing and injection molding. These techniques are commercialized and are applied for the mass production of submicron gratings and micro-lenses in polymers [4,6,14–17]. Replication technology is the key to the one step process and provides a very economical way of producing micro-optical elements in polymers. The key issues for this process are the complex master masks and the limited number of polymers that can be used. An alternative one step process for micromachining of micro-optical elements in polymers is direct exposure of polymers to a focused laser beam or a beam which is imaged through a circular aperture [2,4,7,18–22]. HeCd, ArF, KrF, or XeCl excimer [2,4,20] lasers as well MID IR CO_2 laser [13,23] can be used as irradiation sources.

An alternative one step process for the fast fabrication of complex structures with continuous profiles, e.g. Fresnel micro-lenses in polymers has been demonstrated by David *et al.* [24]. This method will be described in detail later.

UV transparent materials, e.g. quartz, BaF_2, and sapphire have much higher damage thresholds for UV laser irradiation. Therefore, micro-lenses in UV transparent materials, e.g. quartz, are applied in a number of application, such as optical connectors in telecommunication, imaging, wavefront measurements, and beam homogenizing [1,4,7,9,13,25–36].

UV transparent materials, e.g. quartz, can be directly ablated with ultrafast lasers [30,37–48], Vacuum Ultraviolet lasers (VUV) (157 nm, 11 ns) [32,49–53] or by the combination of a VUV laser with a KrF excimer laser [54–57]. MID IR lasers, e.g. CO_2 laser can also be applied for the fabrication of refractive micro-lenses, e.g. on the end of an optical fiber [9]. Another possibility is the fabrication of micro-optics in quartz by using the sequential scanning of a focused ion beam [14,57,58].

An alternative approach for the structuring of micro-optical components in solid state materials is laser assisted etching. The laser light passes through the transparent substrate and is strongly absorbed by a media which is in contact with the material. The highly absorbing media can be in the form of a gas [59], a solid material [60,61], or liquids [62–71], which will be described in detail later. The combination of laser assisted wet etching with the projection of a *Diffractive Gray Tone Phase Mask* (DGTPM) can be applied as an alternative technique for the fabrication of micro-optical components in quartz [31,72,73], which are applied as beam homogenizers for high power lasers.

In this chapter we will give a brief overview of laser ablation of polymers which is the basis for laser fabrication of micro-optical elements in polymers. The focus of this chapter will be on the application of laser ablation and Laser Induced Backside

Wet Etching (LIBWE) combined with the projection of DGTPM for the fabrication of complex three-dimensional structures, such as plano-convex and Fresnel microlens arrays, in polymers, and UV transparent materials.

4.2. Laser Ablation of Polymers

Laser ablation of polymers was first reported in 1982 by Srinivasan *et al.* [74] and Kawamura *et al.* [75]. The discovery of laser ablation of polymers initiated research projects around the world and is today industrially used for the production of nozzles for inkjet printers [76], to prepare the via-holes in multichip modules by IBM [77], and for the fabrication of micro-optical components [2,4,13,18,21,22,78,79].

4.2.1. Mechanisms and Models

The mechanisms of polymer ablation are typically described as photochemical, photothermal, photophysical or as a combination of those, and can be described briefly as follows:

Photochemical: Electronic excitation results in direct bond breaking [80–84].

Photothermal: The electronic excitation is thermalized on a ps timescale, resulting in thermal bond breaking [85–89].

Photophysical: Thermal and non-thermal processes play a role. In this model two independent channels of bond breaking [90,91] or different bond breaking energies for ground state and electronically excited states chromophores are applied [92,93]. This model is most adequate for short laser pulses in the ps and fs range [94].

Another method to describe the different ablation processes is a separation into surface and volume models. The volume model describes the ablation process within the bulk of the material. In the surface models, only a few monolayer of the material are considered. The different models can be described as follows:

Photochemical surface model: Valid for long pulses and higher irradiation fluences [95].

Thermal surface model: The model does not consider the sharp ablation threshold, but can describe the occurrence of an *Arrhenius* tail, which is the linear dependence of the ablation rate with the irradiation fluence at low fluences, that is observed when the ablation rate is determined with mass loss measurements [85,89,90,96].

Photochemical volume model: The model describes a sharp ablation threshold, but the Arrhenius tail is not accounted for [80–82,84,97].

Thermal volume models: These models are often oversimplified by reducing the movement of the solid–gas interface and results therefore in too high temperatures [88,98].

In newer models, several mechanisms of the listed above are combined, e.g. in the volume photothermal model of Arnold and Bityurin [99], that combines features of the photochemical and the thermal surface models. This model includes many material parameters. Several of these parameters are obtained from fitting of experimental data, and have to be adjusted to fit each polymer.

Polymers that show a photochemical ablation behavior at the irradiation wavelength are preferable for structuring, as the damage of the surrounding material due to a thermal processes is minimized. A conversion of the polymer into gaseous product is also of advantage, as no or only minor amounts of ablation products are redeposited on the structured surface, and additional cleaning procedures may not be necessary.

The ablation process is often described by an Lambert–Beer law related Equation (1) [80,100]:

$$d(F) = \frac{1}{\alpha_{eff}} \ln\left(\frac{F}{F_{th}}\right) \tag{1}$$

where $d(F)$ represents the ablation rate per pulse, α_{eff} is the effective absorption coefficient, F is the irradiation fluence, and F_{th} is the ablation threshold fluence.

The dependence of the ablation rate on the irradiation fluence can often not be described by a single set of parameters. In Fig. 1 an example of the dependence of the ablation rate of a polymer on the irradiation fluence is shown. Three fluence regions can be distinguished. From the low fluence range the ablation threshold fluence is defined. The threshold fluence is defined as the lowest fluence, where the onset of ablation can be observed. It is also noteworthy to mention, that in the low fluence range, ablation does not necessarily start with the first pulse, but after multiple pulses. This phenomenon is known as *incubation* and is related to a modification of the polymer by the previous laser pulses [101], which increases the absorption at the irradiation wavelength. Incubation is normally observed only for polymers with low absorption coefficients at the irradiation wavelength.

In the intermediate fluence range an increase of the slope of the ablation rate is often observed. This increase in the ablation rate may be due to an additional or more effective decomposition of the polymer by energy that has been gained from

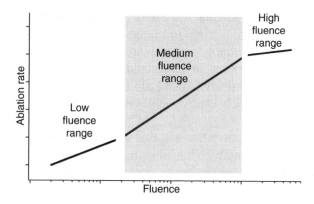

Figure 1: Schematic illustration of the fluence dependence for the ablation rate of a polymer.

decomposing the polymer. In the high fluence range the ablation rates of many polymers are similar, as the incident laser light is screened by ablation products and the plasma which are created during the ablation process [82].

4.2.2. Commercially Available and Designed Polymers

Polymer ablation has been a research field for over 20 years, but its full potential for industrial applications has not yet been explored. One possible reason for this is the fact that commercially available polymers, such as polyimide (PI), poly-methylmethacrylate (PMMA), and polycarbonate (PC) [102] that are applied in many ablation studies have several drawbacks. These include low sensitivity, carbonization upon irradiation, and redeposition of ablation products on the polymer surface [103].

Therefore, novel photopolymers for laser ablation have been designed. The most important criteria for the designed polymers are:

- High absorption coefficients ($\geq 20,000$ cm^{-1}) at the irradiation wavelength.
- Exothermic decomposition at well-defined positions of the polymer backbone.
- Decomposition of the polymer into gaseous products, which are not contaminating the polymer surface [104,105].

An XeCl excimer laser (308 nm, 60 ns) has been selected as irradiation source due to the long lifetime of the laser optics and laser gas fills. Another advantage of this relatively long irradiation wavelength is the possibility to separate the absorption of the photoactive group from other parts of the polymer.

The most promising photopolymers were triazene (-N = N-N<) containing materials, which have high etch rates, low threshold fluences, reveal no surface contamination, and a small heat affected zone [103,106]. The diffractive limited optical resolution which can be achieved by 308 nm irradiation is sufficient for many applications [107].

A typical UV-Vis spectrum (shown in Fig. 2) reveals an absorption band around 196 nm that correspond mainly to the aromatic groups of the polymer. The second strong absorption band at 332 nm corresponds to the triazene chromophore [108]. The chemical structure of a triazene polymer is included as inset in Fig. 2.

An example of the improved ablation quality for the designed polymers is shown in Fig. 3, where the same structure is ablated into a triazene polymer (left) and in polyimide (Kapton™, right). It is noteworthy to mention that both polymers have very similar absorption coefficients at the irradiation wavelength of 308 nm. The structure in the triazene polymer is well defined with no visible debris. For polyimide a pronounced ring of redeposited ablation products can be observed. A closer inspection reveals also contaminations inside the polyimide structure by products that consist mainly of amorphous carbon [106,109–111].

A polyimide with better ablation properties than Kapton™ (Fig. 3 (right)) is Durimid 7020™ (Arch Chemical) (structure shown in Fig. 4) [112]. Durimid belongs

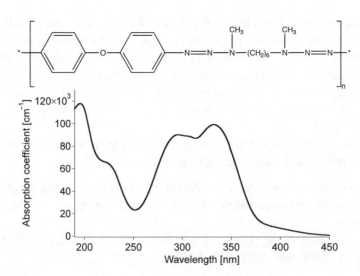

Figure 2: UV-Vis spectrum of a triazene polymer designed for laser ablation at 308 nm and of the chemical structure of the polymer.

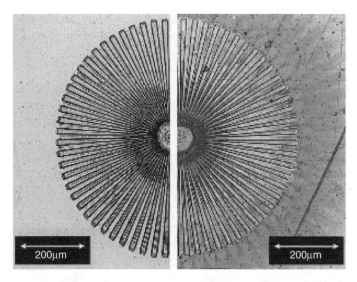

Figure 3: SEM image of a Siemens Star fabricated by laser ablation in a triazene polymer (left) and in polyimide (right) (adapted from [106]).

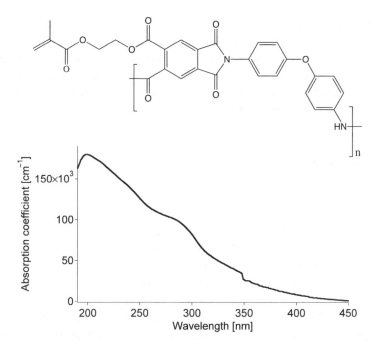

Figure 4: The linear absorption coefficient and chemical structure of Durimid.

to the class of photosensitive polyimides, that can be ablated with all common excimer laser. This polymer and other types, such as Pyrolin[TM] reveals good spin coating properties and have also a linear absorption coefficient (Fig. 4) at 308 nm which is similar to the triazene polymer shown in Fig. 2.

The photosensitive polyimides are after crosslinking, no longer photosensitive and present the typical polyimide properties, i.e. thermal and photo resistance, which makes them suitable materials for optical components in the visible range.

4.3. Methods for the Fabrication of Micro-optical Elements in Polymers

Injection molding is, as discussed in the introduction, one of the key techniques for the low-cost mass production of the complex structures in polymers [4,6,10,15,16]. The principle of these techniques is copying of a surface-relief microstructure of a metal into formable polymers, such as polycarbonate (PC), polymethyl-methacrylate (PMMA), polyvinyl chloride (PVC). Hot embossing is mainly used for reproducing submicron grating structures and micro-optics in PC and PVC polymers. The limitation of this method is the depth of the structures, which is about 1 μm [4,15]. Much deeper microstructures (up to 1 mm) in PC or PMMA can be produced by injection molding [4,15]. The replication technology is capable to achieve nanometer resolution over large areas and is in principle a low-production cost process. The major weaknesses of these replication techniques are the limited aspect ratio and depth of the structures, the complex mask which is necessary for producing micro-lens arrays, and the limited number of polymers which can be applied for the replication techniques.

4.3.1. Laser Beam Writing

Laser beam writing techniques, compared to the replication process are more flexible and do not need the complex masks. This technique is based on the sequential patterning of polymer surfaces by a focused laser beam [2,4,7,18–22]. Various type of lasers e.g. cw HeCd, pulsed excimers, and Nd:YAG can be applied as irradiation sources. The fabrication of circular micro-optical elements e.g. spherical shape micro-lenses in polymer samples is achieved by applying a circular mask, which is imaged onto the sample surface. The polymer sample is positioned on a translation stage, which allows to perform circular movement during laser irradiation [4,7,19,20,78]. For the fabrication of an array of micro-lenses in polymers, a two-dimensional array of a circular apertures can be used [19].

This approach allows a relative fast fabrication of micro-lens arrays in polymers. The advantages of the laser writing technique is its flexibility and that no complex grayscale masks are necessary, but this method is relatively slow compared to replication techniques.

4.3.2. Fabrication of Micro-optics using Laser Ablation and Half Tone or Diffractive Gray Tone Masks

An alternative method to laser writing is the combination of laser ablation with the projection of a half tone (gray tone) mask (as shown in Fig. 5) [25,79, 113–116]. The encoding of the desired structure can be achieved either by a mask with locally variable transmission patterns (the size of the pattern is ≈5 μm) [25,79,113,116].

A half tone (gray scale) mask consists of a number of patterns with chromium layers of various thickness. The modulation of the laser light intensity is based on the local absorption of the laser light by the chromium patterns. One of the major disadvantages of these masks is the low laser damage threshold, which can cause defects in the mask. To overcome this drawback, Smith *et al.* [117,118] invented diffractive phase masks which can be used with high power lasers. Further developments of the DGTPM were performed by David *et al.* [24,119] and Braun *et al.* [120]. A DGTPM is a binary structure fabricated in quartz by *e-beam lithography* and *Reactive Ion Etching* (RIE) which is used to modulate the incoming laser beam intensity [24,120]. No light is absorbed with a DGTPM and, therefore no damage can occur even for very high laser fluences.

The diffractive gray tone masks consist of periodically spaced line structures with various line width. The transmission of the laser light in the zeroth order depends on the ratio between the line width and the grating period, which is also called the duty cycle (DC). Keeping the grating period as a constant and varying the line width allows to continuously change the transmission of the mask (as shown in Fig. 6).

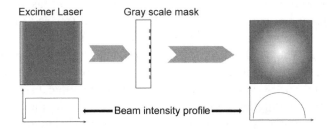

Figure 5: Modulation of the incoming laser beam intensity by a gray scale mask.

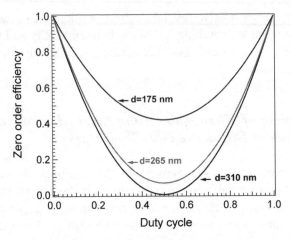

Figure 6: Relation of the Zeroth order efficiency with the duty cycle for a Diffractive Gray Tone Phase Mask.

The transmission (T) of the mask or zero order efficiency can be described by scalar diffraction theory:

$$T = 1 - 2 \cdot DC \cdot (1 - DC) \cdot (1 - \cos \varphi) \qquad (2)$$

where DC is the duty cycle and φ is the phase shift.

The zero order efficiency calculated by Equation (2) is close to zero when the phase shift is equal to π and the DC = 0.5. A π phase shift is obtained when the height (h) of the grating structures is equal to $\lambda/(n-1)$, where λ is the laser wavelength and n is refractive index of quartz. A depth of the structure of 310 nm ensures a complete modulation range for an XeCl excimer laser (308 nm, 60 ns) (shown in Fig. 6 as solid line), which is used as irradiation source for all our experiments. The etch rate and threshold fluence of polymers must be included in the DGTPM design to improve the quality of the complex structures ablated into the polymers with a DGTPM. An optimized DGTPM for the fabrication of various structures in one selected polymer has patterns with a depth of 265 nm (instead of 310 nm). The optimization of the masks yields an improvement of the quality of the complex structures [119]. The depth of the features in the DGTPM was even more reduced ($d \approx 175$ nm) for the application of these masks to obtain plano-convex microlenses in quartz. The reason for using a DGTPM, with a typical size of 5 mm, with only half of the modulation depth (shown in Fig. 6) for the fabrication of three-dimensional structures in quartz is the complex etch rate behavior of quartz which is completely different to the laser ablation of polymers. The experimental setup

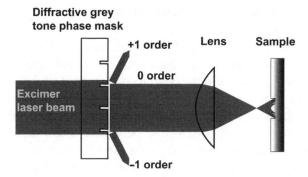

Figure 7: Experimental setup for the fabrication of micro-lenses in polymers.

for the fabrication of complex structures in a photosensitive polyimide (Durimid 7020™) is shown in Fig. 7 and consists of the combination of laser ablation with the projection of a DGTPM.

The intensity of an XeCl excimer is modulated by the DGTPM. The beam is then imaged onto the polymer surface with a 5 times demagnification using a doublet lens. A two-dimensional Fresnel micro-lens array was fabricated in the polyimide (shown in Fig. 8) by moving the sample and repeatedly exposing the DGTPM which encodes the Fresnel lens shape. The size of each micro-lens is 900×900 µm.

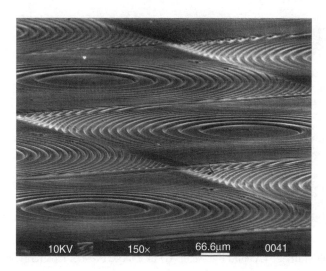

Figure 8: SEM picture of a micro-lens array in Durimid (adapted from [24]).

Such arrays can for example be applied as diffusers, or for beam shaping of white light sources.

One of the advantages of the combination of laser ablation and imaging of the DGTPM compared to the laser writing method is the simple adjustment of the size of the micro-optical elements in the polymers. The same DGTPM can be applied for the fabrication of various sizes of optical components with diameters from 250 μm to several millimeters, by adjusting the demagnification of the DGTPM. A Fresnel lens was transferred in a photosensitive polyimide that was spin-coated onto quartz to test the ability of this method for the fabrication of larger optical components in polymers. The diameter of this lens (shown in Fig. 9A) is 5 mm while the focal length is 32 mm. This lens was fabricated with 40 laser pulses. The depth profile scan of this lens (shown in Fig. 9B) presents the typical features of a Fresnel lens, i.e. parabolic depth profile in the center and triangular shapes on the sides. The size of the optical elements is mainly dependent on the available laser power, which is necessary to induce ablation.

The main disadvantage of the combination of laser ablation with projection of a DGTPM is high cost of the phase mask. The possibility to apply one mask to create various sizes of the micro-optical elements in polymers is a clear advantage to replication techniques, where one mask can be only used for one size of the structure. Another advantage of laser ablation is the fact that most polymers can be structured with UV lasers, especially at 193 nm, while hot embossing and injection molding techniques require thermoplastic or curable polymers that are optimized for the replication process.

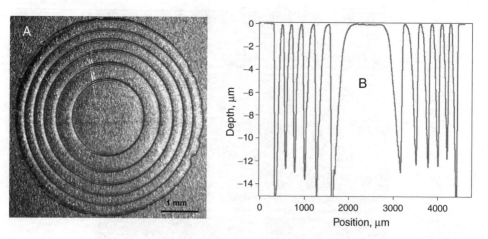

Figure 9: Optical image (A) and line scan (B) of a large Fresnel lens fabricated in Durimid.

4.4. Microstructuring of UV Transparent Materials

UV transparent materials, e.g. quartz, BaF_2, sapphire, etc. are very important in modern optics, mainly due to their wide transparency (from UV to IR), high laser damage threshold (>20 J/cm^2 (at 308 nm)), and high mechanical and chemical stability. One of the commercial applications of micro-lens arrays fabricated in quartz is high power laser beam shaping and homogenizing. However, the fabrication of micro-optics in UV transparent materials is also restricted due to the properties of these materials. Industrial technologies for the fabrication of micro-lens arrays in quartz are based on the combination of photolithography, resist melt-reflow, and RIE [7,34,121–123]. The first step in this approach is the creation of binary cylindrical structures in a photoresist coated on a quartz substrate by photolithography. The two-dimensional features on the sample are converted to three-dimensional by heating the photoresist above the glass transition temperature (>200°C). This results in a flow of the resist to yield the desired spherical shape structures. Finally, the micro-lenses are transferred from the photoresist to the quartz by RIE using a proportional etching process. High-quality refractive micro-lenses with good spherical shape and low etch roughness in quartz can be fabricated by this technique. However, this method has several drawbacks, such as the multiple steps, the necessity for a high control of the proportional etching and resist characteristics, and a relatively slow RIE process (20 nm/min). The drawback of this method resulted in the development of alternative techniques for structuring of UV transparent materials, which are briefly mentioned in the introduction.

Direct laser ablation of quartz can be achieved by ultrafast (femtosecond) lasers. The removal of the material occurs due to the strong multiphoton absorption. A very promising application of femtosecond laser ablation is the three-dimensional microstructuring inside a material [37,44], which can be utilized to create photonic crystals, waveguides, Bragg gratings, etc.

Complex structures with sub-micrometer precision in quartz can be obtained by VUV laser ablation [51,124], where a F_2 excimer laser (157 nm, 11 ns) is used as irradiation source. Well defined structures with the very low etch roughness can be fabricated by this technique in quartz [32,51,52,55]. This method requires transparent gases for the beam path and expensive optics. Fabrication of high quality Fresnel micro-lens arrays in quartz by a focused ion beam was demonstrated by Fu *et al.* [57,125,126]. The drawback of this process is an increase of the absorption in the UV range of the etched areas in quartz due to the exposure of quartz to the Ga^+ ion beam.

Another approach for structuring of UV transparent materials is indirect laser assisted etching using highly absorbing media, which are in contact with the material.

The first application of this approach was performed for the etching of semiconductors materials [59].

Sugioka *et al.* developed a method in which structuring of quartz is assisted by a laser induced plasma [54,60,61,127]. A Nd:YAG laser is applied as an irradiation source, while a metal target, which is in contact with the quartz plate acts as an absorber. The laser light passes through the quartz and is strongly absorbed by the metal target, acts where a plasma is generated. The removal of the quartz is assisted by the plasma process. Well defined gratings in quartz can be prepared by this method, however, the depth of the features is limited to 2 µm.

Another approach for the structuring of sapphire and SiC ceramics with a copper vapor laser (510 nm, 10 ns) laser was demonstrated by Dolgaev *et al.* [64,65,128]. A 0.6 M aqueous CrO_3 solution which is in contact with the sapphire sample was used to achieve the absorption of the laser photons. The key element of the etching process is a temperature jump at the thin sapphire–liquid interface which originates from the strong absorption of the laser light. The rapid rise of the temperature results in heating of the sapphire substrate and thermal decomposition of the solution, which generates non-soluble Cr_2O_3 deposits [129]. These particles form a thin film on the substrate which results in a further increase of the laser induced temperature. The removal of the sapphire is most probably achieved mechanically, resulting from the difference in thermal expansion coefficient between the Cr_2O_3 film and sapphire substrate. A typical etch rate of sapphire with this method is 200 nm per pulse at a laser fluence of 5 J/cm^2. The roughness of the etched features is in the range of several micrometers. The major disadvantage of this method is the high roughness and the deposition of a Cr_2O_3 layer on top of the etched features, which has to be removed by an additional cleaning process.

In 1999, Yabe and coworkers developed a method for precise micromachining of UV transparent materials [69,130]. This technique is similar to the approach of Dolgaev, i.e. in order to increase the absorption of the laser light a strong absorbing organic liquid is used which is in contact with the material. This method has been termed *Laser Induced Backside Wet Etching* (LIBWE) and was applied for the structuring of quartz, CaF_2, BaF_2, and sapphire [62,63,66,69,71,130–136]. Different excimer lasers, i.e. ArF, KrF, XeCl, XeF can be applied as irradiation sources. A number of organic liquids, such as pyrene in acetone (or THF) [31,69,137], aqueous pyranine and naphthalene [63], pure toluene [134,138,139], and naphthalene in methyl-methacrylate [68] can be applied as 'etchant'. An experimental setup for LIBWE is shown in Fig. 10, where an XeCl excimer laser (308 nm, 60 ns) is used as irradiation source with a repetition rate of 5 Hz, while 0.4 M pyrene in acetone and 1.4 M pyrene in tetrahydrofurane were applied as etching media.

A possible mechanism for the LIBWE process is based on the strong absorption of the intense laser light by the organic liquid, which is in contact with the

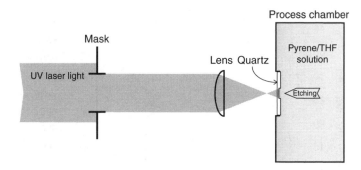

Figure 10: Experimental setup for Laser Induced Backside Wet Etching.

UV transparent material. Rapid relaxation processes of the excited dye molecules generate a fast increase of the temperature at the substrate–liquid interface and may result in softening, melting or even boiling of the UV transparent materials. Secondary processes, which take place at the interphase between the substrate and the liquid, are the creation of a shock wave and the boiling of the solvent. The latter results in the creation of expanding and collapsing bubbles, which generate another strong pressure jump at the interface between the heated material and the liquid. The pressure jump from the shock wave or from the bubble may remove the softened material from the surface [68–70,134,135,137,140].

One characteristic of the structuring of UV transparent materials by LIBWE is the existence of a threshold fluence. The threshold fluence for quartz is defined as the highest fluence, for which no etching can be achieved even if thousands of pulses have to be applied. The threshold fluence is obtained experimentally by measuring the etch rates at various laser fluences (shown in Fig. 11).

The experimentally obtained threshold fluence for quartz using 0.4 M pyrene in acetone solution as etchant is 0.66 J/cm^2, which is much lower than the ablation threshold without solution ($\approx$20 J/cm^2 [142]).

The etching of UV transparent materials by LIBWE is a complex fluence dependent process, which can be divided into several etching regions (marked in Fig. 11 as A, B, C). In the low fluence range the etch rates increases only slowly with an increase of the laser fluence. The roughness of the etched features is in the range of 800 nm and decreases rapidly to 100 nm for higher laser fluences (shown in Fig. 11). Another effect which is observed at these fluences is incubation, which means that the removal of quartz occurs after a number of laser pulses (shown in Fig. 12). Incubation, which is also observed in the case of laser ablation of polymers, is associated with a chemical modification of the surface prior to etching.

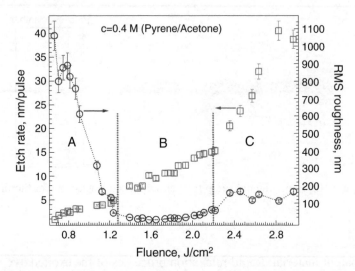

Figure 11: Etch rates and etch roughness of quartz at different laser fluences using a 0.4 M pyrene in acetone solution (irradiation wavelength 308 nm) (adapted from [73]).

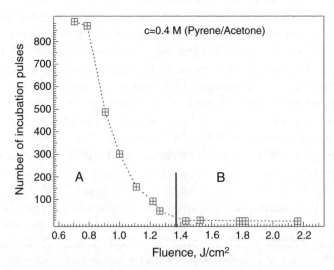

Figure 12: Number of incubation pulses at different laser fluences for a 0.4 M pyrene/acetone solution and 308 nm as irradiation wavelength (adapted from [73]).

On the basis of the experimental results, the following etching mechanism for low laser fluence range can be suggested: the laser induced temperature at the quartz–liquid interface is probably below the melting point of the quartz, but is still high enough to cause boiling of the solution. The pressure waves, generated by the shock wave and collapsing of the laser induced bubbles are also not strong enough to remove the heated material, and no etching occurs. The temperature at the quartz–liquid interface, may however, be high enough to decompose the solution, i.e. solvent or pyrene. This results in the formation of carbon deposits, which strongly adhere to the heated quartz surface [72,131]. The formation of the carbon layer alters the absorption of the laser light, which results in an increase of the temperature jump at the interface between the quartz and the liquid, which is similar to the process described by Dolgaev [65]. The removal of the quartz is possible at the higher temperatures, which may reach the melting point of quartz. Another process which may influence the etching of quartz at the low laser fluences is associated with the mechanical stress, resulting from the different thermal expansion coefficients of quartz and the carbon layer.

The increase of the etch rates with higher laser fluences, i.e. above 1.2 J/cm^2, (marked as B in the Fig. 11), is associated with the generation of higher temperatures at these laser fluences. An increase of the temperature also results in stronger pressure jumps, which remove the softened-molten material yielding smooth etching with a very low etch roughness ($\approx$20 nm). This fluence range is utilized for the fabrication of the micro-optical components in quartz.

A further change of the etch rate is observed for higher fluences (marked as C in Fig. 11) and is accompanied by the formation of a plasma in the solution [73]. The onset of the plasma generation in the solution coincides with the increase of the etch rates but also with an increase of the etch roughness. It is therefore very probable that the plasma assists the etching process but probable in a different way then reported by Sugioka *et al.* [54,60,61,127].

The quality of the etched features in UV transparent dielectrics depends on various etching parameters, such as the laser fluence, pulse number, and concentration of the absorbers (pyrene). In order to obtain the optimum etching conditions for the fabrication of micro-optical elements in quartz different solutions containing various concentrations of absorbers have been studied.

An increase of the pyrene concentration (shown in Fig. 13A) results in the absorption of the laser light in a thinner layer which results in higher temperatures and pressures. The increase of the temperature and pressure jumps is indicated by a shift of the threshold fluence to lower fluences and a reduction of the etch roughness. The number of incubation pulses is also decreasing with an increase of the pyrene concentration [73]. The optimum etching conditions in terms of lowest etch roughness, highest etch rates, and lowest number of incubation pulses for the

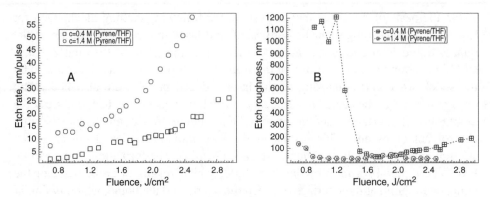

Figure 13: Etch rates (A) and etch roughness (B) of quartz using different concentrations of pyrene in THF (irradiation wavelength 308 nm) (adapted from [73]).

fabrication of micro-optical components in quartz are obtained for a 1.4 M pyrene in THF solution and laser fluences in the range between 0.9–1.6 J/cm^2.

4.4.1. Fabrication and Applications of Micro-optical Elements in Quartz

An array of Fresnel (shown in Fig. 14) and plano-convex (shown in Fig. 15) micro-lenses were fabricated in quartz by the combination of LIBWE and the projecting of a DGTPM. The DGTPM which encodes the Fresnel lens shape was optimized

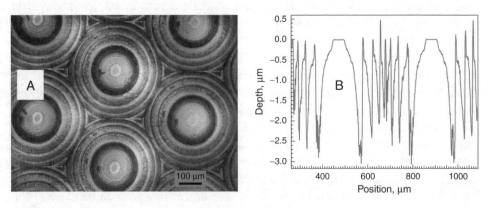

Figure 14: The light microscope image (A) and line scan (B) of a Fresnel micro-lens array fabricated in quartz (adapted from [73]).

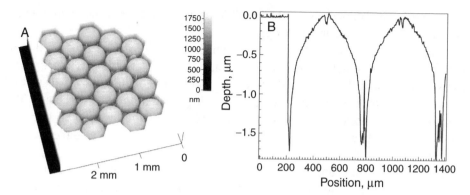

Figure 15: Three dimensional image (A) and horizontal scan (B) of the plano-convex micro-lens array in quartz.

for the fabrication of Fresnel lenses in polymers, which results in the formation of a small plateau in the center of the Fresnel lens in quartz (shown in Fig. 14B). This is caused by the different etch rates and threshold fluences of the photosensitive polyimide compared to quartz.

Another type of micro-lenses, i.e. an array of plano-convex micro-lenses in quartz (shown in Fig. 15) was also fabricated using LIBWE and imaging of a DGTPM. The etch rates and threshold fluence for quartz was included in the design of this DGTPM. The modulation depth of these masks was reduced in order to modulate only the intermediate fluence range (between 0.9–1.6 J/cm^2) where the lowest etch roughness of quartz is observed for a 1.4 M pyrene in THF solution as an etchant.

The focal lengths of the Fresnel and plano-convex micro-lenses are calculated from the depth profile measurements using Equation (3):

$$f = \frac{\left(h_0/2\right) + \left(D^2/8 \cdot h\right)}{2 \cdot (n(\lambda) - 1)} \tag{3}$$

where h_0 is the lens sagitta (the distance from the midpoint of an arc to the midpoint of its chord), D is the diameter of the micro-lens and $n(\lambda)$ is the refractive index of quartz ($n = 1.485@308$ nm) as a function of the wavelength.

The focal length of the Fresnel micro-lens is 5 mm which corresponds well to the measured value [73]. The focal length of the Fresnel lens is however too small for beam homogenizing applications. In order to fabricate Fresnel lenses with larger focal lengths, an optimization of the DGTPM design is necessary, i.e. including the etch parameters into the mask design.

The plano-convex micro-lenses have a much larger focal length ($f = 70$ mm) and are therefore suitable for beam shaping and homogenizing of high-power lasers. Arrays of diffractive and refractive micro-lenses in quartz, as discussed above, are used for the shaping and homogenizing of high-power laser beams. The concept of beam homogenizing is based on splitting of the laser beam intensity into a number of beamlets, which are collected in the focal plane of the collecting lens, as shown in Fig. 16.

The homogenized beam size D obtained at the focal plane of the collecting lens is proportional to the focal length of the collecting lens, the diameter and focal length of the micro-lens, and can be calculated using Equation (4):

$$D = \frac{d_{ul} \cdot F}{f_{ul}} \tag{4}$$

where d_{ul} and f_{ul} are the diameter and the focal length of the micro-lenses, and F is the focal length of the collecting lens.

The homogenizing of a quadrupled Nd:YAG laser beam profile was tested with a quartz micro-lens array which consists of 20×22 plano-convex micro-lenses (same as shown in Fig. 15). The focal length of the collecting lens used for the homogenizing of the Nd:YAG laser was 250 mm. The Nd:YAG laser beam profile with and without the beam homogenizer is shown in Fig. 17. The homogenized beam size obtained at the focal plane of the collecting lens is 1.6 mm, which corresponds well to the value calculated from Equation (4).

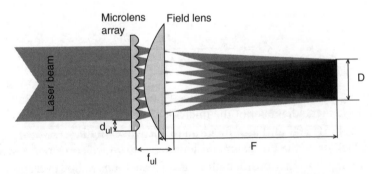

Figure 16: Experimental setup for beam homogenizing (adapted from [72]).

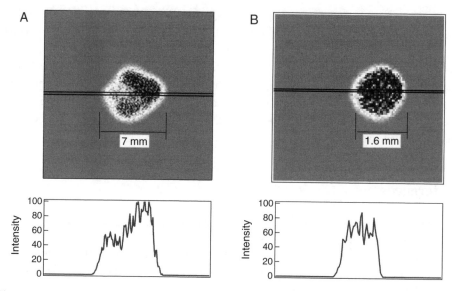

Figure 17: The spatial beam profile of a quadrupled Nd:YAG laser without (A) and with a plano-convex micro-lens array (B).

A clear improvement of the Nd:YAG laser beam profile is obtained when the micro-lens array is applied. However, the line scan profile suggests the existence of energy fluctuation in the intensity of the homogenized laser beam, which is in the range of 30% (RMS). These energy fluctuations are most probably caused by an interference effect, resulting from dividing of the laser beam into a number of beamlets [8,26]. This interference effect is associated with the coherence of the laser and depends on the diameter of the micro-lenses and the focal length of the collecting lens [26]. A further improvement of the micro-lenses is necessary to achieve a more homogenous laser beam profile.

4.5. Conclusions

The development and application of polymers designed specially for laser ablation is a promising approach to optimize the ablation properties, such as reducing the amount of debris and improving the quality of etched surface. Micro-optical elements with dimensions from 250 µm to several millimeters can be fabricated in these polymers by laser ablation and projection of a Diffractive Gray Tone Phase Mask (DGTPM).

The combination of Laser Induced Backside Wet Etching (LIBWE) and the projection of a DGTPM can be utilized for the one step fabrication of plano-convex and Fresnel micro-lenses in quartz. Quartz and other UV transparent materials can be structured with this method with fluences well below the damage threshold of these materials. The roughness of the etched features in quartz is low enough to create micro-optical elements that can be applied as beam homogenizers.

A clear improvement of the Nd:YAG laser beam profile laser is obtained for a plano-convex micro-lens array.

References

[1] Dickey, F.M. and Holswade, S.C., Gaussian laser beam profile shaping, *Optical Engineering,* **1996**, *35*(11), 3285–3295.

[2] Gale, M.T., Rossi, M., Pedersen, J. and Schutz, H., Fabrication of continuous-relief micro-optical elements by direct laser writing in photoresists, *Optical Engineering,* **1994a**, *33*(11), 3556–3566.

[3] Herman, P.R., Chen, K.P., Corkum, P.B., Naumov, A., Ng, S. and Zhang, J., Advanced laser microfabrication of photonic components, *Proceedings of the SPIE - The International Society for Optical Engineering,* **2000a**, *4088*, 345–350.

[4] Herzig, H.P., *Micro-optics elements, systems and applications*, Taylor and Francis Ltd., London, **1997**.

[5] Pachauri, J.P., Baby, A., Chaturvedi, M.N., Kothari, H.S., Singh, A. and Singh, B.R., Deep reactive ion etching of silica for planar lightwave circuits using indigenously developed ecr/rie system, *Proceedings of the SPIE - The International Society for Optical Engineering,* **2001**, *4417*, 267–270.

[6] Pasco, I.K. and Everest, J.H., Plastics optics for opto-electronics, *Optics and Laser Technology,* **1978**, *10*(2), 71–76.

[7] Sinzinger, S. and Jahns, J., *Microoptics*, WILEY-VCH, Weinheim GmbH and Co. KGaA, **2003**.

[8] Williams, S.W., Marsden, P.J., Roberts, N.C., Sidhu, J. and Venables, M.A., Excimer laser beam shaping and material processing using diffractive optics, *Proceedings of the SPIE - The International Society for Optical Engineering,* **1998**, *3343*(1–2), 205–211.

[9] Calixto, S. and Sanchez-Martin, F.J., Spherical fused silica microlenses fabricated by the melting method, *Revista Mexicana de Fisica,* **2003**, *49*(5), 421–424.

[10] Cox, W.R., Hayes, D.J., Chen, T., Ussery, D.W., MacFarlane, D.L. and Wilson, E., Fabrication of micro-optics by microjet printing, *Proceedings of the SPIE - The International Society for Optical Engineering,* **1995**, *2383*, 110–115.

[11] Gale, M.T., Rossi, M. and Schutz, H., Fabrication of continuous-relief microoptical elements by direct laser writing in photoresist, *Proceedings of the SPIE - The International Society for Optical Engineering,* **1994b**, *2045*, 54–62.

[12] Jinwon, S., Pitchumani, M., Brown, J., Hockel, H. and Johnson, E.G., Analog micro-optics fabrication by use of a binary phase grating mask, *Proceedings of the SPIE - The International Society for Optical Engineering,* **2004**, *5347*(1), 62–70.

[13] Calixto, S. and Padilla, G.P., Micromirrors and microlenses fabricated on polymer materials by means of infrared radiation, *Applied Optics,* **1996**, *35*(31), 6126–6130.

[14] Fu, Y.Q. and Bryan, N.K.A., Novel one-step method of microlens mold array fabrication, *Optical Engineering,* **2001**, *40*(8), 1433–1434.

[15] Kluepfel, B. and Ross, F., *Holography marketplace: The reference text and directory of the holography industry,* Ross Books, Berkeley, **1991**.

[16] Palermo, P., Korpel, A., Dickinson, G. and Watson, W., Video disc mastering and replication, *Optics and Laser Technology,* **1977**, *9*(4), 169.

[17] Rossi, M. and Saarinen, J., Replication of micro-optics is cost-effective, *Optoelectronics World,* **2002**, *38*(8), S3–S5.

[18] Beinhorn, F., Ihlemann, J., Luther, K. and Troe, J., Micro-lens arrays generated by UV laser irradiation of doped PMMA, *Applied Physics A-Materials Science and Processing,* **1999**, *68*(6), 709–713.

[19] Naessens, K., Ottevaere, H., Baets, R., Van Daele, P. and Thienpont, H., Direct writing of microlenses in polycarbonate with excimer laser ablation, *Applied Optics,* **2003a**, *42*(31), 6349–6359.

[20] Naessens, K., Ottevaere, H., Van Daele, P. and Baets, R., Flexible fabrication of microlenses in polymer layers with excimer laser ablation, *Applied Surface Science,* **2003b**, *208*, 159–164.

[21] Gale, M.T. and Knop, K., The fabrication of fine lens arrays by laser beam writing, *Proceedings of the SPIE - The International Society for Optical Engineering,* **1983**, *398*, 347–353.

[22] Gale, M.T., Rossi, M., Kunz, R.E. and Bona, G.L., Direct laser writing of planar fresnel elements for optical interconnects, *Optical Computing, Proceedings of the International Conference,* **1995**, 267–270.

[23] Ornelas-Rodriguez, M. and Calixto, S., Mid-infrared optical elements fabricated on polymer materials, *Proceedings of the SPIE - The International Society for Optical Engineering,* **2000**, *4087*, 712–719.

[24] David, C., Wei, J., Lippert, T. and Wokaun, A., Diffractive grey-tone phase masks for laser ablation lithography, *Microelectronic Engineering,* **2001**, *57–8*, 453–460.

[25] Borek, G.T., Brown, D.M. and Shafer, J.A., Gray-scale fabrication of micro-optics in bulk zinc selenide and bulk multispectral zinc sulfide, *Proceedings of the SPIE - The International Society for Optical Engineering,* **2004**, *5347*(1), 28–37.

[26] Dickey, F.M. and Holswade, S.C., *Laser Beam Shaping Theory and Techniques,* Marcel Dekker, Inc., New York, **2000**.

[27] Fu, Y.Q., Kok, N. and Bryan, A., Microfabrication of microlens array by focused ion beam technology, *Microelectronic Engineering,* **2000a**, *54*(3–4), 211–221.

[28] He, M., Yuan, X.C., Ngo, N.Q., Bu, J. and Tao, S.H., Single-step fabrication of a microlens array in sol-gel material by direct laser writing and its application in optical coupling, *Journal of Optics A: Pure and Applied Optics,* **2004**, *6*(1), 94–97.

[29] He, M., Yuan, X.C., Ngo, N.Q., Cheong, W.C. and Bu, J., Simple fabrication of cylindrical sol-gel microlens array by reflowing technique, *Optical Engineering*, **2003b**, *42*(8), 2180–2181.

[30] Ihlemann, J. and WolffRottke, B., Excimer laser micro machining of inorganic dielectrics, *Applied Surface Science*, **1996**, *106*, 282–286.

[31] Kopitkovas, G., Lippert, T., David, C., Wokaun, A. and Gobrecht, J., Fabrication of micro-optical elements in quartz by laser induced backside wet etching, *Microelectronic Engineering*, **2003**, *67–8*, 438–444.

[32] Li, J., Herman, P.R., Wei, X.M., Chen, K.P., Ihlemann, J., Marowsky, G., *et al.*, High-resolution F_2-laser machining of micro-optic components, *Proceedings of the SPIE - The International Society for Optical Engineering*, **2002**, *4637*, 228–234.

[33] Romero, L.A. and Dickey, F.M., Lossless laser beam shaping, *Journal of the Optical Society of America A-Optics Image Science and Vision*, **1996**, *13*(4), 751–760.

[34] Savander, P., Microlens arrays etched into glass and silicon, *Optics and Lasers in Engineering*, **1994**, *20*(2), 97–107.

[35] Shealy, D.L. and Dickey, F.M., Laser beam shaping, *Optical Engineering*, **2003**, *42*(11), 3077–3079.

[36] Yuan, X.C., Yu, W.X., Ngo, N.Q. and Cheong, W.C., Cost-effective fabrication of microlenses on hybrid sol-gel glass with a high-energy beam-sensitive gray-scale mask, *Optics Express*, **2002**, *10*(7), 303–308.

[37] Ashkenasi, D., Varel, H., Rosenfeld, A., Henz, S., Herrmann, J. and Cambell, E.E.B., Application of self-focusing of ps laser pulses for three-dimensional microstructuring of transparent materials, *Applied Physics Letters*, **1998**, *72*(12), 1442–1444.

[38] Bruneau, S., Hermann, J., Sentis, M.L., Dumitru, G., Romano, V., Weber, H.P., *et al.*, Femtosecond laser ablation of materials, *Proceedings of the SPIE - The International Society for Optical Engineering*, **2003**, *5147*(1), 199–203

[39] Campbell, E.E.B., Ashkenasi, D. and Rosenfeld, A., Ultra-short-pulse laser irradiation and ablation of dielectrics, in *Lasers in Materials Science*, Vol. 301., (pp. 123–144), Transtec Publications Ltd., Zurich-Uetikon, **1999**.

[40] Henyk, M., Costache, F. and Reif, J., Femtosecond laser ablation from sodium chloride and barium fluoride, *Applied Surface Science*, **2002**, *186*(1–4), 381–384.

[41] Hertel, I.V., Stoian, R., Ashkenasi, D., Rosenfeld, A. and Campbell, E.E., Surface and bulk ultrashort-pulsed laser processing of transparent materials, *Proceedings of the SPIE - The International Society for Optical Engineering*, **2000**, *4088*, 17–24.

[42] Ihlemann, J., Excimer laser ablation of fused-silica, *Applied Surface Science*, **1992**, *54*, 193–200.

[43] Ihlemann, J., Scholl, A., Schmidt, H. and Wolffrottke, B., Nanosecond and femtosecond excimer-laser ablation of oxide ceramics, *Applied Physics A-Materials Science and Processing*, **1995**, *60*(4), 411–417.

[44] Marcinkevicius, A., Juodkazis, S., Watanabe, M., Miwa, M., Matsuo, S., Misawa, H., *et al.*, Femtosecond laser-assisted three-dimensional microfabrication in silica, *Optics Letters*, **2001**, *26*(5), 277–279.

[45] Nagashima, K., Hashida, M., Katto, M., Tsukamoto, M., Fujita, M. and Izawa, Y., Femtosecond laser ablation of Al_2O_3 ceramics, *Transactions of the Institute of Electrical Engineers of Japan, Part C,* **2004**, *124–C*(2), 388–392.

[46] Perry, M.D., Stuart, B.C., Banks, P.S., Feit, M.D., Yanovsky, V. and Rubenchik, A.M., Ultrashort-pulse laser machining of dielectric materials, *Journal of Applied Physics,* **1999**, *85*(9), 6803–6810.

[47] Varel, H., Ashkenasi, D., Rosenfeld, A., Herrmann, R., Noack, F. and Campbell, E.E.B., Laser-induced damage in SiO_2 and CaF_2 with picosecond and femtosecond laser pulses, *Applied Physics A-Materials Science and Processing,* **1996**, *62*(3), 293–294.

[48] Varel, H., Ashkenasi, D., Rosenfeld, A., Wahmer, M. and Campbell, E.E.B., Micromachining of quartz with ultrashort laser pulses, *Applied Physics A-Materials Science and Processing,* **1997**, *65*(4–5), 367–373.

[49] Dyer, P.E. and Walton, C.D., VUV laser ablation of insulators, *Applied Physics A-Materials Science and Processing,* **2004**, *79*(4–6), 721–727.

[50] Herman, P.R., Marjoribanks, R.S., Oettl, A., Chen, K., Konovalov, I. and Ness, S., Laser shaping of photonic materials: Deep-ultraviolet and ultrafast lasers, *Applied Surface Science,* **2000b**, *154*, 577–586.

[51] Ihlemann, J., Muller, S., Pischmann, S., Schafer, D., Wei, M., Li, J., *et al.*, Fabrication of submicron gratings in fused silica by F_2-laser ablation, *Applied Physics A-Materials Science and Processing,* **2003a**, *76*(5), 751–753.

[52] Mi Li, N., Herman, P.R., Nejadmalayeri, A.H. and Jianzhao, L., F_2-laser microfabrication of efficient diffractive optical phase elements, *Proceedings of the SPIE - The International Society for Optical Engineering,* **2004**, *5339*(1) 127–133.

[53] Sugioka, K., Comparison between DUV/VUV and femtosecond laser processing of dielectrics and semiconductors, *Review of Laser Engineering,* **2002**, *30*(5), 226–232.

[54] Sugioka, K., Obata, K., Hong, M.H., Wu, D.J., Wong, L.L., Lu, Y.F., *et al.*, Hybrid laser processing for microfabrication of glass, *Applied Physics A-Materials Science and Processing,* **2003**, *77*(2), 251–257.

[55] Sugioka, K., Obata, K., Midorikawa, K., Hong, M.H., Wu, D. J., Wong, L.L., *et al.*, Microprocessing of glass by hybrid laser processing, *Proceedings of the SPIE - The International Society for Optical Engineering,* **2002**, *4760*, 230–238.

[56] Sugioka, K., Zhang, J. and Midorikawa, K., Hybrid laser processing of micromachining of hard materials, *Proceedings of the SPIE - The International Society for Optical Engineering,* **1999**, *3822*, 6–17.

[57] Fu, Y. and Bryan, N.K.A., Fabrication of three-dimensional microstructures by two-dimensional slice by slice approaching via focused ion beam milling, *Journal of Vacuum Science and Technology B,* **2004**, *22*(4), 1672–1678.

[58] Fu, Y.Q., Kok, N., Bryan, A. and Shing, O.N., Microfabrication of diffractive optical element with continuous relief by focused ion beam, *Microelectronic Engineering,* **2000b**, *54*(3–4), 287–293.

[59] Karlov, N.V., Lukyanchuk, B.S., Sisakyan, E.V. and Shafeev, G.A., Etching of semi-conductors by products of the laser thermal-dissociation of molecular gases, *Kvantovaya Elektronika*, **1985**, *12*(4), 803–809.

[60] Jie, Z., Sugioka, K. and Midorikawa, K., Direct fabrication of microgratings in fused quartz by laser-induced plasma-assisted ablation with a KrF excimer laser, *Optics Letters*, **1998**, *23*(18), 1486–1488.

[61] Zhang, J., Sugioka, K. and Midorikawa, K., Laser-induced plasma-assisted ablation of fused quartz using the fourth harmonic of a Nd$^+$:YAG laser, *Applied Physics A-Materials Science and Processing*, **1998b**, *A67*(5), 545–549.

[62] Böhme, R., Zajadacz, J., Zimmer, K. and Rauschenbach, B., Topography and roughness evolution of microstructured surfaces at laser-induced backside wet etching, *Applied Physics A-Materials Science and Processing*, **2005**, *80*(2), 433–438.

[63] Ding, X., Kawaguchi, Y., Niino, H. and Yabe, A., Laser-induced high-quality etching of fused silica using a novel aqueous medium, *Applied Physics A-Materials Science and Processing*, **2002**, *75*(6), 641–645.

[64] Dolgaev, S.I., Lyalin, A.A., Simakin, A.V. and Shafeev, G.A., Fast etching of sapphire by a visible range quasi-cw laser radiation, *Applied Surface Science*, **1996b**, *96–8*, 491–495.

[65] Dolgaev, S.I., Lyalin, A.A., Simakin, A.V., Voronov, V.V. and Shafeev, G.A., Fast etching and metallization of via-holes in sapphire with the help of radiation by a copper vapor laser, *Applied Surface Science*, **1997**, *110*, 201–205.

[66] Kopitkovas, G., Lippert, T., David, C., Wokaun, A. and Gobrecht, J., Surface micro-machining of UV transparent materials, *Thin Solid Films*, **2004c**, *453–54*, 31–35.

[67] Niino, H., Ding, X., Kurosaki, R., Narazaki, A., Sato, T. and Kawaguchi, Y., Surface micro-fabrication of uv transparent materials by laser-induced backside wet etching, *CLEO/Pacific Rim 2003, The 5th Pacific Rim Conference on Lasers and Electro-Optics (IEEE Cat. No.03TH8671)*, **2003a**, 742–742.

[68] Vass, C., Hopp, B., Smausz, T. and Ignacz, F., Experiments and numerical calculations for the interpretation of the backside wet etching of fused silica, *Thin Solid Films*, **2004a**, *453–54*, 121–126.

[69] Wang, J., Niino, H. and Yabe, A., Micromachining of quartz crystal with excimer lasers by laser-induced backside wet etching, *Applied Physics A-Materials Science and Processing*, **1999a**, *69*, S271–S273.

[70] Wang, J., Niino, H. and Yabe, A., Micromachining of transparent materials with super-heated liquid generated by multiphotonic absorption of organic molecule, *Applied Surface Science*, **2000**, *154*, 571–576.

[71] Zimmer, K. and Böhme, R., Precise etching of fused silica for micro-optical applications, *Applied Surface Science*, **2005**, *243*(1–4), 415–420.

[72] Kopitkovas, G., Lippert, T., David, C., Canulescu, S., Wokaun, A. and Gobrecht, J., Fabrication of beam homogenizers in quartz by laser micromachining, *Journal of Photochemistry and Photobiology A-Chemistry*, **2004a**, *166*(1–3), 135–140.

[73] Kopitkovas, G., Lippert, T., David, C., Sulcas, R., Hobley, J., Wokaun, A.J., *et al.*, Laser micromachining of optical devices, *Proceedings of the SPIE - The International Society for Optical Engineering*, **2004b**, *5662*(1), 515–525.

[74] Srinivasan, R. and Mayne-Banton, V., Self-developing photoetching of poly(ethylene-terephthalate) films by far ultraviolet excimer laser-radiation, *Applied Physics Letters,* **1982**, *41*(6), 576–578.

[75] Kawamura, Y., Toyoda, K. and Namba, S., Effective deep ultraviolet photoetching of polymethyl methacrylate by an excimer laser, *Applied Physics Letters,* **1982**, *40*(5), 374–375.

[76] Aoki, H., U.S. Patent 5736999, **1998**.

[77] Patel, R.S. and Wassick, T.A., *Paper presented at the Proc. SPIE-Int. Soc. Opt. Eng.,* **1997**.

[78] Gale, M.T., Lang, G.K., Raynor, J.M. and Schutz, H., Fabrication of microoptical components by laser beam writing in photoresist, *Proceedings of the SPIE - The International Society for Optical Engineering,* **1991**, *1506*, 65–70.

[79] Quentel, F., Fieret, J., Holmes, A.S. and Paineau, S., Multilevel diffractive optical element manufacture by excimer laser ablation and halftone masks, *Proceedings of the SPIE - The International Society for Optical Engineering,* **2001**, *4274*, 420–431.

[80] Andrew, J.E., Dyer, P.E., Forster, D. and Key, P.H., Direct etching of polymeric materials using a XeCl laser, *Applied Physics Letters,* **1983**, *43*(8), 717–719.

[81] Deutsch, T.F. and Geis, M.W., Self-developing UV photoresist using excimer laser exposure, *Journal of Applied Physics,* **1983**, *54*(12), 7201–7204.

[82] Lazare, S. and Granier, V., Ultraviolet-laser photoablation of polymers - a review and recent results, *Laser Chemistry,* **1989**, *10*(1), 25–40.

[83] Mahan, G.D., Cole, H.S., Liu, Y.S. and Philipp, H.R., Theory of polymer ablation, *Applied Physics Letters,* **1988**, *53*(24), 2377–2379.

[84] Sutcliffe, E. and Srinivasan, R., Dynamics of UV laser ablation of organic polymer surfaces, *Journal of Applied Physics,* **1986**, *60*(9), 3315–3322.

[85] Arnold, N., Luk'yanchuk, B. and Bityurin, N., A fast quantitative modelling of ns laser ablation based on non-stationary averaging technique, *Applied Surface Science,* **1998**, *129*, 184–192.

[86] Cain, S.R., A photothermal model for polymer ablation - chemical modification, *Journal of Physical Chemistry,* **1993**, *97*(29), 7572–7577.

[87] Cain, S.R., Burns, F.C. and Otis, C.E., On single-photon ultraviolet ablation of polymeric materials, *Journal of Applied Physics,* **1992**, *71*(9), 4107–4117.

[88] D'Couto, G.C. and Babu, S.V., Heat-transfer and material removal in pulsed excimer-laser-induced ablation - pulsewidth dependence, *Journal of Applied Physics,* **1994**, *76*(5), 3052–3058.

[89] Luk'yanchuk, B., Bityurin, N., Himmelbauer, M. and Arnold, N., UV-laser ablation of polyimide: From long to ultra-short laser pulses, *Nuclear Instruments and Methods in Physics Research Section B-Beam Interactions with Materials and Atoms,* **1997**, *122*(3), 347–355.

[90] Schmidt, H., Ihlemann, J., Wolff-Rottke, B., Luther, K. and Troe, J., Ultraviolet laser ablation of polymers: Spot size, pulse duration, and plume attenuation effects explained, *Journal of Applied Physics,* **1998**, *83*(10), 5458–5468.

[91] Srinivasan, V., Smrtic, M.A. and Babu, S.V., Excimer laser etching of polymers, *Journal of Applied Physics*, **1986**, *59*(11), 3861–3867.

[92] Luk'yanchuk, B., Bityurin, N., Anisimov, S., Arnold, N. and Bauerle, D., The role of excited species in ultraviolet-laser materials ablation.3, Non-stationary ablation of organic polymers, *Applied Physics A-Materials Science and Processing*, **1996**, *62*(5), 397–401.

[93] Luk'yanchuk, B., Bityurin, N., Anisimov, S. and Bauerle, D., The role of excited species in UV-laser materials ablation.1, Photophysical ablation of organic polymers, *Applied Physics A-Materials Science and Processing*, **1993**, *57*(4), 367–374.

[94] Bityurin, N., Malyshev, A., Luk'yanchuk, B., Anisimov, S. and Bäuerle, D., *Paper presented at the Proc. SPIE.*, **1996**.

[95] Bityurin, N., UV etching accompanied by modifications, Surface etching, *Applied Surface Science*, **1999**, *139*, 354–358.

[96] Treyz, G.V., Scarmozzino, R. and Osgood, R.M., Deep ultraviolet laser etching of vias in polyimide films, *Applied Physics Letters*, **1989**, *55*(4), 346–348.

[97] Srinivasan, R. and Braren, B., Ultraviolet-laser ablation of organic polymers, *Chemical Reviews*, **1989**, *89*(6), 1303–1316.

[98] Küper, S., Brannon, J. and Brannon, K., Threshold behavior in polyimide photoablation - single-shot rate measurements and surface-temperature modelling, *Applied Physics A-Materials Science and Processing*, **1993**, *56*(1), 43–50.

[99] Arnold, N. and Bityurin, N., Model for laser-induced thermal degradation and ablation of polymers, *Applied Physics A-Materials Science and Processing*, **1999**, *68*(6), 615–625.

[100] Srinivasan, R. and Braren, B., Ablative photodecomposition of polymer-films by pulsed far-ultraviolet (193 nm) laser-radiation – dependence of etch depth on experimental conditions, *Journal of Polymer Science Part A-Polymer Chemistry*, **1984**, *22*(10), 2601–2609.

[101] Srinivasan, R., Braren, B. and Casey, K.G., Nature of incubation pulses in the ultraviolet-laser ablation of polymethyl methacrylate, *Journal of Applied Physics*, **1990**, *68*(4), 1842–1847.

[102] Suzuki, K., Matsuda, M., Ogino, T., Hayashi, T., Terabayashi, T. and Amemiya, K., *Paper presented at the Proc. SPIE*, **1997**.

[103] Lippert, T., Hauer, M., Phipps, C.R. and Wokaun, A., Fundamentals and applications of polymers designed for laser ablation, *Applied Physics A-Materials Science and Processing*, **2003**, *77*(2), 259–264.

[104] Bennett, L.S., Lippert, T., Furutani, H., Fukumura, H. and Masuhara, H., Laser induced microexplosions of a photosensitive polymer, *Applied Physics A-Materials Science and Processing*, **1996**, *63*(4), 327–332.

[105] Lippert, T., Langford, S.C., Wokaun, A., Savas, G. and Dickinson, J.T., Analysis of neutral fragments from ultraviolet laser irradiation of a photolabile triazeno polymer, *Journal of Applied Physics*, **1999a**, *86*(12), 7116–7122.

[106] Lippert, T. and Dickinson, J.T., Chemical and spectroscopic aspects of polymer ablation: Special features and novel directions, *Chemical Reviews*, **2003**, *103*(2), 453–485.

[107] Lippert, T., David, C., Hauer, M., Wokaun, A., Robert, J., Nuyken, O., *et al.*, Polymers for UV and near-IR irradiation, *Journal of Photochemistry and Photobiology A-Chemistry,* **2001**, *145*(1–2), 87–92.

[108] Lippert, T., Bennett, L.S., Nakamura, T., Niino, H., Ouchi, A. and Yabe, A., Comparison of the transmission behavior of a triazeno-polymer with a theoretical model, *Applied Physics A-Materials Science and Processing,* **1996**, *63*(3), 257–265.

[109] Lippert, T., Dickinson, J.T., Hauer, M., Kopitkovas, G., Langford, S.C., Masuhara, H., *et al.*, Polymers designed for laser ablation-influence of photochemical properties, *Applied Surface Science,* **2002**, *197*, 746–756.

[110] Lippert, T., Ortelli, E., Panitz, J.C., Raimondi, F., Wambach, J., Wei, J., *et al.*, Imaging-XPS/Raman investigation on the carbonization of polyimide after irradiation at 308 nm, *Applied Physics A-Materials Science and Processing,* **1999b**, *69*, S651–S654.

[111] Raimondi, F., Abolhassani, S., Brutsch, R., Geiger, F., Lippert, T., Wambach, J., *et al.*, Quantification of polyimide carbonization after laser ablation, *Journal of Applied Physics,* **2000**, *88*(6), 3659–3666.

[112] Brannon, J.H., Lankard, J.R., Baise, A.I., Burns, F. and Kaufman, J., Excimer laser etching of polyimide, *Journal of Applied Physics,* **1985**, *58*(5), 2036–2043.

[113] Braun, A., Zimmer, K. and Bigl, F., Combination of contour and half-tone masks used in laser ablation, *Applied Surface Science,* **2000**, *168*(1–4), 178–181.

[114] Daschner, W., Long, P., Stein, R., Wu, C. and Lee, S.H., Cost-effective mass fabrication of multilevel diffractive optical elements by use of a single optical exposure with a gray-scale mask on high-energy beam-sensitive glass, *Applied Optics,* **1997**, *36*(20), 4675–4680.

[115] Suleski, T.J. and Oshea, D.C., Gray-scale masks for diffractive-optics fabrication:1, Commercial slide imagers, *Applied Optics,* **1995**, *34*(32), 7507–7517.

[116] Zimmer, K., Braun, A. and Bigl, F., Combination of different processing methods for the fabrication of 3d polymer structures by excimer laser machining, *Applied Surface Science,* **2000**, *154*, 601–604.

[117] Smith, A.H. and Hunter, R.O.Jr., High power phase masks for imaging systems, U.S. Patent 5328785, **1992**.

[118] Smith, A.H., Hunter, R.O.Jr. and McArthur, B.B., High power masks and methods for manufacturing same, U.S. Patent 5501925, **1994**.

[119] David, C. and Hambach, D., Line width control using a defocused low voltage electron beam, *Microelectronic Engineering,* **1999**, *46*(1–4), 219–222.

[120] Braun, A. and Zimmer, K., Diffractive gray scale masks for excimer laser ablation, *Applied Surface Science,* **2002**, *186*(1–4), 200–205.

[121] He, M., Yuan, X., Ngo, N.Q., Cheong, W.C. and Jing, B., Reflow technique for the fabrication of an elliptical microlens array in sol-gel material, *Applied Optics,* **2003a**, *42*(36), 7174–7178.

[122] Herzig, H.P., Schilling, A., Stauffer, L., Vokinger, U. and Rossi, M., Efficient beamshaping of high-power diode-lasers using micro-optics, *Proceedings of the SPIE - The International Society for Optical Engineering,* **2001**, *4437*, 134–141.

[123] Nussbaum, P., Volkel, R., Herzig, H.P., Eisner, M. and Haselbeck, S., Design, fabrication and testing of microlens arrays for sensors and microsystems, *Pure and Applied Optics,* **1997**, *6*(6), 617–636.

[124] Ihlemann, J., Muller, S., Puschmann, S., Schafer, D., Herman, P.R., Jianzhao, L., *et al.*, F_2-laser microfabrication of sub-micrometer gratings in fused silica, *Proceedings of the SPIE - The International Society for Optical Engineering,* **2003b**, *4941*, 94–98.

[125] Fu, Y.Q. and Bryan, N.K.A., Investigation of integrated diffractive/refractive microlens microfabricated by focused ion beam, *Review of Scientific Instruments,* **2000**, *71*(6), 2263–2266.

[126] Fu, Y.Q., Bryan, N.K.A. and Zhou, W., Quasi-direct writing of diffractive structures with a focused ion beam, *Optics Express,* **2004**, *12*(9), 1803–1809.

[127] Zhang, J., Sugioka, K. and Midorikawa, K., High-speed machining of glass materials by laser-induced plasma-assisted ablation using a 532-nm laser, *Applied Physics A-Materials Science and Processing,* **1998a**, *A67*(4), 499–501.

[128] Dolgaev, S.I., Lyalin, A.A., Shafeev, G.A. and Voronov, V.V., Fast etching and metallization of sic ceramics with copper-vapor-laser radiation, *Applied Physics A-Materials Science and Processing,* **1996a**, *63*(1), 75–79.

[129] Dolgaev, S.I., Kirichenko, N.A. and Shafeev, G.A., Deposition of nanostructured Cr_2O_3 on amorphous substrates under laser irradiation of the solid-liquid interface, *Applied Surface Science,* **1999**, *139*, 449–454.

[130] Wang, J., Niino, H. and Yabe, A., One-step microfabrication of fused silica by laser ablation of an organic solution, *Applied Physics A-Materials Science and Processing,* **1999b**, *68*(1), 111–113.

[131] Böhme, R., Spemann, D. and Zimmer, K., Surface characterization of backside-etched transparent dielectrics, *Thin Solid Films,* **2004**, *453–54*, 127–132.

[132] Böhme, R. and Zimmer, K., Low roughness laser etching of fused silica using an adsorbed layer, *Applied Surface Science,* **2004**, *239*(1), 109–116.

[133] Ding, X.M., Sato, T., Kawaguchi, Y. and Niino, H., Laser-induced backside wet etching of sapphire, *Japanese Journal of Applied Physics Part 2-Letters,* **2003**, *42*(2B), L176–L178.

[134] Kawaguchi, Y., Ding, X., Narazaki, A., Sato, T. and Niino, H., Transient pressure induced by laser ablation of liquid toluene: Toward the understanding of laser-induced backside wet etching, *Applied Physics A-Materials Science and Processing,* **2004**, *79*(4–6), 883–885.

[135] Vass, C., Smausz, T. and Hopp, B., Wet etching of fused silica: A multiplex study, *Journal of Physics D-Applied Physics,* **2004b**, *37*(17), 2449–2454.

[136] Zimmer, K., Böhme, R., Braun, A., Rauschenbach, B. and Bigl, F., Excimer laser-induced etching of sub-micron surface relief gratings in fused silica using phase grating projection, *Applied Physics A-Materials Science and Processing,* **2002**, *74*(4), 453–456.

[137] Böhme, R., Braun, A. and Zimmer, K., Backside etching of UV-transparent materials at the interface to liquids, *Applied Surface Science,* **2002**, *186*(1–4), 276–281.

[138] Kawaguchi, Y., Ding, X., Narazaki, A., Sato, T. and Niino, H., Transient pressure induced by laser ablation of toluene, a highly laser-absorbing liquid, *Applied Physics A-Materials Science and Processing,* **2005**, *80*(2), 275–281.

[139] Zimmer, K., Böhme, R. and Rauschenbach, B., Laser etching of fused silica using an absorbed toluene layer, *Applied Physics A-Materials Science and Processing,* **2004**, *79*(8), 1883–1885.

[140] Niino, H., Yasui, Y., Ximing, D., Narazaki, A., Sato, T., Kawaguchi, Y., *et al.,* Surface microfabrication of silica glass by excimer laser irradiation of toluene solution, *Proceedings of the SPIE - The International Society for Optical Engineering,* **2003b**, *4977,* 269–280.

[141] Duley, W.W., *Uv Lasers: Effects and Applications in Materials Science,* Press Syndicate of the University of Cambridge, Cambridge, **1996**.

Chapter 5

Ultraviolet Laser Ablation of Polymers and the Role of Liquid Formation and Expulsion

S. Lazare and V. Tokarev

LPCM UMR 5803, Université de Bordeaux 1, 351 cours de la Libération, F-33405 Talence, France

Abstract

Excimer laser ablation of common polymers is introduced and multipulse high aspect (up to 600) ratio microdrilling (diam. 25 µm) is explained. Mechanisms of material removal include new considerations on liquid expulsion driven by the ablation recoil pressure gradient. Experiments of poly(ethylene terephthalate) (PET) ablation with periodic microbeams (period Λ submicronic) are presented and evidence of transient liquid lateral flow is provided. Ablation of polymethylmethacrylate (PMMA) shows a regime of explosive boiling which ejects droplets and millimeter long nanofilaments. As according to our model, liquid expulsion is very dependent on the viscosity of the transient liquid which has to be low enough to allow for a flow.

Keywords: Excimer laser ablation; KrF laser ablation; Laser ablation expulsion; Microbumps; Microdrilling; Micro-Raman spectroscopy; Navier-Stokes equation; Submicron resolution; Transient liquid expulsion; UV polymer ablation; Viscous microflow.

5.1. Introduction

Ablation of polymer surfaces by pulsed ultraviolet lasers [1–3] is studied for more than 20 years [4,5] and is still receiving a growing interest (see a recent review [6]), since it is one of the key processes for future technology and industry

Recent Advances in Laser Processing of Materials
J. Perrière, E. Millon and E. Fogarassy (Editors)

(e.g. laser microdrilling). A full understanding of the complex mechanisms involved in the laser based material removal process is the target of the efforts of the scientific community due to its increasing importance in both academic and application fields. Ultraviolet wavelengths (150 nm $< \lambda <$ 350 nm) match the strong absorption (absorption coefficient $\gamma \sim 10^3$–10^5 cm^{-1}, Table 1) of most polymers so that the incident energy is deposited just below the surface in an absorption length $l_{abs} = 1/\gamma$ of less than ~1 μm. This ensures correspondingly low enough threshold fluences F_t (minimum energy density above which ablation is measured) of the order of ~10–100 mJ/cm^2 for nanosecond pulses (Table 1). Upon absorption of the laser pulse, if the absorbed fluence exceeds the threshold fluence $F > F_t$, the suddenly excited material expands rapidly into the ablation plume, a hot mixture of gas products, particles, liquid microdroplets which are expelled at high speed in a direction roughly normal to the material surface. This direct mechanism of material removal is effective in for instance multipulse laser microdrilling as described below. However as we have demonstrated experimentally with periodic microbeams, a lateral flow of liquid material during the ablation pulse can be obtained owing to the strong recoil pressure which is exerted on the ablating surface. The cleanliness of the impacts left on the surface depends strongly on the transient hot 'liquid' state and on the gradient of recoil pressure. We have shown that some polymers like poly (ethylene terephthalate) (PET) which yields a thin transient liquid layer of low viscosity display a controlled lateral expulsion of liquid by ablation with a moderate fluence (~1 J/cm^2) to form an accumulation of polymer (microbumps) just at the outside of the laser spot, where pressure is lower. In a less controlled expulsion, because in the regime of explosive boiling or phase explosion, KrF laser ablation of poly(methyl methacrylate) (PMMA) forms millimeter long fibers with nanometric diameter. The mechanisms of these liquid formation and expulsion are discussed below. In the following text we will first recall the basics of laser ablation of polymers, then we present the experiments which have lead to the study of high aspect ratio microdrilling, to submicron ablation and then we explain the new experiments which have evidenced the transient liquid expulsion.

Table 1: Polymer absorption coefficient γ at the KrF laser wavelength 248 nm, absorption length $l_{abs} = 1/\gamma$ and ablation threshold F_t

	PET	PEN	PC	PEEK	PI	PS	PMMA
γ (μm^{-1})	16	20	1	~10	22	0.61	0.0063
$l_{abs} = 1/\gamma$ (μm)	0.065	0.05	1	~0.1	0.045	1.6	150
F_t (mJ/cm^2)	30	40	40	50	54	40	250

5.2. Polymers and Lasers: The Situation in 2005

The polymers studied are all of high technological interest: polyimide (PI), poly-(ethylene terephthalate) (PET), poly(ethylene naphthalate) (PEN), polystyrene (PS), poly(ether ether ketone) (PEEK), poly(methyl methacrylate) (PMMA) and bisphenol A polycarbonate (PC). Commercial names are Kapton for PI, Mylar D for PET, Teonex for PEN, Stabar for PEEK. PS samples were taken from petri dish and PC samples from polycarbonate panels (Axxis, DSM Engineering Products). Basic studies are done with this set of usual polymers which have smooth surfaces with properties easily probed by atomic force microscopy (AFM), good optical quality (allowing optical microscope observation) and are like standards for ablation experiments. As seen from Table 1, PI, PET, PEN and PEEK have high absorption coefficient for the KrF wavelength at 248 nm for instance, whereas PMMA, PS and PC are weak absorbers at room temperature for the used laser wavelength. The early work in microdrilling was done with polymer/polymer composite material made by Dassault aeronautics industry. It is composed of Kevlar fibers embedded in an epoxy matrix and because of the presence of the Kevlar fibers it is impossible to microdrill with conventional mechanical tools. Therefore the laser microdrilling offers a good alternative to this problem as seen in Fig. 1. Figure 1 also illustrates laser precision micromachining. From these examples, it is easy to realize the performances of the laser ablation in term of spatial precision and quality of interaction which most of the time leaves a clean ablated surface with negligible damage on it. Most of our experiments are performed with the radiation of the KrF excimer laser since it is the most powerful wavelength from this laser. 248 nm is also a good

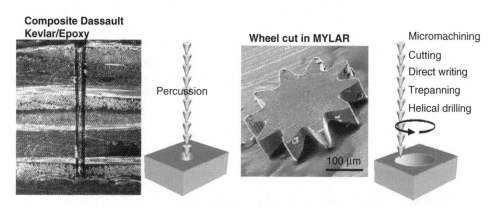

Figure 1: Illustration of the use of ultraviolet laser ablation in 1) multipulse (percussion) microdrilling of kevlar/epoxy composite material (hole diameter of the order of 100 μm) and 2) micromachining of a microwheel in PET by a computer driven ablation process.

compromise in terms of absorptivity of the material (Table 1). Shorter wavelengths ArF/193 nm, F_2/157 nm are good alternatives [7] in the cases of materials essentially transparent at 248 nm, e.g. SiO_2 and poly(tetrafluoro ethylene) (PTFE). In fact for the applications, one might be interested in minimizing the processing time and recent sources like all solid state lasers of Nd/YAG 3 and 4ω (351 and 266 nm) are now proposed with enhanced averaged power by laser manufacturers. As other trend is the use of ultrafast or femtosecond lasers, mainly based on Ti/Sapphire oscillator at 800 nm. This wavelength is usually outside the absorptions bands of most polymers and therefore some laboratories [8] develop tripled Ti/Sa 745 nm femtosecond (200–500 fs) laser systems whose pulses at 248 nm can be amplified in a KrF laser. Such lasers are starting to be experimented in material processing and the studies concerning ablation of polymers are only at the beginning, despite the pioneering publication of Küper and Stuke earlier in 1989 [9].

5.3. From Ablation Curve to Ablation Recoil Pressure

As will be seen below, the liquid expulsion during the ablation phenomenon is due to a strong recoil pressure created by the momentum transfer of the accelerated molecules leaving the surface at high speed. The recoil pressure P is related to the momentum impulse transferred by the mass which leaves the surface and to the leaving velocity. It can be evaluated as follows. For this we first consider the mechanical momentum exerted on the sample during the pulse [10–14]:

$$mw = \rho S \delta w = S \int_0^\infty P(t)dt \tag{1}$$

m is the mass of ejected material, w is the velocity of the center of mass of the plume, S is the area of the irradiated spot, δ is the ablation depth, a function of fluence. With the use of a characteristic duration τ_p of the pressure pulse the integral in Equation (1) is $\bar{P}\tau_p$, where $\bar{P}$ is an averaged pressure. Then from (1) $\bar{P}$ is simply deduced as

$$\bar{P} = \frac{\rho \delta w}{\tau_p} \tag{2}$$

w relates to the kinetic energy of the plume $E_K = \frac{1}{2}mw^2$.

At the late stage of plume expansion, when practically all internal energy of the plume is transferred to the energy of a directed translational motion of plasma cloud from the surface as a result of a so-called adiabatic expansion, $E_k = E_p$ and hence

$$w = \sqrt{\frac{2E_p}{\rho S \delta}} \tag{3}$$

The energy of the plume E_p is related to the laser energy of the pulse for the 'air-plume-target' system by the following relationship:

$$E_p = E - E_R - E_s \tag{4}$$

in which $E = SF$ is the laser energy (F is fluence), E_R is the incident energy loss by reflection on the sample and E_s is the energy staying in the sample after ablation. For polymer with no heat diffusion outside the absorbing volume:

$$E_s \approx AF_t S \tag{5}$$

with $A = 1 - R$ is the absorptivity (R reflectivity) of the sample. The energy lost by reflection on the sample surface is:

$$E_R = SR \left(\int_0^{t_{th}} I(t)dt - \int_{t_{th}}^{\tau} \sigma^2(t)I(t)dt \right) \quad \text{for} \quad F > F_t \tag{6}$$

Here the first term describes reflection losses from the sample during the absence of the plume since t_{th} is the onset of the plume shielding during the pulse. The second term corresponds to the reflection losses form the target-plume system for $t_{th} < t < \tau$ (τ is the laser pulse duration), with $\sigma(t)$ being the attenuation of laser intensity by the plume and $\sigma^2(t)$, because the beam passes 2 times in the plume. It is possible to find explicitly $\sigma(t)$ by using Equation (7) which conveniently approaches the ablation depth [15] of Fig. 2 which displays experimental data obtained with the quartz crystal microbalance [16] and the theoretical model [17] of the moving interface. Equation (7) is different from the model of the moving interface [3,17] and is an analytical function simpler to handle once the three parameters F_t, μ and ζ have been determined by fitting the experimental data (for PET Fig. 2 at 248 nm: $F_t = 0.04$ J/cm^2, $\mu = 0.36$, $\zeta = 1.47$):

$$\delta(t) = \frac{1}{\mu \alpha} \ln \left(1 + \mu \zeta \frac{F(t) - F_t}{F_t} \right) \tag{7}$$

It follows

$$\sigma(t) = e^{-\mu \alpha \delta(t)} = \frac{1}{1 + \mu \zeta (F(t) - F_t)/F_t} \quad \text{for} \quad t_{th} < t < \tau \tag{8}$$

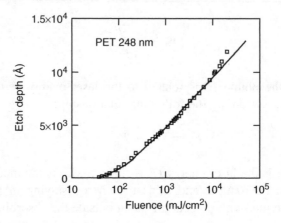

Figure 2: Ablation curve of PET obtained at 248 nm from which the ablated mass is calculated. The points are experimental data obtained with the quartz crystal microbalance and the solid line is a fit with the theoretical model of the moving interface [3,17].

and

$$E_R = SR\left[F_t + \frac{(F - F_t)}{1 + \mu\zeta(F - F_t)/F_t}\right] \qquad (9)$$

By combining Equations (4), (5) and (9) one can obtain the fraction of energy released in the plume:

$$\xi(F) = \frac{E_p}{E} = 1 - R\frac{F_t}{F} - R\left(1 - \frac{F_t}{F}\right)\frac{1}{1 + \mu\zeta(F - F_t)/F_t} - A\frac{F_t}{F} \quad (10)$$

As seen from Fig. 3, calculated for parameters of PET given above, ξ starts from zero at $F = F_t$ and then rapidly increases and levels off to 1. For example, $\xi = 0.583$ for $F = 0.1$ J/cm^2 and $\xi = 0.956$ for $F = 1$ J/cm^2 ($= 25\ F_t$). It means that at high fluences ($F \gg F_t$) practically all laser energy E is released in the plume: $E_p = \xi(F)E \cong E$. The use of $E_p = \xi(F)FS$ in Equations (3) and (2) with $\tau_p \cong 2\tau$ allows finally to find the average recoil pressure:

$$\bar{P} \approx \frac{1}{\sqrt{2\tau}}\sqrt{\rho\delta\xi F} \qquad (11)$$

In fact all plume energy is not converted into kinetic energy concentrated into the translation normal to surface as we assume. The energy dissipation is complex since

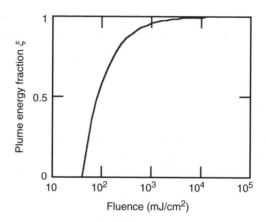

Figure 3: Energy fraction ξ deposited in the plume during the KrF laser ablation of PET.

a part of it is consumed into heat and decomposes the material and some other part is returned to heat when condensation reactions are possible. For polymers, however this chemical exchange energy is of the order of the threshold fluence F_t ($\sim$10–50 mJ/cm^2) and can be neglected against the total pulse energy when it is of the order of 1000 mJ/cm^2 or more. Therefore it is a reasonable approximation for polymers known as thermolabile compounds to replace plume kinetic energy by plume energy.

For example, $\bar{P} = 645$ bar is calculated for PET data ($\rho = 1380$ kg/m^3, $F = 1$ J/cm^2, $\delta = 0.5$ μm [3] and $\tau = 30$ ns) as seen in Fig. 4. This relation is true in a wide range of fluences and is a convenient evaluation of the plume pressure. For high fluences ($F \gg F_t$) it is simplified by putting $\xi \cong 1$. $\bar{P}$ in this case can be calculated directly

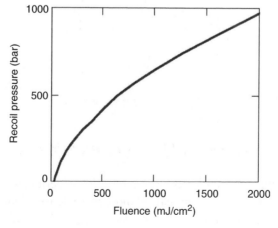

Figure 4: Ablation pressure for PET at 248 nm as calculated from Equation (11).

from the experimental data on ablation depth irrespective to a particular model of ablation. Ablation pressure plays an important role in transient liquid expulsion and will be used in the experimental work presented below.

5.4. Experimental

Although not restricted to UV, polymer ablation takes large advantage of short wavelength radiation of ultraviolet laser beam, thanks to its focusability and the high absorptivity of the polymer in general (Table 1). Transformation of the beam by a single lens (Fig. 5) yields a waist of diameter

$$d = 4\lambda f / \pi D \tag{12}$$

which is proportional to wavelength λ with f being the focal length, in the hypothesis of a perfect beam with $M^2 = 1$. As a consequence, a UV laser of given energy can provide a smaller spot of larger fluence F than a spot made with lasers with larger wavelength and of similar pulse energy. As seen below, microdrilling characteristics like wall angle is strongly dependent on the incident fluence F. Furthermore the ultimate resolution achieved by a light beam is also proportional to wavelength:

$$r = \lambda/2 \times \text{N.A.} \tag{13}$$

(N.A.: numerical aperture of the laser beam)

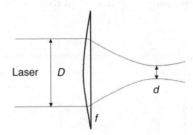

Figure 5: Laser focusing.

5.4.1. General Setup Designed for Laser Microdrilling

The laser beam setup typically used for microdrilling experiments is shown in Fig. 6. It is composed of two silica (Suprasil) plano-convex lenses. One is the condensor ($f_{\text{cond}} = 250$ mm) and the other is the projector ($f_{\text{proj}} = 25$ mm). They are chosen because of their small diameter, in order to keep the thickness at the beam path

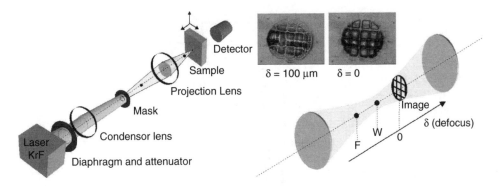

Figure 6: Typical optical setup used for microdrilling or surface irradiation with spot of 100 μm size. Positioning of the sample in the beam waist with the aid of a nickel grid placed on the mask and whose image is looked for in the waist by ablation with one pulse on PC surface. $\delta = 0$ is by definition the image position.

to a minimum and avoid unwanted attenuation of the radiation (they must stand 200 pulses/s for minutes with good stability). The use of a precision lens [18] as projection optics is not completely excluded, but was not preferred for this particular study. The excimer laser is a Lambda Physik LPX 220i (pulse energy $E = 0.4$ J, pulse duration $\tau = 25$–30 ns, pulse peak intensity $I = E/\tau = 1.6 \times 10^7$ W and repetition rate 200 Hz maximum) whose cavity is equipped with regular planar aluminum mirrors and which has a length of about 1.50 m, therefore offering a maximum of energy in the low divergence resonator modes. A circular molybdenum aperture (e.g. $\Phi = 250$ μm) is used to shape the beam at a distance of the condensor equal approximately to f_{cond}. This mask is precisely imaged onto the polymer surface, as illustrated in Fig. 6, with the projection lens so that the demagnification ratio is of the order of 1/5. A fine nickel grid (Goodfellow, period 340 μm) is placed on the mask and the polymer target is moved near the waist position so as to obtain a good position giving the best ablation pattern (which is then called defocus zero $\delta = 0$) as seen in the optical microscopy image in upper left inset in Fig. 6. Polycarbonate is used for this fine adjustment made with one single pulse, since its ablation is clearly not disturbed by liquid formation and lateral flow along the surface. It proved to be a very sensitive technique. Two diaphragms are also used to limit the influence of scattered radiation, which are positioned before each lens. Their adjustment is done on an empirical basis.

5.4.2. Precision Lens Projection Setup for Submicron Ablation

The use of precision lens is necessary when the wanted beam size approaches 1 μm in experiments designed to explore the limits of resolution. The important point

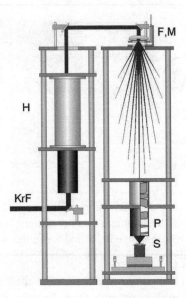

Figure 7: Submicron ablation optical setup using the precision lens (P) to image a mask (M) on the sample (S) surface. (H) Homogenizer, (F) field lens.

also is to have a good knowledge of the intensity profile. With that respect, modelling of the image is required since the experimental determination of the incident fluence is of limited interest. Setup is presented in Fig. 7. Masks are made of Cr on silica binary gratings of period chosen between 2.1 and 3.7 µm. The precision lens is a ×10 fully corrected optical system designed by F. Goodall (R.A.L., U.K.) and contained in a cylinder of 195 mm length. Its numerical aperture is 0.2 and the resolution given by Equation (13) is 0.62 µm. The lens to sample distance is 23.5 mm and the mask to lens distance is 545 mm. The sample positioning is achieved with XYZ microstages and surface tilt is adjusted with the aid of a vertical micrometer head.

5.5. Microdrilling

The science of laser microdrilling is a growing field [19–31] since in the applications, it is a good and effective alternative to mechanical drilling when the hole diameter becomes small (<100 µm). In this framework we have recently [32–36] investigated the ultimate limits of laser microdrilling mainly in polymeric materials. In particular with respect to the two dimensions of interest in microdrilling applications, hole diameter (d) and maximum hole length (l) obtained for a given set of laser parameters (wavelength, beam size, energy density). In the present work, we show that unprecedented high aspect ratio ($R = l/d$, up to 600) can be reached when experimental conditions are adequately chosen.

5.5.1. Rate of Drilling and Stationary Profile with PET Example

A typical microdrilling experiment, done after beam optimization, is illustrated in Fig. 8. It requires usually many pulses. As in Fig. 8A, the rate has an initial value (a) which can be related to surface ablation with 3D expansion characteristic of surface ablation (as opposed to deep ablation), then it displays a constant value (b) in deep keyhole because plume absorption is increased and redeposition sets in, and finally it goes to a 'zero value' (c) when the stationary profile is reached. With an average rate of 0.7 µm/pulse the stationary profile is reached in approximately 7500 pulses. The intermediate pictures **B, C, D** are respectively taken at 3500, 4500, 6000 pulses. We observe the evolution of the main laser front or hole bottom which is rather square at the beginning and which becomes sharper when the stationary state or end of drilling is approached. When the stationary state is reached the tip does not move anymore, the lateral dimensions of the hole do not change as well. The laser energy is therefore trapped within the hole and spread over the hole inner surface so that its surface density is lowered below the material removal threshold.

This result can be better understood with a simplified model of a conical hole in Fig. 9 (we will see below that the real hole profile is not conical). An ideal fully collimated beam however would create the formation of a cone of angle α and

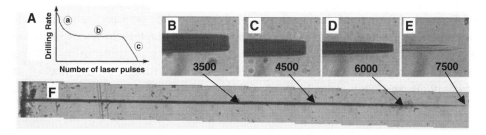

Figure 8: Microdrilling characteristics of a hole with 25 µm diameter and 5 mm length drilled in **PET** with the KrF laser. **A** drilling rate of advancement as a function of time (or pulse number) **B** to **E** shape of the advancing ablation front at various times of the experiment. **E** and **F** hole after stationary profile (end of microdrilling), **E** final sharp tip. The scale of the pictures is given by the hole diameter 25 µm and the hole image results of the assembling of many partial images of the same hole.

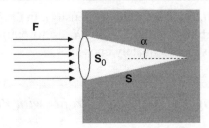

Figure 9: Simplified model of stationary state of drilled conical hole.

aspect ratio $1/\alpha$. The drilling then would stop when the fluence at the wall is a certain threshold F_∞ so that

$$FS_0 = F_\infty S = F_\infty S_0/\alpha \qquad (14)$$

From this aspect ratio can be estimated, for example as follows:

$$R = 1/\alpha = F/F_\infty > F/F_t \qquad (15)$$

since the threshold F_∞ at which the drilling stops is lower than the usual surface ablation threshold F_t. With typical fluence $F = 30$ J/cm^2 and threshold $F_t = 30$ mJ/cm^2, an aspect ratio of the order of 10^3 would be predicted. This is comparable to what experience gives.

5.5.2. Other Polymers

PET under the form of Mylar D, presented above gives the best aspect ratio (Table 2) then there is a group with PC, PEEK and PI which displays lower aspect ratio $R_m \sim 350$ and PMMA, PS have the lower values. It is interesting to understand why, when chemical structure is varied, the microdrilling performance is so strongly affected. This particular point will be discussed below with the presentation of drilling features related to each specific polymer.

5.5.2.1. Polycarbonate (PC)
Polycarbonate also gives high aspect ratio holes (Table 2). The surface after ablation has a black color characteristic of carbon material, already observed in multipulse surface ablation of PC. As we already reported at some point of the drilling, a branching phenomenon can be observed. In Fig. 10 it appears at the tip end when the laser fluence is attenuated to values in the vicinity of the threshold value.

Table 2: Polymer characteristic threshold F_∞ of ablation extinction at the end of drilling provided by the fit with model, Equation (26). Maximum aspect ratio R_m and conicity of the hole $1/R_m$

Material	F_∞ (mJ/cm^2)	R_m	$1/R_m$ (10^{-3})
PET	1	565	1.77
PC	2.2	390	2.56
PEEK	8.5	385	2.60
PI	3	360	2.78
PS	5	315	3.17
PMMA	6.6	255	3.92

The mechanism of formation of the branches [36] is analogous to the formation of cones, reported in surface ablation. They form because the absorbed fluence $FA(i)\cos(i)$ becomes equal to F_t (i is the beam incidence angle). This branching is a specific phenomenon of polycarbonate and other polymers like PI and PEEK. They hold in common the characteristic of yielding by ablation graphite to a large extent, whereas other polymers PET, PS and PMMA much less. It is also thought that expulsed ablation products are gases and nanoparticles as opposed to liquid micro and nano droplets in other cases.

Figure 10: A set of two adjacent holes in polycarbonate of 25 µm diameter and distant of 40 µm showing the precision of the laser drilling. The second hole drilling is not perturbed by the presence of the first one.

5.5.2.2. Poly(methylmethacrylate) (PMMA)
An example of KrF laser drilled hole in PMMA is shown in Fig. 11 with a magnification which allows to reveal several interesting features concerning the drilling mechanism. Its aspect ratio is only $R = 100$ in Fig. 11 and not the maximum aspect ratio $R_m = 250$ (Table 2), because its diameter does not allow to reach the fluence

Figure 11: KrF laser drilled hole in PMMA viewed with optical microscope. Diameter
60 μm and length 6.5 mm (AR = 100). Top left is entrance and bottom right is end tip.
The total image is assembled from several partial pictures.

value high enough to reach R_m. A smaller diameter beam (~30–35 μm) obtained
with a smaller molybdenum mask (30 × 5 = 150 μm) yields the more energetic spot
capable of reaching the maximum aspect ratio R_m of Table 2. One main difference
with the previous case of PC is the shape of the end tip which is now completely
round instead of branched. This is due to the presence of liquid although viscous
PMMA polymer during the laser interaction. The liquid surface tension produces
a meniscus at the bottom of the hole which resolidifies between two consecutive
pulses. Such liquid is not produced in drilling of the other polymers PC, PI, PEEK
and since it forms an ultrathin layer [37,38] in the ablation of PET, it seems to have
a negligible effect on its drilling. SEM imaging in Fig. 12 displays the redeposited
PMMA droplets around the hole entrance. Liquid expulsion may be a significant
mechanism of material removal in the case of PMMA. Recently we also demonstrated the experimental conditions in which nanofilaments [39,40] are formed by
droplet expulsion with a single laser pulse. In these cases, speed of liquid ejection
can be as high as 840 m/s (~Mach 3) according to our recoil pressure model and
depending on the fluence used. Details of the hole surface can be seen in Fig. 11
where in particular some microcracks are visible. It is difficult at this point to explain
their mechanisms of formation but their presence is not surprising, since the

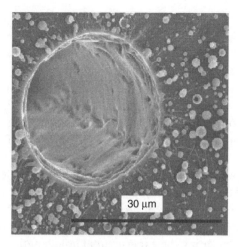

Figure 12: SEM picture of PMMA drilled with the KrF laser. The hole entrance diameter 30 μm. It shows many redeposited polymer droplets originating from the liquid expulsion.

PMMA material is exposed to a large dose of UV and the P and T pulses in the hole during microdrilling have large amplitude and impose some strain to the target. Also due to the large absorption depth ~150 μm involved in PMMA (Table 1), we can imagine that a cylindrical shape affected volume is formed around the hole where molecules receive a lot of photonic and thermal excitation. This volume was measured by micro-raman spectroscopy.

5.5.3. Search for the Optimum Aspect Ratio

As included in Table 2, there exists a maximum aspect ratio R_m. It is mainly a function of polymer type when drilling is performed with the same beam for all polymers. But it is also sensitive to beam diameter since beam fluence is the main controlling parameter in these experiments. We determined that aspect ratio varies with the beam fluence like:

$R = R_m [1 - \exp(-F/F_c)]$ in which F_c is a fluence characteristic of the polymer [36,41]. Now the fluence which can be obtained at the waist center is a very important parameter since it should be as large as possible to drill deeper, as suggested by Equation (15). It is made larger by decreasing the size of the beam but at some point the diffraction tends to spread the beam over a larger solid angle. This is the reason of the maximum aspect ratio R_m that we can demonstrate in these experiments. It is obtained with a beam diameter of the order of 25–30 μm depending on the polymer and it is a limitation due to the laser beam characteristics.

5.5.4. Model

The analytical model presented below was described in detail in previous publications [36,41]. The laser radiation approximated as in Fig. 13a by a collimated beam such that it diverges from the focal point O of the projection lens P and it gives a spot of diameter d at the surface of sample S positioned in the image plane of the mask M. So in the model, O is like a virtual point source (Fig. 13b) shedding some light rays onto the sample and into the hole in a cone of half angle $\alpha_0 = r_0/z_0$. The laser geometrical parameters are determined by the cone of light emitted by O in direction of the hole and defined by distance z_0, the cone angle α_0 (numerical aperture of the beam) and the fluence profile (flat top, gaussian, etc.,) which will be taken as a function of the inclination α of a particular ray (Fig. 13b). We show by the present model that the hole geometry is predetermined by these fixed parameters. The model was validated by fitting the experimental hole length $l(F)$ with the predicted law. For some reason related to hole diameter increase at stationary profile, the aspect ratio R is not suited for this fit.

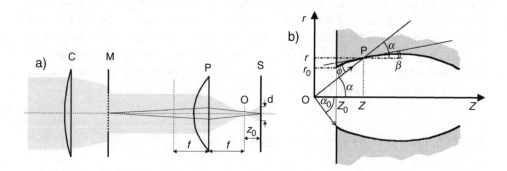

Figure 13: **(a)** Schematic beam setup used in the modelling of laser hole drilling. **(b)** Point source model with ray propagation into the laser drilled hole. O is a virtual point source at distance z_0 of the sample and emitting rays in a cone of angle α_0 into the hole with intensity varying with inclination a (laser profile). Hole diameter is strongly exaggerated for the sake of better viewing.

5.5.4.1. Condition of Stationary Profile

At the end of drilling the hole profile is stationary and therefore does not change significantly since ablation does not produce anymore. If it is so, this means that the laser fluence absorbed at the hole surface F_a has become smaller than the threshold value that we named F_∞, constant over the entire surface. At point P of Fig. 13b, we therefore have the condition

$$F_a(r, z) = F_\infty \qquad (16)$$

the equation which characterizes the stationary profile. F_∞ is a threshold fluence at which the ablation stops, which depends mainly on the polymer and on its modification introduced by the laser dose during the drilling process. We will see that it is fairly different from the well-known ablation threshold. It is simply a threshold below which the disappearance of the laser excited material does not produce anymore.

5.5.4.2. Hole Profile at Stationary Stage

The absorbed fluence at point P with coordinates (r, z) of the profile (Fig. 13b) is given by

$$F_a(r, z) = A_{eff}(\phi(r, z)) \times F(r, z)\sin\phi(r, z) \qquad (17)$$

where $F(r, z)$ is the incident fluence at P and $\phi(r, z)$ is the grazing angle for ray OP. $A_{eff}(\phi(r, z))$ is a local effective absorptivity for the laser radiation at point P. $F(r, z)$ is obtained from the angular fluence distribution $F(\alpha)$ at $z = z_0$ by

$$F(r, z) = F(\alpha)z_0^2/z^2 \qquad (18)$$

in which the factor z_0^2/z^2 describes the attenuation of fluence with distance z. The main two reasons responsible for ablation stop at stationary profile are then increasing attenuation with distance and increasing incidence angle with drilling progress. In Equation (17), $\Phi(r, z)$ can be expressed as a function of the ray inclination α (Fig. 13b):

$$\phi(r, z) = \alpha(r, z) - \beta(r, z) \qquad (19)$$

where $\alpha = \arctan(r/z) \cong r/z$ since r is small and

$$\beta = \arctan(dr/dz) \cong dr/dz = d(\alpha z)/dz = \alpha + z(d\alpha/dz) \qquad (20)$$

is the local inclination of the side wall at point P. Substituting Equation (20) into Equation (19) gives:

$$\sin\phi(r, z) \cong \phi(r, z) \cong -z(d\alpha/dz) \qquad (21)$$

Combining Equations (17), (18) and (21) gives the differential equation determining the keyhole profile:

$$z\frac{dz}{d\alpha} = -A_{\text{eff}}(\alpha)\frac{z_0^2}{F_\infty}F(\alpha) \tag{22}$$

Now in the following A_{eff}, the effective absorptivity will be given the value 1 [36] on the basis of the following arguments. The laser radiation entered in the hole has very little chance to exit again back to the outside and multiple scattering occurs within the hole which makes the absorption of the radiation energy total. So at all point, incident rays are finally absorbed regardless of their origin, since probably scattering dominates the beam propagation inside the hole. Furthermore attempt to fit experimental data with any other values than $A_{\text{eff}} = 1$ does not succeed to predict experimental measurements. With this value 1 the multiple scattering model Equation (22) can be solved in the parametric form $r(\alpha)$, $z(\alpha)$ with α the parameter being the ray inclination: The equation of the hole profile at the final stationary regime is then given by:

$$r(\alpha) = \alpha z(\alpha), \quad z(\alpha) = z_0\left[1 + \frac{2}{F_\infty}\int_\alpha^{\alpha_\infty}F(\alpha')d\alpha'\right]^{1/2} \tag{23}$$

in which α_∞ is the limit angle above which integration is negligible. The hole length or depth can be inferred from Equation (23):

$$l(F) = z(\alpha = 0) - z_0 = z_0\left[\left[1 + \frac{2}{F_\infty}\int_0^{\alpha_\infty}F(\alpha')d\alpha'\right]^{1/2} - 1\right] \tag{24}$$

5.5.4.3. Fit of the Experimental Hole Length for Top-hat Beam Profile
As stated above, the hole aspect ratio was not taken as fitting quantity, since it levels off at high fluence whereas the hole length continues to increase due to slow diameter increase with fluence which cannot be simply predicted by this model. If a constant diameter versus fluence is used besides the predicted length, Equation (24), then theoretical aspect ratio is an increasing function of fluence and does not reproduce the levelling off of the experimental data. On the contrary it was found that the hole length increases with fluence as predicted by Equation (24) as seen in Fig. 14. The experimental data were fitted successfully

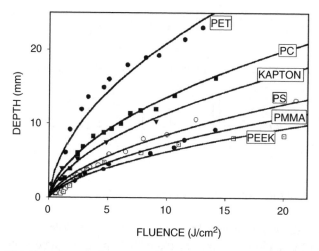

Figure 14: Hole length at stationary profile versus fluence. Points are experimental measurements and solid line are obtained with model Equation (26) fitted by adjusting F_∞.

with the model, Equation (24), for the six polymers studied in this work. The top-hat beam profile is defined by:

$$F = \text{const} = F \tag{25}$$

for $0 < \alpha < \alpha_0$ (see Fig. 13b) and the hole length versus fluence dependence becomes:

$$l(F) = z_0 \left[\left[1 + 2 \left(\frac{F}{F_\infty} \right) \left(\frac{r_0}{z_0} \right) \right]^{1/2} - 1 \right] \tag{26}$$

in which appears the ratio $r_0/z_0 = \alpha_0$ already seen before and which is the predetermined numerical aperture of the beam, and the ratio F/F_∞, an important ratio that can be called overfluence since it must be as large as possible to provide deep hole. Experimental length versus fluence is successfully simulated by Equation (26) with adjusting of parameter F_∞ for the given polymers (Table 2) up to obtaining a good fit as in Fig. 14.

In Equation (26) we see dimensionless overfluence F/F_∞ plays a particular role since a high value provides the good magnitude of drilling length and ensures the high aspect ratio obtained in this work (e.g. for PET it is 3×10^4 with $F_\infty = 1$ mJ/cm^2 and laser fluence $F = 30$ J/cm^2). If the beam profile is taken as a gaussian profile [36]

$$F(\alpha) = F \exp(-2\alpha^2 / w_\alpha^2) \tag{27}$$

the length versus fluence dependence is only slightly modified:

$$l(F) = z_0 \left[\left[1 + 2\left(\frac{F}{F_\infty}\right)\left(\frac{\sqrt{\pi} w_\alpha}{\sqrt{2}}\right) \right]^{1/2} - 1 \right] \tag{28}$$

where w_α is the parameter in the gaussian formula.

5.5.4.4. Hole Profile and Beam Transmission into the Hole

With the help of Equation (24), the calculated hole profile can be plotted as in Fig. 15 for a beam having a flat-top profile with different values of fluence. Fluence can be expressed in unit of $F = F_\infty z_0/r_0$, which is a particular value of fluence at $r = r_0$, providing parallel side walls at the keyhole entrance (Fig. 15 curve (2)). Calculated keyhole profiles of Fig. 15 can be not only convergent but also divergent at the entrance and convergent in a deeper part (profiles 3–5 for example).

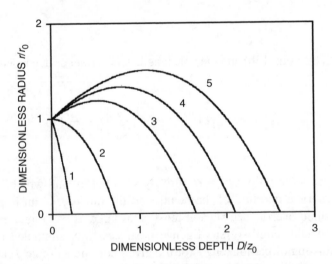

Figure 15: Calculated stationary profiles for a number of laser fluences for top-hat distribution of incident beam, r_0 is beam radius at the entrance plane of the sample, z_0 is the distance between the focal point of the beam and the position of the material front surface, $D = z - z_0$. $F/F_{par} = $ (1) 0.25; (2) 1; (3) 3; (4) 4.3; (5) 6. $F_{par} = F_\infty z_0/r_0$ is a threshold fluence at the spot border, $r = r_0$, providing parallel side walls at the keyhole entrance.

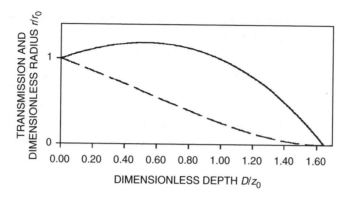

Figure 16: Via hole transmission (dashed) versus dimensionless hole depth for top-hat beam for $F = 3F_{par}$. Here r_0 is the beam radius at the entrance plane of the sample, z_0 is the distance between the beam waist and the position of the material front surface, $D = z - z_0$. The corresponding keyhole profiles are shown with solid curves.

In the multiple-scattering model, all rays propagating from the point source to the side walls are totally absorbed, as we have $A_{eff} = 1$. Therefore the only contribution to the transmission is given by the energy of the rays directly reaching the via hole exit from the point source O. For top-hat beam, the transmitted energy $E(z)$ is then simply directly proportional to the solid angle at which the exit diameter is seen from the point source and the following transmission can be obtained:

$$T(z) = \frac{E(z)}{E(z_0)} = \frac{\pi r(z)^2 z_0^2}{\pi r_0^2 z^2} \tag{29}$$

From Fig. 16 we see that the calculated transmission curve is nearly a linear function of depth and does reproduce well the experimental measurements [36].

5.5.5. Mechanisms and Perspectives

The present model is a good theoretical support for explaining some features of the experimental results, in particular for the prediction of the geometrical dimensions of the laser drilled hole when it has reached its final stationary profile. It gives a reasonable answer to the question which initiated these experiments, i.e. how deep a given laser beam can drill? It defines a contour surface $r(\alpha)$, $z(\alpha)$ where $F_a(\alpha) = F_\infty$, and inside which the polymer is removed by ablation (ablative volume). It is important to note that it is independent of the pulsed character of the laser radiation. It is equally applicable to continuous wave (cw) laser drilling, provided the mechanism of material removal involves only a solid-to-gas transition in the absorbing volume when fluence is larger than the minimum for material removal F_∞. But it is not intended to explain how the drilling performs in the intermediate steps. In the present pulsed laser drilling, the process is repetitive with a period consisting of the following sequence of events: absorption, excitation, ablation, expansion and cooling. At the frequency of our laser (200 Hz) it is considered that complete cooling takes place between two pulses. It means that temperature goes back to initial one because heat diffuses fast enough (the surface cooling time is of the order of 1–10 ms) around the hole in a volume strictly limited to what can be reached by the energy content of a single pulse, that is to say only ~1–10^2 μm in most cases. The absorption step is then in general done on a new cold surface for each pulse because transmitted energy decreases with depth, the end of drilling is reached when on each point of the keyhole surface $F(r, z) = F_\infty$. After some time of drilling, when ablation is deeper in the material, redeposition of the ablated products on the side walls comes into play but to a more or less extent depending on the nature of the polymer and its ablation mechanisms. For instance PMMA is typically ablating by a thermal mechanism with the formation of a substantial amount of liquid which can be expulsed under the form of droplets or redeposited on the walls. Redeposition and 're-ablation' is one of the two main reasons which reduce the drilling rate in Fig. 8A. The second reason is the increased plume shielding due to a more 1D expansion for deep ablation rather than the 3D for surface ablation. Furthermore at this level of fluence this strongly absorbing and expanding plume, probably a transient plasma, plays also the role of storage of the incoming energy and can release it to the material surface during expansion. After each pulse the plasma energy transfer to the surface can have a cleaning or ablative effect on the redeposited products. It is also likely that if a liquid film is formed on the inner wall during drilling, it is accelerated toward the exit of the drilled hole by momentum transferred from the outcoming gas plume. This may also contribute to an increase of hole diameter. The nature of the laser/matter interaction is mainly laser-surface but can be laser-plasma and plasma-surface with the later one being more delocalized over the entire surface of the hole and mainly over the side walls. Up to now little is known on the intimate mechanisms of material transport to the outside when the ablation front is deep in the material.

A good control on the laser beam leading to deep ablation or machining is suitable in many laser machining approaches. An experimental setup identical to ours can be used for precise or complicated cutting materials over a large thickness just by a computer controlled moving of the target relative to the beam: see the Mylar wheel example in Fig. 1 [32], helical drilling in reference [23]. Extensive microdrilling of multi-hole arrays is also needed for some applications. Higher frequency lasers [42] (up to ~5–50 kHz, 10 ps–30 ns, $I = 10$–20 MW for instance) are becoming available and might be of interest if one looks for higher processing speed. The short excimer laser wavelengths (193 and 248 nm) are preferable in most cases, but 4 and 3ω harmonics of the Nd/YAG laser are interesting sources with the highest frequencies. Such higher frequency laser may in principle offer a larger speed of processing. Nevertheless the new problem of target temperature increase may rise with increasing repetition rate of the laser pulse [42].

Other materials like metals, ceramics and glasses can also be treated by the KrF laser beam in a similar way [43]. However due to larger thresholds of ablation (F_t and F_∞) only much lower aspect ratio can be reached: 10–50. Furthermore the interest for using an ultraviolet laser is less obvious in this case. Many other sources of visible wavelength are being used for these particular materials like high power copper vapor lasers [44]. Also interesting work is done with femtosecond lasers since virtually liquid-free ablation in metal [45], silica [46], quartz [47] and polymers [48] drilling is reported. Ultraviolet femtosecond laser microdrilling of polymers is of basic interest and remains a perspective for the future.

5.5.6. Microdrilling Summary

We have presented recent experimental and theoretical work on high aspect ratio microdrilling of polymers with the pulsed KrF laser radiation. Good laser beams are obtained by condensing the laser on a circular molybdenum mask which in turn is imaged with a short focal projection lens onto the surface of the target. The best aspect ratio ~600 is obtained in PET for a 25–30 μm hole diameter and 18 mm length. Six different polymers have been studied. Hole length versus fluence dependence is predicted successfully with our model, which also provides a new threshold fluence F_∞, minimum of energy density required for material removal. An important parameter for deep drilling is the overfluence ratio F/F_∞ which determines the depth at which the drilling profile becomes stationary. In this model F_∞ is the threshold of material removal and not the usual threshold of ablation. It is indeed close to minimum fluence necessary for melting of the polymer at the wall of the microhole. With the rapid plume expansion, a strong gas flow to the exit of the hole can induce the expulsion of liquid droplets to the outside of the microhole. It is therefore realized that liquid expulsion phenomena presented below are important in laser processing of materials.

5.6. Submicron Resolution

Exploring the potentially high spatial resolution of laser ablation is a challenging scientific goal and is of great technological interest. In a recent paper [18] we have shown that KrF laser ablation on the surface of various polymers is a process of submicron resolution if a precision lens system is used to project the image of a mask on the surface of the treated material [49]. Using micro- or submicron beams is an interesting way to try gaining some more detailed knowledge about the material removal by looking at ablation profile, nanoparticles, liquid formation, etc. With the submicron ablation setup, we have evidenced and modelled new features of the ablation mechanisms, like transient liquid flow (see below) on the surface of PET (Mylar) during a single pulse KrF laser ablation [38,50]. In this part are presented two new experiments using the precision lens laser ablation system that are aimed at doubling and quadrupling the resolution of the ablation patterns (essentially gratings). In the previous experiments of reference [18] we used chrome-on-quartz grating masks which projected an image grating of period 3.7 µm and 2.1 µm. In these new experiments we use the same masks but use two different experimental procedures: 1) the first one uses 'defocus adjustment' in which the defocus is purposely varied in order to look for the appropriate and distorted image intensity profile and 2) the second one uses 'diffraction order filtering' in order to select two conjugated orders ($\pm n$) and let them interfere at the image level to get a sine type intensity modulation with increased spatial frequency. We have shown that for a number of usual polymers (PC, PET, PEN, PI-Upilex), chosen for their original surface quality, gratings with submicron period are successfully obtained by ablation with a single pulse of the KrF laser. The perspectives opened by such results are discussed.

5.6.1. Two Beams Imaging Experiments by Filtering

The incident laser light is diffracted by grating mask into many beams identified by their diffraction order $\pm n$. The larger value of n corresponds to beams of larger inclination with respect to the optical axis. Due to increasing inclination with n, only a limited number of beams (1–3 in our case) are entering the imaging lens and as a consequence after transmission reconstitute the projected image with loss of some details. In theory all diffracted beams are necessary to reconstitute a good image, but in reality the lens acts as a filter which transmits only low inclination beams, but looses high inclination beams responsible for the information on small details of the mask. By analogous filtering in the following experiment, the idea is to allow the transmission by the imaging lens of only two beams of conjugated diffraction orders $\pm n$ by placing blocking paper masks directly on the lens. The image in this case is simply due to the interference of these two beams and the period of the interference modulation depends only on ±*n* and the grating mask parameters. The fluence at the ablated sample surface [51] is given by:

$$F(x) = F_0 \frac{2\alpha_n^2}{\alpha_0^2 + \sum_{i=1}^{\infty} 2\alpha_i^2} \left(1 + \cos\left(2\pi \frac{2n}{p} x \right) \right) \tag{30}$$

where F_0 is the fluence without mask, x is the position coordinate in the image plane, p is the period of the non-filtered image of the mask and α_n^2 are the intensities of the diffraction orders, defined as:

$$\alpha_n^2 = \left(\frac{\sin\left(\pi \frac{p-a}{p} n \right)}{\pi \frac{p-a}{p} n} \right)^2 \tag{31}$$

in which the mask parameter a is the spacing between black stripes, for instance: 1.4 μm for $p = 3.7$ μm and 0.7 μm for $p = 2.1$ μm. Equation (30) tells us that the spatial frequency of the interference modulation in the filtered image is now $2n/p$ or equivalently the new period is $p/2n$. By these experiments the resolution can be improved by a factor equal to $2n$ (Fig. 17).

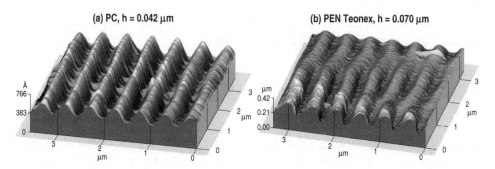

(a) PC, h = 0.042 μm   (b) PEN Teonex, h = 0.070 μm

Figure 17: KrF laser ablation patterns obtained with the 2 beams filtering experiments, by using diffracted beams of orders ±2. The obtained period is 0.53 μm which corresponds to $p/4$ where $p = 2.1$ μm is the period of the normal grating image (the spatial period is quadrupled).

5.6.2. Defocus Adjustment Experiments (3 Beams)

In this experiment the $n = 0, \pm 1$ orders are selected by filtering, similarly to the previous case, and now the idea is to explore the intensity profile around the good image position (defocus $z = 0$) by varying the defocus or the axial coordinate z. z is positive when the lens to sample distance is increased by definition. In practice the z variation is in the range of ~ 0–100 μm. It is shown by a Fourier optics calculation [51] that the intensity profile at the sample surface can be expressed by:

$$I(n,x,z) = 2\alpha_n^2 \cos\left(2\pi \frac{2n}{p} x\right) + 4\alpha_n \alpha_0 \cos\left(2\pi \frac{n}{p} x\right) \cos\left(\pi \frac{n^2 z \lambda}{p^2}\right) + \alpha_0^2 + 2\alpha_n^2$$

(32)

where λ is the laser wavelength (248 nm), n is equal to 1 (not ± 1) and which holds true for small values of z ($|z| < 100$ μm). Equation (32) is of the form $I_1(x) + I_2(x, z) + $ Const. where the two first terms are periodic functions of spatial coordinates x and the second also of z. The term of interest is the first one since its spatial frequency is $2n/p$, that is to say double of that of the mask image n/p which is represented by the second term. Due to the dependence on z the second term vanishes for a certain positions of defocus called $z_{1/2}$ and given by:

$$Z_{1/2} = \frac{p^2}{2\lambda}$$

(33)

which takes for instance for the mask period $p = 3.7$ μm the value $z_{1/2} = 27.6$ μm. The two defocus positions $z = \pm z_{1/2}$ providing an intensity modulation in x with the doubled frequency were found experimentally. It is shown (Fig. 18) that the ablated surfaces successfully reproduce these predictions.

5.6.3. Discussion and Perspectives of Submicron Experiments

These experiments can be viewed as two and three 'in phase' beams interference experiments. Interference experiments are classical optics and conventional 2 beams interference laser ablation approaches have used various beam splitting elements like phase gratings [52], splitter and mirrors [53,54], prisms, cubes [55], etc. Our defocusing experiments (6.2) in reality use the interference field of 3 selected beams of the mask. Less conventional and more recent multibeam (more than two)

NORMAL IMAGES   **DEFOCUSSED IMAGES**

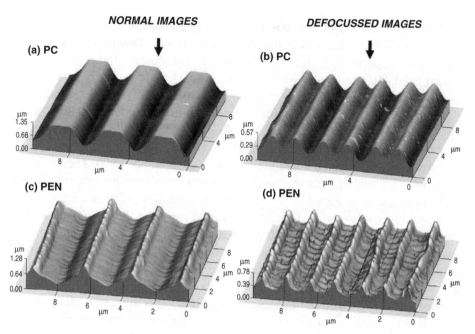

Figure 18: KrF laser ablated patterns by projection with a precision lens measured by AFM. Defocusing experiments ($z_{1/2}$) images displaying a double spatial frequency for PC and PEN (b,d) are displayed in the right-hand column. In the left-hand column (a,c) normal grating images for comparison. Normal period are 3.7 μm for PC and 3.0 μm for PEN. Irradiations were done with a single pulse of fluence $F = 1$ J/cm² in air.

interference experiments [56] using diffracting beam splitters present some analogous features. The use of the present filtered imaging with a projection lens offers several advantages among which is the large working distance of lens to sample. It is of the order of 30 mm in our setup, a distance which permits the use of vacuum or environmental chamber, for instance. It is interesting to note that the simple variation of the defocus distance z revealed the doubling of the lateral modulation in the ablation patterns. To our knowledge there is no literature report of this z dependence of the laser intensity in such 3 beams interference. The best example of this effect is given by PC in Fig. 18a and b, since its ablation is known to be 'dry' i.e. not producing a transient liquid film like PET and PEN, that could erase the wanted pattern by flowing before solidification. Therefore PC ablation profile is sharp (Fig. 18a) and predictable with simple model [51] by knowing the laser intensity profile and the ablation curve. For other polymers PET and PEN (Fig. 18c–d), although transient liquid is formed and displaced during ablation [38] the spatial frequency doubling is clearly seen but the profile is strongly affected by the transient liquid effect. It is thought that this will be a main limitation for smaller grating periods. In the other type of experiment with the 2 filtered beams, we have reached smaller grating period 0.53 μm with orders ±2 selected (see Fig. 17). PC is still giving a good ablation pattern with high regularity (Fig. 17a). The other two polymers PET and PEN have similar ablation behaviors which depend on their crystalline morphology. They can be totally amorphous or partially crystalline like in many commercial films (Mylar, Teonex). We have observed for a long time [57] that amorphous PET and PEN are less prone to transient liquid formation for reason of polymer solubility in the ablation products (see reference [38]). Therefore they give profiles not too far from the expected one that is to say with very few defects. On the contrary as in Figs 17b and 18c and d, crystalline PET and PEN display profiles strongly distorted by the transient liquid (see next figures in part for PET results). The valleys are filled with resolidified liquid and the lines are not fully straight anymore, they show some lateral waviness. The dynamics of the transient liquid layer has been modelled in reference [38]. Because of its low transient viscosity [58], it is accelerated by the lateral recoil pressure (several hundred bars) developed by the departure of the ablation products. This induces a lateral flow of the liquid toward the regions exposed to lower fluence. During the timescale of one laser pulse 25 ns, the flow distance is of the order of ~4 μm [38] a value close to the period of the grating image used in this work (see below). If the pressure modulation induced by the grating ablation has a period lower than this value, then the liquid accumulates on the top of the dark stripes as in Figs 18 and 19 and its level rises above that of the original surface. This phenomenon is easily evidenced by these experiments. We can further imagine that on the top of a hill made by a dark stripe, the two incoming flows of transient liquid collide and therefore acquire some momentum in the upward direction. This is the onset of the formation of a

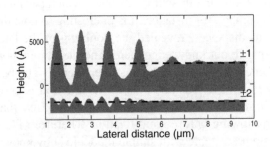

Figure 19: AFM profile of the edge of the KrF ablated patterns $p = 2.1$ μm on PEN showing the rise of transient liquid above the original surface level (horizontal dotted line). Upper profile is for the experiments with ±1 order beams and lower one for ±2 beams.

liquid jet. In other experiment of laser fibering of the PMMA surface [59–61], we study the ejection of nanofilaments resulting of the acceleration of microdroplets in a similar way but at much higher fluence. They can be explained by a similar type of mechanism.

KrF laser ablation of polymers has been done with micron and submicron period grating-like patterned beams, created with a precision projection lens and by selective filtering of the diffracted orders of the grating mask. The two new approaches (with 2 and 3 beams) which have been developed lead to grating patterns of submicron periods as measured by AFM. New features of the ablation behavior of polymers can be measured in these patterns, like transient liquid flow in the case where irradiation tends to melt the target or to produce liquid products. These submicron laser ablation experiments are promising for the future research development.

5.7. Viscous Microflow on Polymers Induced by Ablation

A microscopic flow of transient liquid film produced by KrF laser ablation is evidenced on targets of PET and PEN. Experiments were done by using single pulses of the excimer laser beam micropatterned with the aid of submicron projection optics and grating masks. The samples of various crystalline states, ablated with a grating forming beam (period $\Lambda = 3.7$ μm) were precisely measured by AFM to find any deviation from the ablation behavior predicted by the existing 'dry ablation' model, where the liquid flow effects are neglected and ablation spot profile is obtained from combination of ablation curve and beam profile. Poly(carbonate) (PC), poly(ethylene terephthalate) (PET) and poly(ethylene naphthalate) (PEN) exhibit different results.

PC is a 'normally' ablating polymer in the sense that ablated profile can be predicted with such a 'dry ablation' model. However, a crystalline PET deviates strongly from this simple model: during ablation a film of transient liquid, consisting of decomposition products, can flow under the transient action of the ablation plume pressure gradient and re-solidify at the gratings tops after the pulse termination forming a characteristic bump with a form close to a hemicylinder of radius R_b – a parameter easily measured by AFM (see Fig. 20).

At the conditions of a *stationary melt flow*, defined as

$$\tau_p \gg h^2/\nu \tag{34}$$

viscosity has a strong retardation effect on the melt motion in the spot [38,50,62]. The hydrodynamical model of viscous flow based on the solution of Navier-Stokes equation (see details in [38]) allows in this case to derive the following relationship

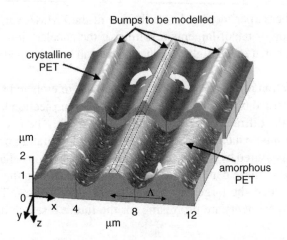

Figure 20: KrF laser ablation profiles obtained with one pulse on poly(ethylene terephthalate) surfaces in air (fluence = 1 J/cm² period of pattern is $\Lambda = 3.7$ μm) and measured by contact mode AFM. See discussion in the text for different behaviors of crystalline and amorphous PET.

between the fluid layer viscosity v during the action of the ablation pressure pulse and this bump radius R:

$$v = \frac{16\tau P h^3}{3\pi\rho\Lambda R_b^2} \qquad (35)$$

Here ρ and h are the density and depth of the fluid layer, P is an averaged ablation pressure having an effective duration $\tau_p = 2\tau$, $\tau = 30$ ns is laser pulse duration. Taking into account a laser radiation absorption in ablation plume, P is estimated from Equation (11) (see details in [38]). As we mentioned above in Section 5.3, $P = 645$ bar is calculated for PET data ($\rho = 1380$ kg/m³, $F = 1$ J/cm², $\delta = 0.5$ μm [3] and $\tau = 30$ ns) as in Fig. 4.

The substitution of the measured value by AFM bump radius $R_b = 0.3$ μm (as shown in Fig. 20), of other experimental values for liquefied layer depth $h = 0.16$ μm [63] and period $\Lambda = 3.7$ μm alongside with above mentioned $P = 645$ bar, $F = 1$ J/cm² and $\tau = 30$ ns into Equation (35) gives the value of kinematic viscosity of polymer melt during the action of ablation plume pressure gradient: $v = 0.28$ cm²/s. It appears to be more than order of magnitude higher than the viscosity of water at room temperature which is around 10^{-2} cm²/s. On the other hand, the same experiment done on the amorphous PET does not show any microflow, which looks as a 'dry ablation'. We explain this difference in our model not by the absence of any liquefied layer in this case, but in terms of a higher viscosity of the film composed of ablation products. In the case of amorphous PET, this viscosity is expected to be much higher than in the case of the crystalline one. Indeed, the relation (35) can be rewritten as

$$R_b = [16\tau Ph^3/(3\pi\rho\Lambda v)]^{1/2} \tag{36}$$

This formula shows that the bump radius R_b decreases with viscosity and for $v \gg 0.28$ cm²/s becomes practically negligible compared to the reported above value 0.3 μm.

Thus a high viscosity value in the considered conditions of the stationary melt flow stabilizes surface relief with respect to lateral redistribution of a polymer along the surface and hence is an important factor for obtaining quality and precision in laser processing.

KrF laser ablation of PET has been shown to produce a fragmentation in the molten layer of big polymer molecules into relatively small ones – ethylene terephthalate cyclic oligomers (ETCs) [64]. The dynamic viscosity of ETCs measured in non-laser experiments has been reported to be $\eta = 30$ cP at 295°C [65], which corresponds to the kinematic viscosity $v = \eta/\rho = 0.22$ cm²/s. This value is very close to the value obtained above by us, $v = 0.28$ cm²/s (taking into account that the uncertainty in measured polymer viscosities is usually great enough). It proves the adequacy of the proposed technique of polymer viscosity measurement at the conditions of nanosecond laser ablation.

5.8. Phase Explosion and Formation of Nanofibers on PMMA

A new form of material removal in laser ablation – an expulsion from the irradiated spot of long (up to 1 mm) nanofibers with a radius 150–200 nm during ablation of poly(methyl methacrylate) (PMMA) target in air by a single pulse of KrF laser (see Fig. 21) is explained in this section by using the concept of a *non-stationary viscous melt flow*, when the opposite to Equation (34) inequality is fulfilled

$$\tau_p \ll h^2/v \tag{37}$$

and the viscosity has practically no retardation effect on the melt motion in the spot [38,50,62]. In the experiment two types of fibers were observed: Type 1 were across a border of the laser spot and Type 2 were freestanding long fibers attached

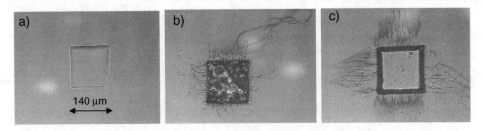

Figure 21: KrF excimer ablation of PMMA with a single pulse in air at three characteristic fluences: (a) 2.4 J/cm^2, above ablation threshold; (b) 5.2 J/cm^2, above fiber formation threshold (3 J/cm^2); and (c) 13.1 J/cm^2, at high fluence. Above 10 J/cm^2 the final surface after ablation is smooth again.

inside the spot (from the ablation spot to the upper right corner of Fig. 21b). At higher fluence shown in Fig. 21c, when the turbulence in inner part of the spot disappears, only fibers of Type 1 are seen. In contrast to Type 2, Type 1 fibers lay on the surface and therefore their cross sections were measured by AFM (Thermomicroscope CP Research system, using the non-contact mode with NCL tips from Nanosensors). These measurements give elliptical cross sections, but in the model below we put these fibers to be circular with a radius about 150 nm. Freestanding fibers may have similar diameters, as judged by the optical micrographies.

Confocal micro-Raman spectroscopy measurements performed with a Jobin-Yvon Labram spectrometer (with a He–Ne laser at 632 nm as a laser probe with the resolution close to 2 μm^3) on the freestanding nanofiber of Fig. 21b revealed no visible degradation of PMMA in the fiber, i.e. no double bond formation (opposite to previous measurements [66] at lower fluence but higher dose irradiated surface).

5.8.1. Possible Mechanisms

Thermal expansion of PMMA due to laser heating induces a surface elevation in the spot (Fig. 22). Liquefied polymer layer comes in motion as a result of acceleration in the peripheral regions A and B, where fluence distribution goes to zero and therefore there is a pressure gradient dP/dx, created by ablation plume and by an explosive decomposition of the metastable, superheated fluid layer, which is a mixture of decomposed gaseous products like H_2, C_2H_6, C_2, monomers, and liquid PMMA. We consider that highly energetic droplets of molten material are expelled to the exterior of the spot in the directions where pressure is not counterbalanced, i.e. in the direction (1) along the surface that corresponds to the formation of observed Type 1 fibers, and in the direction (2) practically parallel to the side

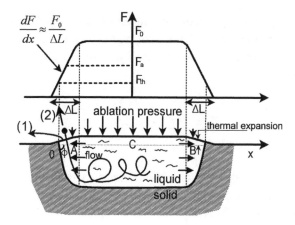

Figure 22: Schematic of droplets and jets expulsion from PMMA surface in the KrF laser ablation due to a collision of the fast liquid flow with a solid wall of the melt bath at the spot border.

wall of melt bath, corresponding to the formation of long Type 2 fibers. These droplets provide the heads for the jets of viscous liquid, a slightly degraded polymer with a viscosity adequate for fibering. Fibers are drawn from the depth of the melt bath and then re-solidified in ambient gas. It becomes evident from such scheme that Type 1 jets cannot propagate too long (compared to Type 2) due to their stronger friction with the surface.

On the other hand, a relatively high angle of expulsion of Type 2 jets with respect to the surface (hence providing a relatively long distance of their propagation) takes place when an inclination of side wall of the melt bath with respect to the original sample surface is not small:

$$\tan \Phi = \frac{dh}{dx}(x = 0) = \frac{dh}{dF}(F = F_t) \times \frac{dF}{dx}(x = 0) > 1 \qquad (38)$$

Here $h(x)$ is a melt depth versus coordinate x dependence in the vicinity of point O, where melting or glass transition starts, but vaporization is absent, F_t is a threshold fluence for melting or glass transition. Using that [67]

$$h(F) = \frac{1}{\gamma} \ln \frac{F}{F_t} \qquad (39)$$

(γ is absorption coefficient), and estimating dF/dx as $\cong F_0/\Delta L$, where F_0 is maximum fluence in the inner part of the spot and ΔL is a characteristic distance of fluence distribution, one can rewrite the high side wall inclination condition Equation (38) as

$$\Delta L \; < \; \frac{1}{\gamma} \frac{F_0}{F_t} \qquad\qquad (40)$$

For a moderate ratio (F_0/F_t), such condition is easy to fulfill for a sharp focusing when ΔL is small and for a relatively small absorption coefficient, as in the case of PMMA, where $\gamma \cong (1-3) \times 10^2$ cm^{-1} and $1/\gamma \cong 30-100$ µm. For comparison, for highly absorbing polymers, e.g. with $\gamma \cong 2 \times 10^5$ cm^{-1} it demands to have $\Delta L < (0.05-0.1)$ µm, which is not possible to fulfil even with sharp focusing.

The maximum value of liquid velocity u at the border of the spot is evaluated from the Bernoulli's law

$$u \; = \; \left(2P/\rho\right)^{1/2} \qquad\qquad (41)$$

where the time-averaged ablation plume pressure P on the surface is given by Equation (11). In the experiments, the formation of long fibers is observed at $F = 3$ J/cm^2 and above. For such F the experimental ablation depth is $\delta = 4$ µm [68]. The substitution of these F, δ and $\rho = 1200$ kg/m^3, $\xi = 1$ into Equation (2) gives $P \cong 4.2$ kbar. One can evidently expect that an explosive decomposition pressure created in the metastable superheated fluid layer is about the same, at least by order of magnitude. The use of such P in Equation (41) gives the liquid velocity $u \cong 840$ m/s (~Mach 3). Such high velocity is valid for a so-called 'non-stationary' melt flow defined by inequality, Equation (37), when viscosity has no retardation effect on polymer liquid flow in the spot. This regime of flow is opposite to the considered above in Section 5.7. case of a stationary flow, Equation (34), where viscosity had a strong retardation effect on melt motion in the spot. Indeed, the depth h of laser-induced liquid layer for our case of high-intensity irradiation of PMMA is estimated as $30-100$ µm and for melt bath with a not too high v (< 100 cm^2/s) the inequality, Equation (37), is really fulfilled.

The Type 2 jet (which becomes a fiber after a re-solidification) is considered below for simplicity as a long cylinder of a variable length $L(t)$, coming out of the melt bath, with a constant across length radius r, a variable mass $m(t) = \rho\pi r^2 L(t)$, and a velocity of all its points being the same: $w(t) = dL/dt$. Let $t = 0$ correspond to the end of pressure pulse acceleration of the melt. For $t > 0$, the Newton's second law for the jet motion has the form:

$$\frac{d}{dt}\left[\rho\pi r^2 L(t)\frac{d}{dt}L(t)\right] = -F_s, \quad L(t = 0) = L_0, \quad dL/dt(t = 0) = u \quad (42)$$

$F_s = 2\pi r\sigma$ is a surface tension force retarding the exit of the fiber from the melt bath, air resistance force is neglected. σ is a surface tension coefficient. L_0 and u are the fiber initial length and initial velocity acquired at $t \leq 0$ at the accelerating stage of the ablation pressure pulse of an effective duration τ_p. Therefore one can evidently put $L_0 \cong h_{A,B}$, where

$$h_{A,B} = \frac{1}{\gamma}\ln\left(\frac{T_{abl} - T_i}{T_g - T_i}\right) \quad (43)$$

is a maximum melt depth in the regions of melt expulsion A and B in Fig. 22. Here $T_{abl} = 225°C$ is a threshold temperature for PMMA ablation [69], $T_g = 105°C$ [70] is glass transition temperature, $T_i = 20°C$ is initial (room) temperature. For example, for $\gamma = 3 \times 10^2$ cm^{-1} gives $L_0 \cong 29$ μm. This relatively short but energetic initial part of few microns pulls behind it from the melt bath the remaining much longer part of the fiber of the length of the order of 1 mm, whose initial velocity in the liquid layer is practically zero. Such situation is similar to the mechanism of operation of harpoon for chasing whales, which is expelled from the gun and pulls behind a long string. The solution of Equation (42) for $L(t)$ is:

$$L(t) = \left(L_0^2 + 2L_0 ut - \frac{2\sigma}{\rho r}t^2\right)^{1/2} \quad (44)$$

$$L_{max} \cong L_0 u\left(\frac{\rho r}{2\sigma}\right)^{1/2}, \quad t_{max} = \frac{L_0 u\rho r}{2\sigma} \quad (45)$$

here L_{max} is a maximum length reached at time $t = t_{max}$, when the jet velocity dL/dt drops to 0. The calculation for $\sigma = 39$ mN/m [70], $r = 150$ nm and mentioned above ρ, u and L_0 gives $L_{max} = 1.17$ mm and $t_{max} = 56$ μs. This L_{max} agrees well with the above mentioned experimental maximal fiber length, about 1 mm. The variation in initial velocities and radii of jets can explain, according to

Equation (45), the variation in their lengths L_{max}. One can expect that a reverse motion of the fiber back to the melt, i.e. a decrease of the fiber length in time at $t > t_{max}$, is not possible due to the fast fiber solidification (which is faster than the bulk melt bath solidification).

A liquid cylinder diameter instability could be another factor limiting the evolution of $L(t)$. It might be effective and could cut-off the jet into shorter than L_{max} pieces, if a characteristic time for instability development (or for cutting), estimated approximately as $t_{inst} = 8v\rho r/\sigma$ [71] (here v is kinematic viscosity of the jet material), appears to be much smaller than the time of fiber solidification t_{sol}^f. On the other hand, a formation of continuous long fiber means that the condition is fulfilled: $t_{sol}^f, t_{max} < t_{inst}$. For example, the use $v = 16$ cm^2/s (with the above r, ρ and σ) gives $t_{inst} = 59$ μs, which is comparable with the obtained above $t_{max} = 56$ μs. Thus a jet viscosity increase up to $v >$, or $>> 16$ cm^2/s could make such instability not effective. The mechanism of a fast increase of jet viscosity might be a fast cooling down during jet propagation in ambient air. In this case t_{inst}, considered as the time of cut-off, can significantly or even practically infinitely increase. Therefore the mechanism of cut-off due to diameter instability seems to be more relevant for the fibers obtained by ablation in vacuum, where the cooling down due to convective heat losses is absent. The comparison of irradiation results in air and in vacuum really exhibits much better continuity of fibers obtained in air.

In the above modelling, the time for solidification of the liquid layer in the spot t_{sol} was assumed to be much greater than t_{max} and t_{inst}. Indeed, t_{sol} can be estimated as $t_{sol} \cong h^2/\chi$ where $\chi = 7.5 \times 10^{-4}$ cm^2/s is thermal diffusivity of PMMA [70]. For the liquid layer thickness $h = 20$ μm, it gives $t_{sol} \cong 5.2$ ms, and for $h = 50$ μm one obtains $t_{sol} \cong 33$ ms. As we see, in both cases, $t_{sol} >> t_{max} = 56$ μs, therefore the melt bath solidification in the spot cannot limit the fiber growth. Finally, charging phenomena due to laser photoemission of electrons may also to some extent play a role in liquid-phase explosion.

5.9. Comparison with Ablation of Metals (Titanium)

In this section we present some of the recent results concerning the KrF excimer laser ablation of titanium alloy Ti-6.8Mo-4.5Fe-1.5Al (Timet, USA) performed in vacuum – see details in [72–74]. The same concept of non-stationary flow as used above for PMMA describes well the surface relief features in ablated spot related to lateral melt flow, although the physical and chemical properties of these materials are strongly different. The original polished surface of this alloy appears as a patchwork of α and β nanocrystals. For fluences beyond the ablation threshold $F_{th} = 0.8$ J/cm^2, the ablation spot has a form of a crater whose geometry is defined

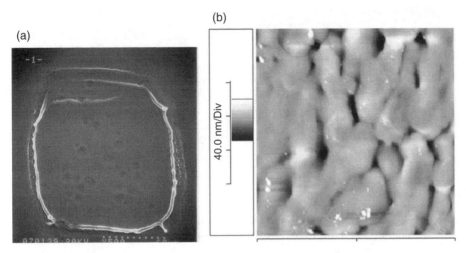

Figure 23: Ti-alloy surface KrF laser ablated in vacuum at the fluence 5 J/cm^2 (beyond the ablation threshold 0.8 J/cm^2) as observed by (a) SEM and (b) AFM.

by the shape of the beam, as shown in Fig. 23a obtained by SEM using secondary electrons. Ablation depth of this crater is about 0.35 μm for laser fluence 5 J/cm^2, as follows from ablation curve in Fig. 24.

At the border of the spot a bump of re-solidified transiently molten metal is formed. Its width and height, measured by AFM (Fig. 25), were found to be equal to 2 and 1 μm, respectively. This liquid is expelled to the periphery of the crater during

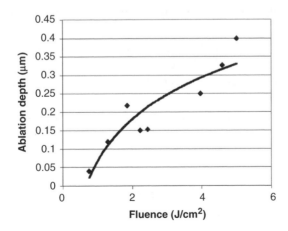

Figure 24: Ablation depth versus fluence measured by AFM for the studied Ti-alloy for single pulse KrF laser irradiation in vacuum.

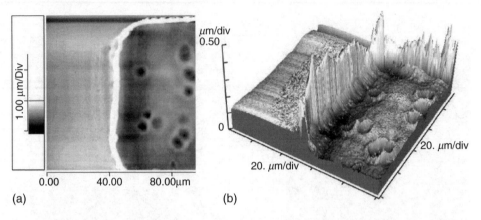

Figure 25: AFM images of the laser ablated spot on Ti-alloy showing the details of the rim and the bottom. KrF laser fluence is 3.9 J/cm². (a) View from above and (b) 3D view.

the ablation phase owing to the high pressure created by the ablation plume in such a way that the lateral melt flow along the surface is directed towards a lower pressure zone, i.e. to the exterior of the ablating region. AFM profilometry of the irradiated spot exhibits an effect of a significant smoothing of the surface in the peripheral region adjacent to the spot border, i.e. an erasing of a non-uniform coarse relief structure of a β-phase coverage onto a residual α–β network observed in the inner part of the spot. The width of this region is about 8 μm. Besides, in this region the ablation depth is about 1.5–2 times greater compared to the remaining inner part of the irradiated spot.

We can propose the following explanation of these observations. For irradiation in vacuum, the ablation plume pressure on the irradiated zone has a characteristic duration about $\tau_p = 2\tau$ and its amplitude averaged over the time is approximated by Equation (11). For laser fluence $F = 3.9$ J/cm² used to irradiate the spot in Fig. 25 the corresponding ablation depth δ as seen from Fig. 24 is about 0.30 μm. Then the use of these F and δ in Equation (11) gives ablation pressure $P = 2.1$ kbar. The difference of pressures from $P = 2.1$ kbar inside to 0 bar outside the spot is capable of accelerating the melt to a maximum velocity u at the border of the spot given by Equation (41). The substitution of this P and $\rho = 4.5$ g/cm³ into Equation (41) gives $u = 3.0 \times 10^4$ cm/s. Then a characteristic distance passed by the melt along the surface can be estimated as

$$L \cong \left\langle u \right\rangle \tau_p \cong u\tau \tag{46}$$

where we used that $\left\langle u \right\rangle \cong 0.5u$ is the average velocity for time $\tau_p = 2\tau$. The calculation for the obtained above u gives $L \cong 7.5$ μm. This value gives an order of magnitude of the width of the ring region immediately adjacent to the spot border, where the surface relief is modified by the lateral melt motion to the exterior of the spot. In particular, such melt flow along the surface can give rise to an efficient smoothing of the coarse relief structure of a β-phase coverage onto a residual α–β network. As we see, the calculated width is in a satisfactory agreement with the above reported experimentally observed width of peripheral smoothed region. The above mentioned enhanced material removal in the considered peripheral region can be explained by the fact that material here is removed not only by a direct ablation (vaporization), but also by the additional mechanism of melt displacement along the surface. Besides, the same mechanism could explain the appearance on the ablated bottom surface of a number of relatively small circular craters with diameters about $d = 13$–16 μm, clearly seen in Fig. 25b. Indeed, we suppose that they are created by the inhomogeneities (spikes) of pressure on the molten surface, induced, for example, by accelerated evaporation of some absorbing point inclusions (i.e. with a characteristic size $<< d$). In this case the radial washing out of the melt from the epicenter of such a pressure spike can give the circular craters with a diameter estimated as $2L$, i.e. around ≈ 15 μm, which corresponds well to the observed d in Fig. 25b. Thus a lateral melt flow that accompanies the ablation can provide a new surface topography with potentially useful properties.

5.10. Conclusions and Perspectives

Now, laser ablation of polymers has developed into many applications, one of them is microdrilling of high aspect ratio holes. We have presented in this review recent drilling experiments and theoretical work designed at understanding how far a laser beam can drill in materials. It is also understood from drilling characteristics which depends on the chemistry of the polymer that the decomposition pathway plays an important role in the drilling mechanisms. Droplets expulsion is clearly observed in the case of PMMA microdrilling. In a special experimental approach using periodic microbeams, we have studied and explained the liquid expulsion phenomena. From these studies, we have learned that upon laser irradiation a thin layer of liquid can be formed on polymer surface and depending on its viscosity its expulsion behavior can be strongly different. Two particular manifestations of viscous melt flow in laser ablation [75] corresponding to a so-called 'stationary' and 'non-stationary' liquid layer flow, respectively, have been studied here:

(i) A microperiodic single pulse KrF laser ablation of crystalline PET was measured precisely by AFM, and a model to interpret it was constructed. It was shown that a liquid film of thickness 0.16 μm, which is formed on the ablated surface, is expelled laterally toward the non-irradiated areas. According to the model, the liquid flow is due to the plume pressure gradient and the low viscosity of the ablation products film in the case of the crystalline polymer. The amorphous PET does not reproduce this behavior because its surface remains highly viscous.

Thus a high viscosity value in the considered 'stationary flow' conditions stabilizes surface relief with respect to lateral flow along the surface and hence is an important factor for obtaining higher precision and resolution in laser processing.

(i) For the first time the kinematic viscosity of laser melt layer on the surface of poly(ethylene terephthalate) at extreme conditions of KrF laser ablation is found: $v = 0.28$ cm^2/s. It appears to be very close to the kinematic viscosity value found recently [65] for ethylene terephthalate cyclic oligomers (ETCs) [64] $v = 0.22$ cm^2/s, giving an indication on the likely composition of the transient melt on PET at laser ablation conditions.

(ii) A new form of material removal in laser ablation is explained – expulsion of long (up to 1 mm) nanofibers with a radius of about 150–200 nm when a PMMA target is irradiated with a single pulse of a KrF excimer laser. The model suggests an emission from the spot of energetic droplets by intense pressure of the plume. Such droplets pull behind them the jets of polymeric viscous liquid from the transient melt bath created by UV laser excitation, giving rise, after re-solidification, to nanofibers. Relatively small absorption coefficient and small thermal diffusivity of PMMA provide a long time of liquid layer solidification, allowing an exit of a long jet. Fast cooling of the jet stabilizes it and prevents from a capillary collapse into smaller pieces allowing the formation of a long continuous nanofiber.

Acknowledgments

The research is partly funded by Région Aquitaine and FEDER program. Authors thank CNRS (Centre National de Recherche Scientifique) of France for a research position for VNT, who is on leave from the General Physics Institute, Moscow.

References

[1] Srinivasan, R. and Mayne-Banton, V., *Appl. Phys. Lett.*, **1982**, *40*, 374.
[2] Kawamura, Y., Toyoda, K. and Namba, S., *Appl. Phys. Lett.*, **1982**, *40*, 374.
[3] Lazare, S. and Granier, V., *Laser Chem.*, **1989**, *10*, 25.
[4] Dyer, P.E., *Appl. Phys. A.*, **2003**, *77*, 167–173.
[5] Bäuerle, D., *Laser Processing and Chemistry*, Springer, Berlin, 3rd edition, **2000**.
[6] Georgiou, S. and Hillenkamp, F. (Eds.), Laser ablation of molecular substrates, *Chem. Rev.*, **2003**, 103(*2*).

[7]   Dyer, P.E. and Walton, C.D., *Appl. Phys. A*, **2004**, *79*, 721.
[8]   Békési, J., Klein-Wiele, J.H., Schäfer, D., Ihlemann, J. and Simon, P., *Proc. SPIE*, **2003**, *4830*, 497.
[9]   Küper, S. and Stuke, M., *Appl. Phys Lett.*, **1989**, *54*, 4.
[10] Phipps, C.R., Turner, T.P., Harrison, R.F., York, G.W., Osborne, W.Z., Anderson, G.K., Corlis, X.F., Haynes, L.C., Steele, H.S., Spicochi, K.C. and King, T.R., *J. Appl. Phys.*, **1988**, *64*, 1083.
[11] Fabbro, R., Fournier, J., Ballard, P., Devaux, D. and Virmont, J., *J. Appl. Phys.*, **1990**, *68*, 775.
[12] Zweig, A.D., Venugopalan, V. and Deutsch, T.F., *J. Appl. Phys.*, **1993**, *74*, 418.
[13] Bennett, T.D., Grigoropoulos, C.P. and Krajnovich, D.J., *J. Appl. Phys.*, **1995**, *77*, 849.
[14] Oraevsky, A., Jacques, S.L. and Tittel, F.K., *J. Appl. Phys.*, **1995**, *78*, 1281.
[15] Tokarev, V.N., Lunney, J.G., Marine, W. and Sentis, M., *J. Appl. Phys.*, **1995**, 78, 1241.
[16] Lazare, S., Granier, V., *Amer. Chem. Soc. Sympos. Ser. Polym. in Microlithography*, **1989**, 412, Chap. 25, 411.
[17] Lazare, S. and Granier, V., *Appl. Phys. Lett.*, **1989**, *54*, 862.
[18] Weisbuch, F., Lazare, S., Goodall, F.N. and Débarre, D., *Appl. Phys. A*, **1999**, *69*, S413.
[19] Braren, B. and Srinivasan, R., *J. Vac. Sci. B*, **1985**, *3*, 913.
[20] Olson, R.W. and Swope, W.C., *J. Appl. Phys.*, **1992**, *72*, 3686.
[21] Olfert, M. and Duley, W.W., *J. Phys. D: Appl. Phys.*, **1996**, *29*, 1140.
[22] Basting, D., *Lambda Highlights*, **2003**, *62*, 4.
[23] Dausinger, F., Hügel, H. and Konov, V., *Proc. SPIE*, **2003**, *5147*, 106.
[24] Körner, C., Mayerhofer, R., Hartmann, M. and Bergmann, H.W., *Appl. Phys. A*, **1996**, *63*, 123.
[25] Paterson, C., Holmes, A.S. and Smith, R.W., *J. Appl. Phys.*, **1999**, *86*, 6538.
[26] Luft, A., Franz, U., Emsermann, A. and Kaspar, J., *Appl. Phys. A*, **1996**, *63*, 93.
[27] Chen, Y., Zheng, H.Y., Wong, K.S. and Tam, S.C., *Proc. SPIE*, **1997**, *3184*, 202.
[28] Rykalin, N., Uglov, A. and Kokora, A., *Laser Machining and Welding*, MIR, Moscou, **1978**.
[29] Lehane, C. and Kwok, H.S., *Appl. Phys. A*, **2001**, *73*, 45.
[30] Lu, X., Yao, Y.L. and Chen, K., *J. Mat. Sci. Eng.*, **2002**, *124*, 475.
[31] Lu, J., Xu, R.Q., Chen, X., Shen, Z.H., Ni, X.W., Zhang, S.Y. and Gao, C.M., *J. Appl. Phys.*, **2004**, *95*, 3890.
[32] Lazare, S., Lopez, J. and Weisbuch, F., *Appl. Phys. A*, **1999**, *69*, S1.
[33] Lopez, J. and Lazare, S., *J. Phys. IV*, **1999**, *9*, 153 (in french).
[34] Lopez, J., Lazare, S. and Weisbuch, F., *Proc. SPIE*, **1999**, *3822*, 77.
[35] Tokarev, V.N., Lopez, J. and Lazare, S., *Appl. Surf. Sci.*, **2000**, *168*, 75.
[36] Tokarev, V.N., Lopez, J., Lazare, S. and Weisbuch, F., *Appl. Phys. A*, **2003**, *76*, 385.
[37] Lazare, S. and Benet, P., *J. Appl. Phys.*, **1993**, *74*, 4953.
[38] Weisbuch, F., Tokarev, V.N., Lazare, S. and Débarre, D., *Appl. Phys. A*, **2003**, *76*, 613.
[39] Weisbuch, F., Tokarev, V.N., Lazare, S., Belin, C. and Bruneel, J.L., *Appl. Phys. A*, **2002**, *75*, 677.
[40] Weisbuch, F., Tokarev, V.N., Lazare, S., Belin, C. and Bruneel, J.L., *Thin Solid Films*, **2004**, *453–454*, 394.

[41] Tokarev, V.N., Lopez, J. and Lazare, S., *Appl. Surf. Sci.*, **2000**, *168*, 75.
[42] Gamaly, E.G., Rode, A.V. and Luther-Davis, B., *J. Appl. Phys.*, **1999**, *85*, 4213.
[43] Lopez, J., PhD thesis, *Université de Bordeaux, 1*, **1997**.
[44] Knowles, M.R.H., *Opt. Expr.*, **2000**, *7*, 50.
[45] Kamlage, G., Bauer, T., Ostendorf, A. and Chichkov, B.N., *Appl. Phys. A*, **2003**, *77*, 307.
[46] Shah, L., Tawney, J., Richardson, M. and Richardson, K., *Appl. Surf. Sci.*, **2001**, *183*, 151.
[47] Varel, H., Ashkenasi, D., Rosenfeld, A., Wähmer, M. and Campbell, E.E.B., *Appl. Phys. A*, **1997**, *65*, 367.
[48] Zhang, Y., Lowe, R.M., Harvey, E., Hannaford, P. and Endo, A., *Appl. Surf. Sci.*, **2002**, *186*, 345.
[49] Srinivasan, R. and Braren, B., *Chem. Rev.*, **1989**, *89*, 1303.
[50] Weisbuch, F., Tokarev, V.N., Lazare, S. and Débarre, D., *Appl. Surf. Sci.*, **2002**, *186*, 95.
[51] Weisbuch, F., PhD thesis, *Université de Bordeaux, 1*, **2000**.
[52] Anderson, E.H., Komatsu, K. and Smith, H.I., *J. Vac. Sci. Technol. B*, **1988**, *6*, 216.
[53] Fedosejevs, R. and Ilcisin, K.J., *Appl. Opt.*, **1987**, *26*, 396.
[54] Dyer, P.E., Farley, R.J. and Giedl, R., *Optics Comm.*, **1996**, *129*, 98.
[55] Sauerbrey, R., *Appl. Phys. Lett.*, **1991**, *58*, 2761.
[56] Kondo, T., Matsuo, S., Juodakis, S., Mizeikis, V. and Misawa, H., *Appl. Phys. Lett.*, **2003**, *82*, 2758.
[57] Lazare, S., Granier, V., Lutgen, P. and Feyder, G., *Revue Phys. Appl.*, **1988**, *23*, 1065.
[58] Weisbuch, F., Tokarev, V., Lazare, S. and Débarre, D., *Appl. Surf. Sci.*, **2002**, *186*, 95.
[59] Weisbuch, F., Tokarev, V.N., Lazare, S., Belin, C. and Bruneel, J.L., *Appl. Phys. A*, **2002**, *75*, 677.
[60] Weisbuch, F., Tokarev, V.N., Lazare, S., Belin, C. and Bruneel, J.L., *Thin Solid Films*, **2004**, *453–454*, 394.
[61] Tokarev, V.N., Lazare, S., Belin, C. and Débarre, D., *Appl. Phys. A*, **2004**, *79*, 717.
[62] Tokarev, V.N. and Kaplan, A.F.H., *J. Phys. D,: Appl. Phys.,* **1999**, *32*, 1526.
[63] Lazare, S. and Benet, P., *J. Appl. Phys.*, **1993**, *74*, 4953.
[64] Watanabe, H. and Yamamoto, M., *J. Appl. Polym. Sci.*, **1997**, *42*, 1203.
[65] Youk, J.H., Kambour, R.P. and MacKnight, W.J., *Macromolecules*, **2000**, *33*, 3594.
[66] Lazare, S., Lopez, J., Turlet, J.M., Kufner, M., Kufner, S. and Chavel, P., *Appl. Optics*, **1996**, *35*, 4471.
[67] Tokarev, V.N. and Kaplan, A.F.H., *Lasers in Eng.*, **1998**, *7*, 295.
[68] Srinivasan, R., Braren, B., Seeger, D.E. and Dreyfus, R.W., *Macromol.*, **1986**, *19*, 917.
[69] Hare, D.E., Franken, J. and Dlott, D.D., *J. Appl. Phys.*, **1995**, *77*, 5950.
[70] Brandrup, J. and Immergut, E.H., (Eds.), *Polymer Handbook*, Wiley, New York, 3rd edition, **1989**.
[71] Funada, T. and Joseph, D.D., *J. Non-Newt. Fluid Mech.*, **2003**, *111*, 87.

[72] Guillemot, F., Tokarev, V.N., Lazare, S., Belin, C., Porté-Durrieu, M.C. and Baquey, C., *Appl. Surf. Sci.*, **2004**, in press.

[73] Guillemot, F., Tokarev, V.N., Belin, C., Porté-Durrieu, M.C., Baquey, C. and Lazare, S., *Appl. Phys. A*, **2003**, *77,* 899.

[74] Guillemot, F., Tokarev, V.N., Belin, C., Porté-Durrieu, M.C., Baquey, C. and Lazare, S., *Appl. Phys. A*, **2004**, *79,* 811.

[75] Tokarev, V.N., Lazare, S., Belin, C. and Debarre, D., *Appl. Phys. A*, **2004**, *79,* 717.

Chapter 6

Nanoscale Laser Processing and Micromachining of Biomaterials and Biological Components

D.B. Chrisey[a], S. Qadri[b], R. Modi[b], D.M. Bubb[c], A. Doraiswamy[d], T. Patz[d] and R. Narayan[d]

[a]*Center for Nanoscale Science and Engineering, Fargo, ND;* [b]*US Naval Research Laboratory Washington, DC;* [c]*Rutgers University, Department of Physics, Camden, NJ;* [d]*Georgia Institute of Technology, Materials Science and Engineering, Atlanta, GA*

Abstract

Lasers are increasingly proving to be an enabling approach to process biomaterials and biological components on the nanometer length scale. While nanotechnology encompasses an array of enabling technologies that utilize the fact that matter at length scales less than 100 nanometers have distinctly different physical and chemical properties than the same matter at larger length scales, biology and biological components are a special subset since the fundamental building blocks are almost all less than 100 nanometers in size, i.e. biological molecules are constantly being used in a directed self-assembly manner to communicate and build new materials. There are already several successful examples of nanotechnology starting from the somewhat mundane sharper scalpels and protective sunscreens, to more effective drug delivery, and even biomolecular motors, gene therapeutics, tissue engineering, and improved medical diagnostics. The unique capabilities of laser processing for these applications is based on tuning the laser-material interaction to create novel structures, i.e. the laser wavelength, pulse width, and power can be varied over a wide range as can the composition and state of the material. This chapter will summarize several successful examples of the laser processing of biomaterials and biological components.

Keywords: Angiogenesis; Calf thymus DNA; Extracellular matrix (ECM); Hematoporphyrin; Laser micromachining; Matrix assisted pulsed laser evaporation (MAPLE); Nanotechnology; Vascular endothelial growth factor (VEGF).

Recent Advances in Laser Processing of Materials
J. Perrière, E. Millon and E. Fogarassy (Editors)

6.1. Introduction

6.1.1. Nanotechnology in Biology and Medicine

Nanotechnology encompasses an array of enabling technologies that utilize the fact that matter at length scales less than 100 nanometers have distinctly different physical and chemical properties than the same matter at larger length scales [1]. In general, it is still an emerging area of science and most of the breakthrough advances are only in the early development stage. Biology and medicine are a special subset of nanotechnology since the fundamental building blocks are almost all less than 100 nanometers in size, i.e. biological molecules are constantly being used in a directed self-assembly manner to communicate and build new materials. Engineers and physicians are just beginning to apply the concepts of nanotechnology to biology and medicine, yet there are already several successful examples starting from the somewhat mundane sharper scalpels and protective sunscreens, to more effective drug delivery, and even biomolecular motors, gene therapeutics, tissue engineering, and improved medical diagnostics. Especially in medicine, some of the more remarkable applications of nanotechnology may be even a decade away because of the long approval process for new therapeutics. But the potential applications for nanotechnology and nanoparticles in the body is fascinating and possibly even revolutionary as they are metabolized in different ways, e.g. they can be excreted in the urine, broken down in the liver, or recycled in the body, thus they have the potential to be utilized in almost any region of the body. Several traditional drugs have used the term nanotechnology since they necessarily meet the 100-nanometer criterion. The difference between these traditional drugs and nanotechnology is that traditional drugs are based on small molecule chemistry. Nanotechnology is more related to the design and structuring of complex arrays of molecules and these arrays can be the therapeutic agents and/or the ability to deliver those agents to specific regions or tissues in the body and even to specific cells. One of the major reasons for considering synthetic nanomaterials in medicine is to avoid immune responses, i.e. several synthetic nanosystems have been tested and they do not induce antibody responses and as such are a major medical advance. In addition to improved capabilities, there is also the hope that nanomedicines, nano-based health monitoring systems, and nanodiagnostics can reduce the cost of healthcare. Still, nanotechnology is an entirely new area of science and laboratory safety and the standard approval process for therapeutics by regulatory agencies may need to be reconsidered. One area that has not been explored significantly is how tools developed in nanoscience could be used to learn how biological systems function. Many structures in cells and organelles function on the nanoscale, therefore, understanding them from a bottom-up approach using molecular biology will further the understanding of the fundamental building blocks at these dimensions.

6.1.2. *Laser Material Interactions for Biological Materials*

The study of laser interactions with materials ranges from low-temperature heating to explosive, plasma-forming reactions. Figure 1 shows a qualitative representation of the laser material interaction and the different types of possible physical events that could occur. Compared to other methods to physically process materials, e.g. electrons, ions, solution-based methods, organics and high voltages, and lasers have several novel features:

- Vacuum is not required
- Can process metals, polymers, ceramics, and biological materials
- Selective absorption (photolytic as well as pyrolytic)
- Wide range of laser fluences (<100 J/cm^2) and powers (<10^{12} W/cm^2)
- Safe and clean, no impurities
- Small laboratory footprint
- Wide range of feature size (submicron – 100 cm^2)

Moreover, researchers are becoming savvier in utilizing this interaction to process materials. Shortly after the discovery of lasers, researchers were, for the most part, studying the modification of mostly solid targets irradiated under varying laser conditions. This included, amongst many other areas, laser micromachining, laser annealing, and laser spectroscopy. It took some 25 years before what could be termed a second generation of laser material interactions was developed. At that point, researchers now utilized the primary laser interaction to do subsequent material processing. Examples of this secondary interaction to do material processing would include pulsed laser deposition (PLD) and narrow band excimer laser lithography [2]. Today there are many exciting new laser-based techniques to process materials

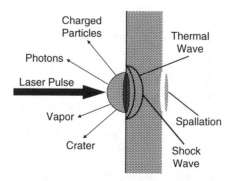

Figure 1: Schematic diagram of the different possible physical events during the laser interaction with a solid is shown.

that could be categorized as the secondary interaction, but they use a novel combination of lasers and materials. Examples of this would include laser capture microdissection, matrix assisted pulsed laser evaporation direct write, and resonant infrared pulsed laser deposition.

Biomaterials and biological components can be categorized loosely as organics (composed of tissues, cells and proteins), ceramics, and metallic alloys [3]. The laser interaction with ceramics and metallic alloys is well understood and treated elsewhere.

Lasers are able to produce a range of biological responses in tissue that are determined by the various processes of energy conversion within the biomolecules [4]. The laser parameters that are important in determining its interaction with tissue are wavelength, power density, and pulse duration. There are four basic types of laser tissue interactions that can occur (Fig. 2):

- Electromechanical
- Photoablative
- Thermal
- Photochemical

Electromechanical interactions occur at extremely high power densities at extremely short pulse widths so that the electromagnetic fields induce dielectric breakdown. This interaction is currently being used for laser assisted shock-wave lithotripsy for calculi in the biliary tree. In addition, this mechanism is used in ophthalmology for photodisruption of the posterior capsule when the posterior

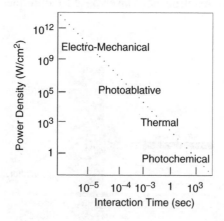

Figure 2: Qualitative description of the regions in power density and pulse width space that the four different types of laser material interactions that can occur.

membrane becomes opaque – secondary cataracts. Photoablation involves the direct breaking of intramolecular bonds in biopolymers caused by absorption of incident photons and subsequent desorption of biological materials. Ultraviolet laser radiation is strongly absorbed by biomolecules leading to small absorption depths of a few microns. This interaction is used in medicine, e.g. radial keratotomy and in orthopedic surgery for shaping the bone surface. The thermal mechanism is the conversion of laser energy into heat. This process occurs first with the absorption of photonic energy producing a vibrational excited state in the biomolecules and then in elastic scattering with neighboring molecules increasing their kinetic energy or a temperature rise. Laser heating is controlled by molecular target absorption in tissue components, such as free water, hemoproteins, melanin, and other macromolecules such as nucleic acids. The depth of penetration into the tissue varies with wavelength of the incident radiation and this determines the amount of tissue removal and bleeding control. The photochemical interaction involves the uptake by target cells of chromophores capable of causing light-induced reactions in molecules to produce a cytotoxic effect. Using chromophores sensitive to a narrow wavelength band and that are introduced and selectively retained in cancer cells at specific sites, the photochemical interaction can offer an effective therapy if irradiated by a laser in this band, especially if the chromophore is delivered to deep target cells by penetrating wavelengths. This principle is used in psoralen ultraviolet therapy in treatment of psoriasis and to treat malignant tumors using a hematoporphyrin derivative.

Cells are very sensitive to laser-produced changes in temperature. The normal human core temperature is 37°C and protein coagulation begins to occur in the cytoplasm at only 45°C. In addition, thermal damage to the cell can occur in the plasma membrane where lipid peroxidation from free radical production may be initiated. The DNA in the nucleus will begin to have random deletions induced and if these become permanent in the cell they will likely be transmitted to future cells.

6.2. Nanoscale Laser Processing and Micromachining of Biomaterials and Biological Components

The unique capabilities of laser processing for these aforementioned applications are based on tuning the laser parameters for constructive interactions with natural biological materials. This process can be optimized further by varying this interaction to create novel structures, i.e. the laser wavelength, pulse width, and power can be varied over a wide range as can the composition and state of the material. This chapter will summarize several successful examples of the laser processing of biomaterials and biological components on the nanoscale.

6.2.1. Matrix Assisted Pulsed Laser Deposition of Novel Drug Delivery Coatings

While lasers have been used to process materials in a multitude of different ways almost since their discovery, it is only recently that they have been successfully employed for the processing of more fragile organic materials, e.g. biological materials. In particular, new methods have been developed to exploit the laser material interaction in a novel way to process these fragile biological materials into films and patterns for numerous electronic, optical, and sensing applications. These new films and prototype devices have performance metrics that far exceed what can be done by conventional coating techniques.

One unique area of application for laser processed biological films and coatings are in medical implants and drug delivery. When a foreign object is implanted in the body there is typically an inflammatory encapsulation response. This response must be suppressed in order for an implantable object or device such as a biosensor or arterial stent to function for extended period of time. Inflammatory mediators (e.g. anti-inflammatory drugs, cytokines, growth factors, or hormones) can be used to control inflammation and encapsulation in tissues surrounding the implant. For example, steroidal agents can exert an anti-inflammatory and immunosuppressive action through release of vaso-active and chemo attractive factors, changes in the circulatory kinetics of leukocytes, alterations in the function of inflammatory cells, and modification of soluble mediators. These anti-inflammatory actions must be provided during acute (24–48 h) and chronic (1–2 weeks) phases of the inflammatory response in order to prevent encapsulation and ensure normal tissue growth. Long-term systemic use of steroids is not desirable. Individuals on systemic steroid therapy are more susceptible to viral, bacterial, and fungal infections. In addition, many side effects, including posterior sub-capsular cataracts, glaucoma, and peptic ulcers, may result from systemic use of corticosteroids. Local, continuous, controlled release of steroids reduces systemic side effects and improves the therapeutic response at the implant site. This approach is currently being used to deliver the steroid dexamethasone from small reservoirs at the tips of pacemaker leads. For arterial stents and other implantable structures conventional coating techniques have significant problems in terms of producing the desired coating thickness, homogeneity, and multilayers.

To overcome these problems, we have used laser processing to deposit the anti-inflammatory layer, for example, on an arterial stent. The coating technique used is called matrix assisted pulsed laser evaporation (MAPLE). In this technique, the fragile organic or biological material to be desorbed is diluted to concentration of ~2–5%. The target is then frozen to liquid nitrogen temperatures and mounted in a specially designed target carousel. Under ideal circumstances, the laser interacts with the volatile solvent and only minimally with the solute. The absorbed laser energy causes the solvent to evaporate. The collective action of the evaporating solvent results in gentle collisions with the solute and it desorbs softly to produce a homogeneous and chemically identical thin film on a receiving substrate placed opposite to it. In this case, the desorbed vapor contains a dilute amount of anti-inflammatory agent dexamethasone in an ethyl acetate solvent. The ethyl acetate matrix is relatively volatile, possesses a high vapor pressure, and preferentially absorbs the laser energy. While the laser-produced plume contains both solvent and solute, the evaporated solvent molecules do not deposit on the substrate because of their low sticking coefficients. The less volatile solute molecules deposit on the substrate and forms a continuous and uniform film. Biomaterial films of simple sugars, complex carbohydrates, proteins, albumin, insulin, and hydrogels have all been deposited using this technique.

Matrix assisted pulsed laser evaporation has several advantages over conventional techniques for making biological and organic thin films (e.g. conventional dip coating, spin coating, inkjet methods, Langmuir–Blodgett dip coating, and pin arraying). Most importantly, MAPLE is a line-of-sight process, and will not alter pore diameter, geometry, or chemistry. Matrix assisted pulsed laser evaporation provides excellent control over process parameters, including thickness, roughness, homogeneity, and reliability. Furthermore, this process allows one to deposit an organic material with the precision and control offered by a 'dry' physical vapor deposition technique. In contrast, film thickness in conventional processing is not well controlled, e.g. dispensed droplet size and surface wetting may significantly affect film thickness. For multilayers, the solvent containing the material of interest may dissolve underlying layers. Matrix assisted pulsed laser evaporation is a PVD process and the coatings produced have many of the same attributes as conventional PVD methods, such as sputtering, MBE, and e-beam evaporation. Yet the target material is conveniently handled in a manner similar to conventional wet processing techniques. MAPLE is thus often referred to as a pseudo-dry process. The benefits of MAPLE processing of organic and biological materials is more clear when compared to conventional methods that have limitations in resolution, scale up, speed, and film quality. Finally, membrane contamination is minimized when PVD of the diamond-like carbon layer and MAPLE of the anti-inflammatory layer are performed using the same deposition system.

For the experiments in this section, the 5% dexamethasone/ethyl acetate target was frozen and quickly mounted inside the deposition chamber. A Lambda Physik LPX 305 excimer laser was used for target ablation at 193 nm (ArF) laser wavelength. The incident laser energy was maintained within the range of 10.2–21.7 mJ, and the laser spot area was maintained within the range of 2.5–3.5 mm^2. During deposition, the target was rotated at 5 rpm, and is kept at ~173 K using copper braids connected to a liquid nitrogen reservoir. The deposition was performed under vacuum of ~10^{-4} Pa. Fourier transform infrared (FTIR) spectra X-ray photoelectron spectra, mass spectra, and dynamical light scattering data from MAPLE-deposited dexamethasone thin films demonstrate that these films exhibit nearly identical structural properties to as-prepared dexamethasone. Figure 3 compares the FTIR of the MAPLE film (A) with a drop cast film of the starting material in (B).

As mentioned above, the ideal release profile for dexamethasone should be continuous and constant over one to two weeks in order to best prevent the acute and chronic inflammatory encapsulation response. A single thin film of MAPLE-deposited dexamethasone provides initial burst release, but does not provide the required long-term continuous release. We have developed bioresorbable polymer/ dexamethasone nanoscale composite films that provide sustained release of dexamethasone for significantly longer periods of time (see Fig. 4). Poly (D, L) lactic acid (PDLLA) (Boehringer Ingelheim Chemicals, Petersburg VA) provides reproducible degradation rates, and allow for continuous, controlled dexamethasone release. Release of dexamethasone from this composite layer occurs due to one of the following mechanisms: (1) diffusion of dexamethasone through the polymer matrix and (2) release of dexamethasone via polymer degradation. These processes are controlled

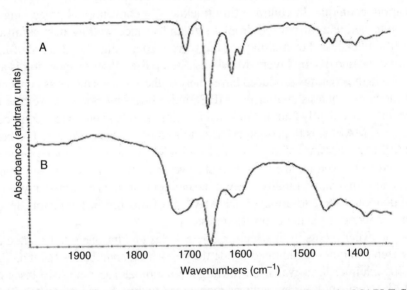

Figure 3: Fourier transform infrared spectra of dexamethasone from the MAPLE film in (A) and with a drop cast film of the starting material in (B). These spectra demonstrate that the deposited films exhibit nearly identical structural properties to as-prepared dexamethasone.

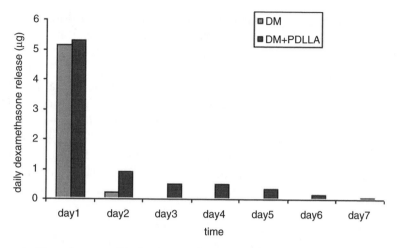

Figure 4: The release profile for dexamethasone, and PDLLA and dexamethasone compared over one week. There is improved drug release performance for the combination of dexamethasone and PDLLA over a single thin film.

by drug physiochemistry and polymer physicochemistry (e.g. molecular weight). This nanostructured composite film can be tailored to provide extremely well-defined drug release properties eliminating the undesirable release properties observed in conventional drug coatings (e.g. the initial burst release).

Matrix assisted pulsed laser evaporation processing of time-released drug coatings is a unique area of application and we have shown that the technique is extremely effective in transferring the fragile dexamethasone steroid and polymer composite. Further optimization of the thickness, composition, and multilayers will result in improved treatment for encapsulation response and other drug delivery related problems.

6.2.2. Resonant Infrared Pulsed Laser Deposition of Drug Delivery Coatings

The MAPLE modification to conventional PLD was and is truly enabling, but its potential for UV photochemical modification of the target highlights how PLD could be further improved for these fragile materials. To avoid irreversible photo-chemical modification of the target by electronic transitions caused by conventional UV lasers, resonant infrared PLD modifies the laser-material interaction by tuning the laser to the vibrational modes in the target. The intense IR radiation imparts enough kinetic energy to the target material to desorb the constituents by exciting the irradiated target to a highly vibrationally excited state (Fig. 5 shows a schematic diagram comparing the energy level diagrams for UV and IR excitation). Using the mid-infrared (2–10 μm) to excite many common functional groups in biopolymers, resonant infrared PLD has demonstrated the deposition of several polymer thin films whose material properties such as chemical structure and functionality also have been maintained. While the light source typically used to tune over this wavelength range is best achieved with a free electron laser, *ad hoc* resonant infrared PLD systems can be realized with some commercial lasers. It is important to note that the IR tuning to the absorption band is critical. When the IR laser is not tuned to a vibrational mode of the polymer, charring results and subsurface defects are created (see Fig. 6). In an analogous way to optimize the solvent/solute system with MAPLE, the deposition results for resonant infrared PLD can vary depending on the particular vibrational mode excited. Resonant infrared PLD does have the advantage of high deposition rates compared to MAPLE because a bulk target is used and not a diluted one. Moreover, the efficacy of the interaction is greater with the resonant infrared approach and it is a dry deposition technique. However, molecular entanglement of the polymeric target poses quite a difficult challenge and we expect that for some polymers segmentation will always occur with this approach.

Poly(DL-lactide-co-glycolide) (PLGA) is a polymer, which degrades by hydrolysis and can be used to controllably release therapeutic levels of drugs. For example, in the lungs, optimized drug delivery would involve the coating of 1–5 μm drug

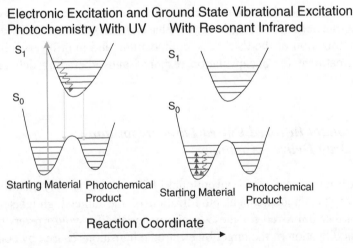

Figure 5: A qualitative schematic drawing is given comparing the energy level diagrams for UV and IR excitation of a target for evaporation.

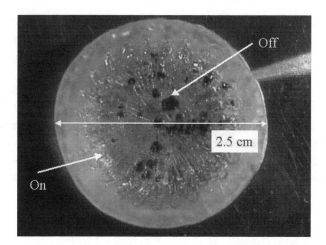

Figure 6: A polymer target irradiated with infrared laser radiation. When the infrared laser is not tuned to a vibrational mode of the polymer, charring results and subsurface defects are created. In an analogous way to optimizing the solvent/solute system with MAPLE, the deposition results for resonant infrared PLD can vary depending on the particular vibrational mode excited.

particles with a thin porous time-release coating of PLGA. The polymer coating is semi-permeable to water and degrades slowly with time. The net result is a delay in the release of the drug and an extension of the time over which the drug is active. Using an ultraviolet laser in a coating process such as with PLD or MAPLE could initiate substantial photochemical rearrangement of the polymer. Also, one must be concerned about potential toxicity that may arise if the chemical structure of PLGA is altered, since the byproducts of UV photolysis may not break down into biodegradable components. Resonant infrared PLD is an exciting alternative.

The starting material was PLGA 50:50 obtained from Birmingham Polymers, Inc. and the light source used in the experiments is the W.M. Keck Foundation Free Electron Laser at Vanderbilt University. Briefly, the laser is tunable between 2 and 10 μm and operates at a repetition rate of 30 Hz. The 4 μs macropulses in turn comprise some 10^4 1 ps micropulses that are separated by 350 ps. In these experiments, films were deposited using 2.90 and 3.40 μm light at fluences of 7.8 and 6.7 J/cm^2, respectively. The spot size was 0.0008 cm^2, measured by placing burn paper on the surface of the PLD target and examining the paper in an optical microscope. The depositions were carried out in vacuum ($\sim 10^{-5}$ torr). Both Si wafers and NaCl plates were used as substrates. The Si wafers were used in order to collect material for GPC analysis and the NaCl plates were used for infrared absorbance measurements with a Bruker IFS 66 spectrometer. The deposition rate was determined by gravimetric analysis of the Si wafers to be ~10 ng/cm^2 × macropulse. Using the bulk density supplied by the manufacturer (1.34 g/cm^3), this corresponds to a films thickness of 750 nm. Therefore, using these conditions, a 0.5 μm thick film can be grown in less than 5 min.

Figure 7 displays the infrared absorbance spectrum of a film deposited using 3.40 μm (resonant with C-H stretch) light and is compared with the native polymer. Not only is there a one-to-one correspondence between the features in the native polymer spectrum and the resonant infrared PLD film, but the ratios of band intensities reproduce as well. The film deposited with the laser tuned to 2.90 μm (O-H stretch) likewise shows virtually perfect correspondence between the native polymer spectrum and that of the deposited film.

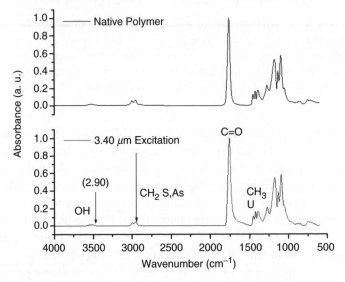

Figure 7: Fourier transform infrared spectra of PLGA 50:50 deposited by resonant infrared PLD compared to the native starting material. These spectra demonstrate that the deposited films exhibit nearly identical structural properties to native PLGA.

6.2.3. Resonant Infrared Matrix Assisted Pulsed Laser Evaporation

With a full understanding of the advantages of the two modifications to conventional PLD above (MAPLE and resonant infrared PLD), a natural, if not required, next step would be to combine the two approaches. In this case, the target would be a frozen composite target of mixed volatility excited by a laser tuned to the vibrational excitation of only the frozen solvent molecules. This combined approach would utilize the convenience of wet target handling as well as eliminate the possibility of photochemical modification.

There is a high level of interest in using thin films of large organic molecules such as DNA as the active element in sensors. One such example uses calf thymus DNA immobilized on electrodes in order to detect pollutants. Previous work has been done with resonant infrared MAPLE of salmon sperm DNA and pBluescript DNA. Here, we describe the deposition and characterization of laser-deposited calf thymus DNA. Highly polymerized fibrous calf thymus DNA was purchased from Sigma-Aldrich (D-1501) and reconstituted in 10 μM potassium phosphate buffer (pH = 7.0, 1 μM EDTA, 100 μM K$^+$). The concentration of DNA in the MAPLE solutions varied from 0.47 to 1.88 mg/mL. The depositions were carried out in a vacuum chamber that was kept at a pressure below 10 mTorr. The spot size was between 0.042 and 0.063 cm^2, yielding a fluence range from 8.6 to 10.8 J/cm^2. We have characterized the deposited films with UV and circular dichroism (CD) spectroscopy. In addition gel electrophoresis was performed in order to determine if the DNA has been fragmented.

The UV spectra of films deposited at three different laser energies are shown in Fig. 8. The concentration of the MAPLE solution was 0.47 mg/mL. 20,000 laser pulses were used to deposit each film. It is interesting to note that at the highest

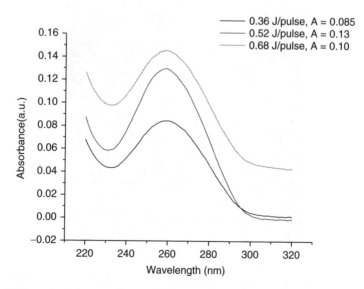

Figure 8: UV spectra taken on three calf thymus DNA films deposited at differing pulse energies by resonant infrared MAPLE.

energy (0.68), the absorbance (difference between absorbance at 260 and 320 nm), and hence the deposition rate is lower than that at 0.52 J/pulse. There are a number of possible reasons for this. It may be that we are destroying some of the DNA when the laser is operated at the highest pulse energies. Another possibility is that the shape of the plume changes with increasing fluence in such a way as to reduce the deposition rate. However, at this point, the reasons why the deposition rate is lower at 0.68 J/pulse as compared with 0.52 are uncertain, and are the subject of further sodium silicate glass microscope slides and washed off in buffer solution after deposition. The DNA was extracted from solution and it was placed in 10 μL of a solution containing dye (Xylene–Cyanol Blue 1X TBE) and injected into one of eight wells in a 1% agarose gel. The results, shown in Fig. 9, clearly show that the DNA has not been fragmented in the deposition. If fragmentation occurred, bright banding would be seen in intermediate positions throughout the gel. In contrast, our

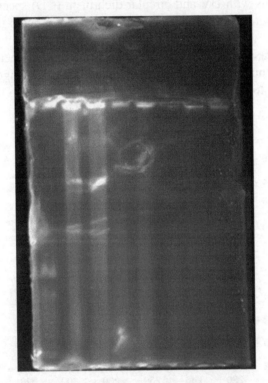

Figure 9: Gel electrophoresis of DNA films clearly show that the DNA has not been fragmented in the deposition. There are eight wells and the leftmost well contains a molecular ladder. The second well contains native calf thymus DNA and the rest are from deposited films.

DNA samples remained localized in their respective wells, thus indicating that they had not been appreciably reduced in size. The first (leftmost) well contains a molecular weight ladder. Next is the native calf thymus DNA. The remaining six wells are populated with DNA that had been extracted from deposited films, UV Circular dichroism measurements were performed in order to verify the structure and conformation of the deposited DNA films, confirming both the primary and secondary structures are preserved in the deposition process. This is a crucial point in the development of sensors.

6.2.4. Laser Micromachining of Differentially Adherent Substrate for Three-Dimensional Myoid Fabrication

New tissue engineering techniques must be developed to create a three-dimensional, functional replacement for natural tissue. Current tissue engineering techniques [5], such as homogeneous seeding of synthetic scaffolding, is limited by bioreactor tissue growth, which cannot adequately supply oxygen and nutrients nor expel waste from cells deep within the tissue. It is important to note that cellular self-assembly occurs between the nano- and macro-scale in natural environments. On the other hand, when small groups of adherent cells (e.g. about 40 myoblast cells) are initially deposited in relative proximity (about 400 µm spacing) onto a two-dimensional substrate, they grow together to form a multinucleated myotube in only about 24 h [6]. Unfortunately, during the next 24 h they will continue to grow in all directions. To control cell attachment, movement, and expansion, the cells must be coupled with a differentially adherent substrate. We have explored the use of micromachined agarose, a hydrophilic polymer, to prevent cell adhesion. Within the micromachined channels, we added polymerized extracellular matrix (ECM), a hydrophobic basement membrane mixture, which has been known to induce cellular adhesion, growth, and proliferation [7].

A 2% agarose gel was prepared by hydrating 1 g of electrophoresis-grade agarose powder with 50 mL of water and warming to 80°C. 17 mL of liquid agarose was poured into a 3″ petri dish and allowed to set. The agarose filled petri dish was mounted on a computer controlled X–Y transition stage and micromachined using an ArF excimer laser ($\lambda = 193$ nm) maintained with a spot size of 20 µm and an energy per pulse of 0.9–1.0 µJ. Channels were milled from 60–400 µm wide and 1 cm long. Extracellular matrix solution (ECM) (ATCC, Manassas, VA) was micropipetted into the channels and polymerized [8].

C2C12 mouse myoblasts (ATCC, Manassas, VA) were grown in tissue culture flasks, trypsinized, centrifuged, and resuspended in culture medium (90% Dulbecco's Modified Eagle's Medium (DMEM) supplemented with 10% fetal bovine serum and 1% penicillin–streptomycin–glutamine) [9–11]. 1 mL from a 1×10^4 cell/mL myoblast suspension was pipetted over the ECM filled, micromachined agarose channels and 10–15 mL of culture media was added. The number of cells per 10×10 μm area was equal to ~0.22. The cells were maintained in a 37°C, 5% CO_2 incubator.

A live/dead cell staining kit (Biovision, Mountain View, CA) was used to stain the templated myoblasts. Staining solution, containing propidium iodide and Live-dye™ stained dead cells red and live nuclei green, respectively. The stained myoblasts were viewed with an inverted optical microscope and a digital still camera with an epi-fluorescence attachment.

Upon the initial addition of the myoblasts to the micromachined channels, less than 20% of the cells contacted the ECM. This small percentage ensured that the adhered cells would form non-confluent surfaces within the channels and their interactions with neighboring cells could be easily observed. Within the first 2 h, adhered myoblasts migrated 5–10 μm along the channel length, and, sometimes, came into contact with other myoblasts. Two hours post-deposition, the myoblasts maintain 50–60% coverage along the ECM-lined channel.

Between 2 and 24 h, the myoblasts continue to grow and proliferate within the channel (Fig. 10). In that time, two-dimensional confluence is reached and cells reorient to align parallel to the channel and form the lowest energy, most highly packed alignment within the channel. Figures 11 and 12 reveal that nearly 90% of the myoblasts have elongated parallel to the channel after 24 h growth. Cellular alignment along channels or groove features greater than 2 μm has been previously noted. However, after 24 h growth, only the channels 60–150 μm in width were able to induce myoblast alignment. The 400 μm width channel was not able to induce cellular alignment. Myoblasts randomly aligned within the channel, while maintaining

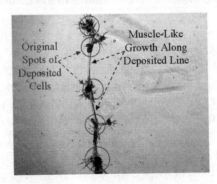

Figure 10: The growth of mouse myoblasts is shown after 24 h in culture. The individual spots have grown together to form a muscle fiber.

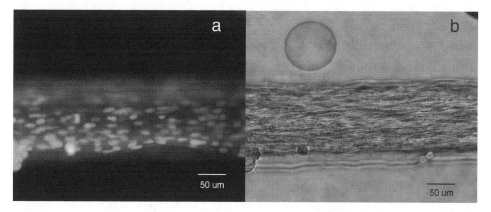

Figure 11: Live/dead and optical photograph, (a) and (b), respectively, of the myoblasts in the 150 μm wide channel 24 h post-deposition.

some alignment with neighboring cell. It is believed that there is no cooperative signaling across the channel width and cell networks greater than 10 cells wide are not able to effectively communicate over the channel distance.

In the 24–72 h growth period, the myoblasts continue to fill the micromachined channel through supplemental growth and vertical stacking of daughter cells. After 72 h, the myoblast density has dramatically increased. Over an area spanning roughly

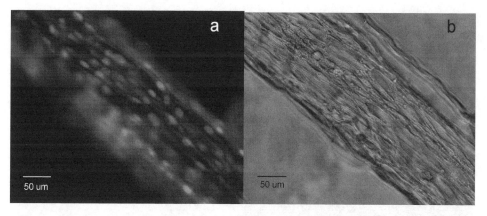

Figure 12: Live/dead and optical photograph, (a) and (b), respectively, of the myoblasts in the 150 μm wide channel 3 days post-deposition. Cellular density has increased dramatically, alignment is maintained throughout the channel. The increase in the number of nuclei indicates myoblast fusion and differentiation into multinucleated myotubes.

40,000 μm^2, the average number of nuclei has nearly doubled from 90 ± 10 to 160 ± 15 for the 24- and 72-h samples, respectively. The abundance of nuclei in such a densely packed space suggests that some myoblasts have differentiated into multinucleated myotubes. After 72 h, the myoblasts have nearly formed a micro-scale three-dimensional cell organization called a muscle organoid. For the 1 cm long 100 μm channel, the aspect ratio for the fully grown organoid was 100:1 (see Fig. 13).

These results have shown that controlled two-dimensional adherence can promote single layer myoblast alignment and three-dimensional myoblast organoid growth when the channel dimension is on the order of the elongated cell width. Specifically, for C2C12 myoblasts, we have observed channel alignment along widths between 60–150 μm and misalignment in a channel width of 400 μm. A micro-scale, three-dimensional, densely aligned channel of cylindrical, elongated cells was observed in these channels after 72 h growth. A schematic summary of the time progression of cell growth, alignment, and differentiation is shown in Fig. 14. This section has shown that controlled two-dimensional adherence can promote single layer myoblast alignment and three-dimensional myoblast organoid growth when the channel dimension is on the order of the elongated cell width. A micro-scale, three-dimensional densely aligned channel of cylindrical, elongated cells was observed in these channels. The combination of differential adherence and micromachined topological features will aid in the development of sophisticated heterogeneous tissues by direct write methods [12].

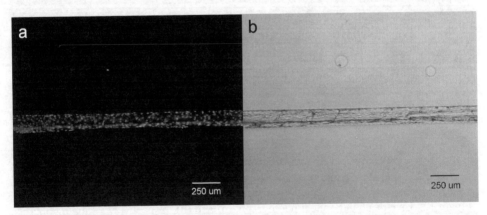

Figure 13: Live/dead and optical photograph, (a) and (b), respectively, of the myoblasts in the 100 μm wide channel 24 h of cellular growth. It is important to note that the channel growth is occurring over a macroscopic scale. Although the entire channel is not shown, the cells completely fill the 1 cm long channel.

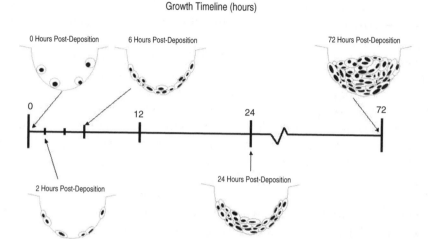

Figure 14: This schematic diagram reveals the cross-sectional adhesion, growth, reproduction and differentiation of the 100 mm channel. The timeline units are given in hours.

6.2.5. Matrix Assisted Pulsed Laser Evaporation Direct Write Applied to the Fabrication of Three-Dimensional Tissue Constructs

Tissue engineering integrates cell biology and engineering with novel materials for the development of next-generation medical technologies [1]. Tissue engineering and regenerative medicine can be defined as the use of living cells that are manipulated genetically or through their environment with bioresorbable materials as a matrix to develop biologically functional tissue substitutes for implantation into the body. Construct technology for tissue engineering is generally approached in either a cell-seeded method using various types of specially functionalized matrices (e.g. hydrogels and degradable polymer scaffolds) or a cell-mediated self-assembly system. In this section we present a novel laser-based approach for depositing cell-ceramic composites that allows the integration of cells to the matrix whilst the construct is being created.

Selection of biodegradable or bioactive materials and processes for developing three-dimensional constructs are important steps in tissue construct technology. Traditional scaffold materials include natural materials, e.g. collagen and ECM, and synthetic degradable organic polymers such as PCL (Poly-caprolactone), PLA (Poly-lactic acid), PGA (Poly-glycolic acid), hydrogels, and other non-biodegradable materials. For organic polymers such as PGA and PLA, the degradation rate has been well studied and is established and has been discussed earlier with respect to drug delivery. Apart from orthopedic applications such as for total hip replacements and filling bone voids, bioceramics have not been employed much in tissue engineering as scaffolds. Inorganic materials such as bioceramics are known to be osteoconductive and offer excellent structural integrity with varying bioresorbability, both of which are vital for any tissue construct.

Recently, technologies have been developed to direct write materials including micropen dispensing, inkjet, and laser-based methods. Originally developed to satisfy the need for conformal and mesoscopic passive electronic components [13], these processing techniques have been extended to soft and biological materials. There are several new direct writing techniques that attempt to achieve precise cell growth and differentiation. Direct writing techniques include dispensing (e.g. inkjet printing, and micropen), stamping (e.g. microcontact printing), proximal probe (e.g. AFM dip pen nanolithography), and laser forward transfer approaches [14–19]. The beauty of these direct writing approaches is that customized heterogeneous biomaterial deposition and even three-dimensional engineered tissue constructs can be fabricated to meet an existing need [20]. For example, a replacement tissue substitute can be fashioned so that it corresponds to the excised tissue site. Moreover, direct writing techniques can be used to precisely fashion conventional scaffold biomaterials or even develop new biomaterials.

The direct write approach used in this chapter is termed matrix assisted pulsed laser evaporation – direct write (MAPLE DW). The experimental MAPLE DW setup is shown in Fig. 15. The material to be transferred is solvated and evenly coated on the ribbon surface. The solvent absorbs the incident laser light and the solute material is driven forward to a receiving substrate. The distance between the ribbon and substrate, the fluence, the spot-size, and the movement of the stage controls the resolution and pattern. In the past, we have successfully used this approach to write mesoscopic patterns of passive electronic pastes or inks and developed them at sintering and laser annealing temperatures commensurate with plastic substrates [2]. Imaging of the MAPLE DW transfer process showed it to be a highly forward-directed technique and it was applied to soft biomaterials of increasing complexity. The MAPLE DW process has clear advantages over other patterning processes in providing a versatile technique to transfer a wide range of materials, with great accuracy, resolution, speed, and efficiency [21–24].

Two-dimensional patterning of a single cell-type has been demonstrated previously using various techniques. Surface recognition techniques have been used frequently to achieve cell patterning. Laminar flow of the media in capillary systems has been used to create simple patterns of different types of 'cells and proteins' in the same substrate [25]. Laborious steps are used to create elastomeric stencils using photoresists that are then used as masks to pattern cells and proteins [26].

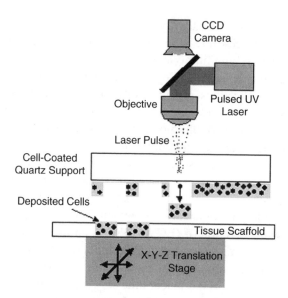

Figure 15: A schematic diagram highlighting the potential of MAPLE DW of bioceramics in developing tissue scaffolds.

Soft photolithographic methods such as microcontact printing, which involve the fabrication of stamps using masking techniques, have been demonstrated to create cell and protein patterns [27,28]. Rapid patterning techniques were proposed by using dielectrophoretic force to transfer cells from the electrode to a glass substrate, however with relatively poor resolutions and accuracy [29]. Inkjet technology, termed organ printing, has been used to demonstrate patterning of proteins and cells [30–32]. The technology uses a CAD setup with cell transfer through jet-based cartridge [33]. Laser Induced Forward Transfer (LIFT) has been shown to success-fully transfer various dry films with good resolutions. Recently, cells and biomol-ecules have also been transferred using the same basic experimental setup, however, it only works with the use of an intermediate film to partially absorb the incident laser [34]. MAPLE DW is recognized as a simple technique to achieve well-defined high-resolution patterns with relativistic ease and short times, involving no masks, screens or preparation of stamps. MAPLE DW has been used successfully to transfer a variety of materials including inorganic materials for electronics, synthetic organic polymers, and various types of viable cells, providing foundation for the subsequent development of *in vitro* three-dimensional tissue constructs. In this section we describe the co-deposition of viable cells and bioceramics into mesoscopic patterns with direct write methods like MAPLE DW. With rapid deposition, the need for immediate structural integrity and accelerated tissue maturation is critical while constructing a three-dimensional network by MAPLE DW.

Bioceramics such as alumina, zirconia, calcium phosphates (hydroxyapatite, α-TCP), silica-based Bioglass®, and pyrolitic carbons have been frequently used in various biomedical applications, including orthopedic restorations. These bioceramics can serve as excellent matrices providing the structural integrity required by direct write techniques and the variable bioresorbability desired for the construct. The deposited layers could then be integrated to serve as a useful mechanical, bioresorbable, and cell-conductive scaffold for tissue engineering.

Hydroxyapatite has been demonstrated to stimulate osteoinductive response, where the cell deposits mineral until a homeostasis is reached. Hott *et al.* have shown that human trabecular osteoblastic cells attach proliferate, and differentiate on hydroxyapatite, with secretion of (ECM) [35]. In an interesting study, Park *et al.* have shown that bioceramics such as tricalcium phosphate and hydroxyapatite can stimulate osteoblastic differentiation of human bone marrow stromal cells [36]. This is beneficial in both ways, as the bioceramic serves as a structural component and as a differentiation factor for the construct. In this section we show the integration of the bioceramic with the cells prior to fabricating the construct. By creating a composite that contains osteoblast cells, ECM, and hydroxyapatite we can address important issues of cell–matrix integration and subsequent functional differentiation. Our findings of depositing mixtures of viable cells, ceramics, and ECM into mesoscopic patterns was the first step towards developing cell-integrated mechanically robust three-dimensional tissue constructs; in this case the application would be bone related, but the goal is to suggest a more general approach to integrating ceramic construct materials. Figure 16 summarizes the potential of this approach, highlighting the effectiveness of using MAPLE DW of bioceramics in developing a bioceramic scaffold. This concept can be further extended to a variety of biomaterials both organic and inorganic that can be deposited by MAPLE DW. The integration of viable cells with the choice of material (with controlled porosity, variable bioresorbability and differential structural integrity) makes MAPLE DW unique and promising for developing tissue constructs.

A schematic diagram of a MAPLE DW is shown in Fig. 15. A pulsed excimer UV laser (operating at 193 nm at 30 ns) is used as the source light and is focused onto the ribbon plane. A 1″ quartz disc is used as a ribbon with the material coated on one side. The ribbon sits on a holder that is placed directly above the substrate. The gap between the ribbon and substrate is controlled using a Z-stage translation. The X–Y translation stage is controlled by CAD setup. In our experiments, the laser was operated at a constant repetition rate of 10 Hz at fluences appropriate to the material for transfer. The laser spot size was determined prior to transfer. MAPLE DW experiments are performed in air and at room temperature.

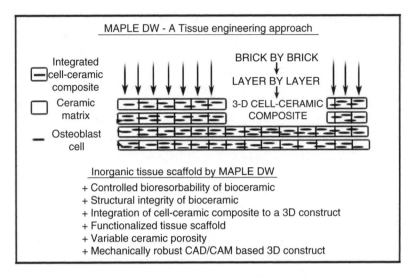

Figure 16: A schematic diagram of a MAPLE DW setup is shown.

Preparation of the ribbon was critical in achieving an effective transfer of the material. HA and zirconia were solvated to 50–70% solutions in a 50:50 glycerol: water solvent to establish an appropriate viscosity for ribbon fabrication and successful transfers. The pastes were spin coated separately on 1″ clean quartz ribbons at 1000 rpm for 10 s, prior to transfer. Clean borosilicate glasses were used as substrates. Using the preprogrammed definitions for the X–Y translation, the patterns were achieved for both HA and ZrO_2, separately at fluences of 0.22 and 0.18 J/cm^2, respectively. An optical microscope was used to image the patterned ceramics. Scanning electron microscopy (SEM) was used to study the structural pattern and imperfections of the deposited ceramics.

MG63 human osteosarcoma (ATCC, Manassas, VA), were started and sub-cultured in a growth media (containing 89% minimum essential Eagle's medium with 2 μM L-glutamine and Earle's BSS adjusted to contain 1.5 g/L sodium bicarbonate, 0.1 μM non-essential amino acids, and 1.0 μM sodium pyruvate, 10% heat-inactivated fetal bovine serum, and 1% Antibiotic Syrup) [37]. The cells in media were stored in a culture incubator at 37°C containing 95% air and 5% CO_2. ECM (Extracellular Matrix, ATCC, Manassas, VA), containing laminin, collagen IV, and various growth factors, was used as the matrix in the ribbon and as the receiving base in the substrate for all cell transfers. 0.5 mL of ECM was spin coated at 1000 rpm for 10 s on a sterilized 1″ quartz ribbon and glass substrate. The ribbon and substrates were left in the incubator for 30 min to allow cross-linking. Meanwhile, sub-confluent

MG63 osteoblast cells were counted, trypsinized, and centrifuged at 5000 rpm for 3 min. The concentrated cell pellet was reconstituted with 0.25 mL media, which was pipetted onto the ECM containing ribbon and allowed to settle. Fluence slightly above the determined threshold, at approximately 0.15 J/cm^2 was used for the cell transfers. A similar protocol was developed for the cell-ceramic composites. HA paste (HA in Phosphate buffer solution) was obtained at varying proportions with the ECM solution and spin coated onto the ribbon at 1000 rpm for 10 s. Known amount of cells were added to the ribbon and allowed to settle. Cell-ceramic composites were transferred at approximately 0.22 J/cm^2. After the transfer, both the ribbon and the substrate were stored in separate petri dishes containing pre-warmed media. The cells on the substrate were then studied for viability and proliferation over days using an optical microscope and later fixed for a live/dead assay for fluorescence imaging.

On the MAPLE DW of HA, we find that the consistency of the paste was crucial in determining the success of the transfer. Transfer of HA was achieved just above the threshold fluence, at 0.22 J/cm^2. Figure 17a shows the mesoscopic view of the transferred HA into a box pattern, with well-defined boundary and relatively low splashing. SEM images at higher magnifications show a porous like continuous network (Fig. 17b). Figure 18 shows the X-ray diffraction spectrum for this film. The diffractions peaks correspond to HA and are further evidence that no material medication has been done in the transfer. The porosity is dependent on the amount and type of solvent used to prepare the mix. Thus, the porosity of the HA could ideally be controlled by changing the slurry mix for developing scaffold structures for bone implants. In a more general way, a porous network could be considered advantageous for vascularization and thus for the cells to proliferate in the scaffold. Angiogenesis (birth of blood vessels for vascularization) is known to be expressed either by using chemical growth factors such as VEGF (Vascular Endothelial Growth Factor) or by controlling the matrix micro-architecture. Brauker *et al.* have shown that pore sizes large enough to allow complete penetration by host cells (0.8–8 μms pore size) favors angiogenesis leading to more vascular structures. As observed from the SEM images the pore size (particle size distribution) is roughly in the range of the critical size [38]. Since changing the solvent amount can control the microscopic porosity of the bioceramic, type and drying conditions, it is possible to deposit specific bioceramic or polymeric pores within the above range. The threshold porosity for vascularization can be tailored to permeate the entire scaffold, a critical benefit for any tissue-engineering construct.

A similar box pattern is used for zirconia and shown with well-defined edges in Fig. 19a. Unlike in HA where a continuous mesh-like structure is observed, SEM images at high magnifications (Fig. 19b) show isolated agglomerated particles of zirconia. Studies on surface uniformity and thickness as a function of the ribbon

a)

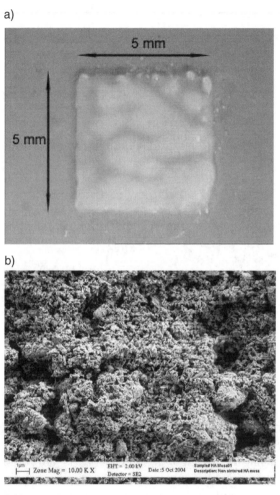

b)

Figure 17: (a) Shows the mesoscopic view of the transferred HA into a box pattern, with well-defined boundary and relatively low splashing and (b) shows the SEM images (at 10 KX) at higher magnifications indicating a porous-like continuous network.

thickness, separation, fluence, and the slurry consistency will be the next step to develop better surface and pattern definitions. The choice of solvent used was important for the fabrication of a uniform ribbon of the appropriate thickness and for the laser-produced evaporative desorption mechanism. More importantly, unless the solvent chosen is predominately water-based, the ribbon fabrication protocol could preclude the possible transferring of living cells by MAPLE DW.

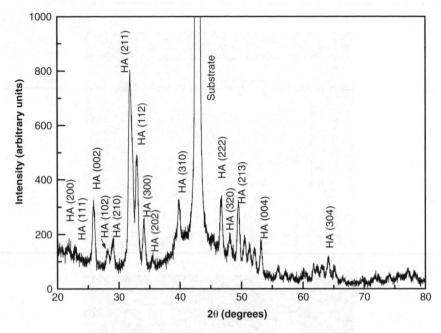

Figure 18: X-ray diffraction spectrum of the transferred HA film. The diffraction peaks observed correspond to those of HA.

The concept of transfer of live cells combined with inorganic materials is a novel approach in developing bioactive, bioresorbable scaffolds. Figure 20a shows an optical microscopic image of the ribbon, showing etched out pattern from the cell-ceramic layer. The resolution of the material transferred can be effectively controlled by changing the laser spot size and fluence. Figure 20b shows the substrate containing the transferred osteoblastic cell and hydroxyapatite particles. The cell can be observed to extend and adhere in the presence of transferred hydroxyapatite. Figure 21a and b shows the optical microscopic image of the transferred cells-ceramic composite at 3 and 96 h post transfer, respectively. It can be clearly noted that the cell-density (in the presence of the transferred ECM and ceramic) increases over time, indicating good viability and proliferation of cells post transfer. The cells can be seen alive *in vitro* even after 5 days. The live-dead stains followed by fluorescent imaging (Fig. 22a,b) shows viable cells 5 days post transfer. At this point, the osteoblasts can be seen at a near confluent state. Dissolution of calcium and phosphate ions into the medium is known to simulate osteoblastic differentiation, which can only aid the sustenance of the tissue scaffold.

For the cell-ceramic composite transfer, the absorbing matrix for the transfer was a mixture of PBS (solvent for HA) and ECM (for cells). ECM was used with

a)

b)

Figure 19: (a) Shows the mesoscopic view of the transferred zirconia into a box pattern, with well-defined boundary and relatively low splashing and (b) shows the SEM images (at 10 KX) at higher magnifications indicating isolated agglomerated particles of zirconia.

HA to facilitate the transfer of viable cells. Recently, it was shown that native ECM matrix proteins induce the nucleation and growth of hydroxyapatite [39]. The nucleation and crystallization of hydroxyapatite was facilitated by ECM-coated stainless steel substrate. This can be shown to be beneficial in more than one way. The presence of ECM aids the cell growth and proliferation and induces an increase in HA growth and crystallization, improving the structural integrity of

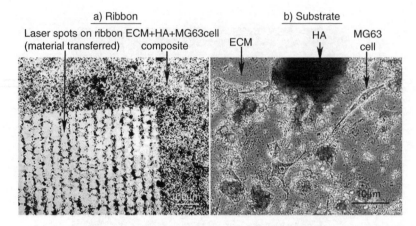

Figure 20: Optical microscopic image of: (a) the ribbon showing the etched out spots of the HA-osteoblast composite layer (at 40X); (b) the substrate where the live cells and ceramic are deposited (at 200X).

the scaffold. By using varying proportions of ECM, bioceramic, and viable cells, the mixture can be deposited to tailor functionalized scaffolds.

The growth rate of the cells in the transferred composite was recorded, and was compared with that of cells transferred with no ceramic and with as-seeded cells (no laser interaction). These results are shown in Fig. 23. The growth rate can be observed to be statistically similar in all cases, suggesting co-deposition has no adverse effect on the cell-proliferation. As with the previous transfer, the cells are

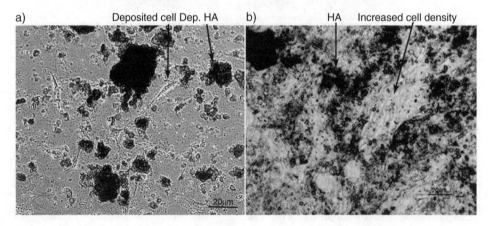

Figure 21: Optical microscopic image (100X) of the transferred cell-ceramic composite taken 3 and 96h post-transfer, showing the proliferation of the cells after transfer.

a)   b)

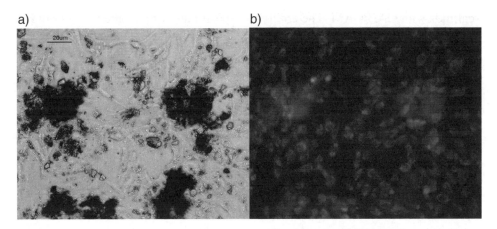

Figure 22: Optical light and fluorescence microscopic image (100X) of the live-dead
stained cell-ceramic composite, 72 h post-transfer.

well adhered to the substrate and show no morphological changes when compared
to the cells before transfer. This suggests that the cells have undergone no mutations
or apoptosis after transfer, and it also signifies the biocompatibility of HA.

The results of this section show that osteoblast cells and HA can be transferred at
the same time using MAPLE DW to form mesoscopic patterns. This result is encouraging for the development of heterogeneous biodegradable and bioresorbable

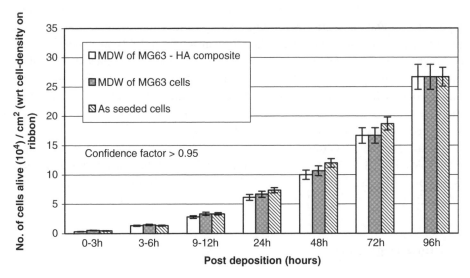

Figure 23: The growth profile of MG63 cells with and without HA is compared.

scaffolds with CAD/CAM. The ability to deposit mesoscopic patterns of both HA and MG63 cells is the first step towards using two-dimensional layers of inorganic materials as tissue scaffolds and for the growth of mechanically robust three-dimensional heterogeneous tissue constructs. The versatility of transferring cells alongwith ceramics to achieve well-defined patterns makes MAPLE DW unique and very promising for heterogeneous scaffolding.

The concept of heterogeneous transfer can be extended to a variety of combinations of materials and living cells, with varying degrees of bioresorbability and functionality. Such heterogeneity could be employed to a recipe of various organic polymeric materials, bioceramics, viable cells, and ECM and deposited as a whole unit. The versatility of such a process offers good potential for engineering functionalized tissue constructs.

6.3. Conclusions

This chapter has demonstrated the unique capabilities of lasers for the processing and micromachining of biomaterials and biological components to produce novel biological structures. MAPLE and resonant infrared PLD were shown to be a new route to deposit thin films for controlled release of drug coatings and biological sensors. The combination of laser micromachining and biomaterials was used to produce a differentially adherent surface for cell growth. Using mouse myoblasts we demonstrated the growth of high aspect ratio (100:1) three-dimensional myoids. MAPLE DW was demonstrated to be a versatile technique to develop novel, multi-layer, heterogeneous scaffolds for engineering tissue constructs. Using a brick-by-brick to a layer-by-layer approach a three-dimensional scaffold can be developed. The ability to deposit mesoscopic patterns of cell-ceramic composites offer structural integrity in using two-dimensional layers to build mechanically robust three-dimensional heterogeneous tissue constructs. Additionally control of microporosity in bioceramics can be used to promote vascularization, while varying degrees of bioresorbability can be used to develop functionalized scaffolds. The versatility of transferring cells with ceramics to achieve well defined, CAD/CAM-based patterns, makes MAPLE DW unique and very promising for heterogeneous scaffolding. Based on the functionality and structural requirements, this concept can be extended to integrate a variety of cell-lines with most organic/inorganic materials.

Acknowledgments

The authors gratefully acknowledge financial support from the Office of Naval Research for this work.

References

[1] *Springer Handbook of Nanotechnology*, Springer, **2004**.
[2] Chrisey, D.B. and Hubler, G.K., *Pulsed Laser Deposition of Thin Films*, Wiley, **1994**.
[3] Niemz, M.H., *Laser-Tissue Interactions: Fundamentals and Applications (Biological and Medical Physics)*, Springer, New York, **2002**.
[4] Von Allmen M. and Blatter, A., *Laser-Beam Interactions with Materials: Physical Principles and Applications*, Springer-Verlag, **1995**.
[5] Lanza, R.P., Langer, R. and Vacanti, J.P., *Principles of Tissue Engineering*, 2nd edition, Academic Press, New York, **2000**.
[6] Chromiak, J.A., Shansky, J., Perrone, C. and Vandenburgh, H.H., *In Vitro Cell. & Dev. Bio.-Anim.*, **1998**, *34*, 694–703.
[7] Maley, M., Davies, M.J. and Grounds, M.D., *Exp. Cell Res.*, **1995**, *219*, 169–179.
[8] www.atcc.org - EC Matrix Solution, Catalog No. 30-2501.
[9] Yaffe, D. and Saxel, O., *Nature*, **1977**, *270*, 725–727.
[10] Lawson, M.A. and Purslow, P.P., *Cel. Tis. and Org.*, **2000**, *167*, 130–137.
[11] Bardouille, C., Lehmann, J., Heimann, P. and Jockusch, H., *Appl. Microbiol. Biotech.*, **2001**, *55*, 556–562.
[12] Wilkinson, C., Riehle, M., Wood, M., Gallagher, J. and Curtis, A., *Mat. Sci. and Eng.*, **2002**, *C 19*, 263–69.
[13] Chrisey, D.B. *et al.*, Direct writing of conformal mesoscopic electronic devices by MAPLE DW, *Applied Surface Science,* **2000**, *168*, 345–352.
[14] Turcu, F., Tratsk-Nitz, K., Thanos, S., Schuhmann, W. and Hieduschka, P., *J. Neurosci. Meth.*, **2003**, *131*, 141–148.
[15] Fan, H.Y., Reed, S., Baer, T., Schunk, R., Lopez, G.P. and Brinker, C.J., *Micropo. and Macropo. Mat.*, **2001**, *44*, 625–637.
[16] James, C.D., Davis, R.C., Kam, L., Craighead, H.G., Isaacson, M., Turner, J.N. and Shain, W., *Langmuir*, **1998**, *14*, 741–744.
[17] Li, Y., Maynor, B.W., Liu, J., *J. Am. Chem. Soc.*, **2001**, *123*, 2105–2106.
[18] Hopp, B., Smausz, T., Antal, Z., Kresz, N., Bor, Z. and Chrisey, D., *J. Appl. Phys.*, **2004**, *96*, 3478.
[19] Karaiskou, A., Zergioti, I., Fotakis, C., Kapsetaki, M. and Kafetzopoulos, D., *Appl. Surf. Sci.*, **2003**, *208*, 245.
[20] Pique, A. and Chrisey, D.B., *Direct Writing Technologies for Rapid Prototyping Applications: Sensors, Electronics, and Integrated Power Sources*, Academic Press, New York, **2002**.
[21] Chrisey, D.B. *et al.*, Laser Deposition of Polymer and Biomaterial Films, *Chem. Rev.*, **2003**, *103*, 553–576.
[22] Wu, P.K. *et al.*, Laser transfer of biomaterials: Matrix-assisted pulsed laser evaporation (MAPLE) and MAPLE Direct Write, *Rev. of Sci. Instr.*, **2003**, *74*(4), 2546.
[23] Ringeisen, B.R. *et al.*, Generation of mesoscopic patterns of viable Escherichia coli by ambient laser transfer, *Biomaterials*, **2002**, *23*(1), 161–166.
[24] Ringeisen, B.R. *et al.*, Laser printing of Pluripotent embryonal carcinoma cells, *Tissue Engineering,* **2004**, *10*, 3–4.

[25] Takayama, S. *et al.*, Patterning cells and their environments using multiple laminar fluid flows in capillary networks, *Proc. Natl. Acad. Sci. US,* **1999**, *96*, 5545–5548.

[26] Folch, A. *et al.*, Microfabricated elastomeric stencils for micropatterning cell cultures, *J. Biomed. Mater Res.*, **2000**, *52*, 346–353.

[27] Kane, R.S. *et al.*, Patterning proteins and cells using soft lithography, *Biomaterials,* **1999**, *20*, 2363–2376.

[28] Patel, N. *et al.*, Printing patterns of biospecifically adsorbed protein, *J. Biomater. Sci. Polymer Edn.,* **2000**, *11*(2), 319–331.

[29] Matsue, T. *et al.*, Rapid micropatterning of living cells by repulsive dielectrophoretic force, *Electrochimica Acta.*, **1997**, *42*(20–22), 3251–3256.

[30] Wilson, W.C. *et al.*, Cell and organ printing 1: Protein and cell printers, *The Anatomical Record Part A,* **2003**, *272A*, 491–496.

[31] Mironov, V. *et al.*, Organ printing: Computer-aided jet-based 3D tissue engineering, *Trends in Biotechnology,* **2003**, *21*, 4.

[32] Barron, J.A. *et al.*, Biological laser printing of three dimensional cellular structures, *Appl. Phys. A,* **2004**, *79*, 1027–1030.

[33] Kantor, Z. *et al.*, Metal pattern deposition by laser-induced forward transfer, *Applied Surface Sciences,* **1995**, *86*, 196–201.

[34] Fernandez-Pradas, J.M. *et al.*, Laser-induced forward transfer of biomolecules, *Thin Solid Films,* **2004**, *453–454*, 27–30.

[35] Hott, M. *et al.*, Proliferation and differentiation of human trabecular osteoblastic cells on hydroxyapatite, *J. Biomed. Mater Res.*, **1997**, *37*, 508–516.

[36] Park, E.K. *et al.*, Cellular biocompatibility and stimulatory effects of cmp on osteoblastic diff of human bone marrow derived stromal cells, *Biomaterials,* **2004**, *25*, 3403–3411.

[37] Billaiau, A. *et al.*, Human Interferon: Mass production in a newly established cell line, MG 63, *Antimicrobial Agents and Chemotherapy,* **1977**, *12*(1), 11–15.

[38] Brauker, J.H. *et al.*, Neovascularization of synthetic membranes directed by membrane microarchitecture, *J. Biomed. Mat. Res.*, **1995**, *29*, 1517.

[39] Pramatarova, L. *et al.*, Hydroxyapatite growth induced by native extracellular matrix deposition on solid surfaces, *European Cells and Materials,* **2005**, *9*, 9–12.

Chapter 7

Direct Transfer and Microprinting of Functional Materials by Laser-Induced Forward Transfer

K.D. Kyrkis[a], A.A. Andreadaki[a], D.G. Papazoglou[b] and I. Zergioti[a]*

[a]*National Technical University of Athens, Physics Department, Iroon Polytehneiou 9, 15780 Zografou, Athens, Greece;* [b]*Foundation for Research and Technology – Hellas, Institute of Electronic Structure and Laser, P.O. Box 1527, Heraklion 711 10, Greece*

Abstract

Laser-induced forward transfer (LIFT) is a technique which enables the controlled transfer of a thin film of a material, from a transparent carrier to a receiving substrate. In most cases, the transfer of the thin film is achieved by using single laser pulses, but it has also been demonstrated by the use of continuous-wave lasers. Usually, the receiving substrate is placed in parallel and at a close proximity to the thin film source under air or vacuum conditions. Realizing the transfer while the supporting carrier and the receiving substrate can be moved with respect to each other, it allows the production of micrometric scale patterns of many different materials and components on diverse substrates. The resolution and speed should be more than adequate, due to the small spot size and the precision scanning systems available. Alternatively, the pattern can be produced on the substrate by using an appropriate projection mask. This method has been applied to fabricate mesoscopic patterns of biological, organic and inorganic materials and this will be presented in the next paragraphs.

Keywords: Laser and micro-capacitors; Laser printing of biomaterials; Laser printing of organic optoelectronics; Laser thermal imaging.

*Corresponding author, Tel. 0030 210 7723345, *zergioti@central.ntua.gr

Recent Advances in Laser Processing of Materials
J. Perrière, E. Millon and E. Fogarassy (Editors)

7.1. Introduction to the Laser-Induced Transfer Methods

7.1.1. Overview of the Laser-Induced Forward Transfer Process

The general principle of the LIFT process [1] is outlined in Fig. 1. The thin film of the material to be transferred is deposited on a transparent carrier. The receiving substrate is placed parallel and facing the thin film at a very short distance. This distance can vary from near contact to several micrometers. The laser beam is then focused through the transparent carrier at the thin-film/carrier interface. The laser wavelength is selected to the range of minimum carrier absorption and maximum film absorption.

The mechanisms and dynamics of the transfer of material from the carrier to the receiver substrate are complex. The type of laser (i.e. its wavelength, intensity, focal spot size and pulse length or scanning speed), the type of the material (i.e. optical absorption coefficient, thermal diffusivity) and the geometry of all components determine the quality of the transferred material arrives onto substrate.

Although laser-induced forward transfer technologies were originally used to transfer metals, the technique was quickly extended to the transfer of other materials such as metal oxides, semiconductors, superconductors and, more recently, bio-materials and organics. The process has been shown to be viable for the fabrication of micro-devices such as capacitors, resistors, chemosensors, biosensors, power sources, thin-film transistors (TFTs), organic TFT arrays and organic light-emitting diode (OLED) displays. An overview of the research efforts made in this field is outlined in Table 1.

Following to a historical account of laser-induced forward transfer methods, selected examples of current developments with technological interest will be presented in this review article. The potential of these techniques will be discussed together with their advantages. Finally, aspects of theoretical investigations over LIFT and their results will be discussed.

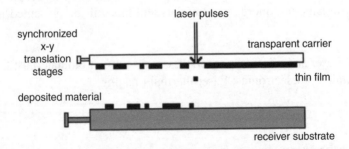

Figure 1: Scheme of the laser-induced forward transfer (LIFT) process.

Table 1: Selected works on laser-induced forward transfer materials deposition

Transferred material	Substrate	Feature size (μm)	Laser type (λ)	Applications	References
Cu	Silicon, silica	50	ArF (193 nm)		Bohandy (1986), Adrian (1987)
Cu, Ag	Fused silica	15	2ω-Nd:YAG (532 nm)		Bohandy (1988) [15]
Al	Copper, silica		Nd:YAG (1.06 μm)		Alexander (1988)
Au, Cr, Cu, Ti, Ta, Pt, Ni	Polyimide, ceramics, glasses, metals, silicon		KrF (248 nm)		Baseman (1988)
Al_2O_3	Quartz, glass		Nd:YAG (1.06 μm)		Greer (1988)
Au, Cr	Quartz, glass		KrF (248 nm)		Baseman (1989)
$YBa_2Cu_3O_{7-x}$, BiSrCaCuO	MgO, silicon	100	ArF (193 nm), Nd:YAG (1.06 μm, 5 ns), 2ω-Nd:YAG (530 nm, 100 ns)	Patterning of thin film superconductors	Fogarassy (1989, 1992) [20–22]
V, Cr, Ti, Ge, Sn	Glass, quartz		Ruby (694 nm), XeCl (308 nm)		Mogyorosi (1989)
Au	Quartz, glass		2ω-Nd:YAG (532 nm, 15 ns), KrF (248 nm, 25 ns)		Baseman (1990) [72]
Ti, Cr, Ge/Se	Glass		Ruby (694 nm)		Tóth (1990)
Al	Silicon	200	Nd:YAG (1.06 μm, 120 ns)		Schultze (1991) [73]
Ti, Cr	Glass	10	Ruby (694 nm)		Kántor (1992) [16,17]
W, Cr	Glass	10	Ar+ (515 nm)		Tóth (1993) [18]
Au, Al	Silicon, quartz	7–10	ArF (193 nm, 18 ns)		Lätsch (1994) [74]
PMMA	KBr, Si, CaF_2		4ω-Nd-YAG (266 nm, 10 ns)	Deposition of thin polymer films	Blanchet (1995) [37]

Continued

Table 1: Selected works on laser-induced forward transfer materials deposition—cont'd

Transferred material	Substrate	Laser type (λ)	Feature size (μm)	Applications	References
Pd	Quartz, ceramics, polymers	ArF, KrF, KrCl, XeCl, XeF (193 nm, 248 nm, 222 nm, 308 nm, 351 nm)	150	Metal patterns	Esrom (1995) [75]
Diamond	Silicon	KrF (248 nm, 15 ns), Cu (510 nm, 20 ns)	10		Pimenov (1995) [76]
Al	Doped p-Si, silica	XeCl (308 nm, 30 ns)		Fabrication of top-gated TFTs	Sameshima (1996) [19]
C, Cr	Glass	Nd:YLF (1047 nm)	20		Poon (1996) [77]
Cr, In$_2$O$_3$	Glass, silicon	KrF (248 nm, 500 fs)	3	Fabrication of diffractive microstructures	Zergioti (1998) [23]
Pyrene molecules	PBMA	KrF (248 nm, 30 ns)	Pyrene molecule	Implantation of molecules	Karnakis (1998) [38]
Cr, In$_2$O$_3$	Glass	KrF (248 nm, 500 fs)		Diffractive microstructures	Zergioti (1999) [78]
Cr	Glass	KrF (248 nm, 500 fs)	2	Fabrication of microstructures on non-planar and high-curvature surfaces	Mailis (1999) [25]
Au	SiO$_2$	Dye (440 nm, 9 ns)			Nakata (1999) [60]
Pt, Cr, In$_2$O$_3$	Glass	KrF (248 nm, 500 fs)	3	Diffractive microstructures	Papakonstantinou (1999) [79]

Material	Substrate	Thickness/Size	Application	Laser	Reference
Ag, Au, NiCr, $BaTiO_3$, Y_3FeO_{12}	Glass, alumina, silicon FR-4 and RO4003 circuit boards	~25	Fabrication of electronics and sensors	KrF (248 nm, 20 ns)	Piqué (1999) [28]
p-Si, Al	Silicon		Fabrication of top-gated TFTs	XeCl (308 nm, 35 ns)	Toet (1999, 2000) [26,27]
Dense oxide phosphor powders of Y_2O_3:Eu and Zn_2SiO_4:Mn	Alumina, polymers		Fabrication of high-definition displays	KrF (248 nm, 25 ns)	Fitz-Gerald (2000) [32]
In_2O_3	Glass		Microprinted gratings	KrF (248 nm, 500 fs)	Koundourakis (2001) [24]
Au/Sn	Silicon	30	Die/flip-chip bonding	Ti:sapphire (775 nm, 0.1–8 ps)	Bähnisch (2000) [80]
Pyrene-doped PMMA	PMMA	~100	Patterning of thin film materials	KrF (248 nm, 30 ns)	Mito (2001) [36]
Rhodamine 610 dye	SiO_2	~350	Controlled transfer of sensitive materials	Nd:YAG (532 nm, 8 ns)	Nakata (2002) [39]
Au, Ni	Si		Metal patterns	KrF (248 nm, 30 ns)	Sano (2002) [71]
Ni	Si		Organic electronics	KrF (248 nm, 30 ns)	Yamada (2002) [81]
Pentacene	Si wafers			4ω-Nd–YAG (266 nm, 10 ns)	Blanchet (2003) [40]
DNNSA-PANI/ SWNT	Glass resin	5 × 2.7	Organic TFTs	Diode 780 nm	Blanchet (2003) [41]
Blend of LEP and small molecule hosts	PEDOT: PSS or PFO–TPD on PEDOT	35	Organic LED displays	Nd:YAG (CW)	Chin (2003) [43]

Continued

Table 1: Selected works on laser-induced forward transfer materials deposition—cont'd

Transferred material	Substrate	Feature size (μm)	Laser type (λ)	Applications	References
Lambda phage DNA dissolved in Tris-HCl, EDTA solution	Glass	100	KrF (248 nm, 500 fs)	Research on genome functions	Karaiskou (2003) [53]
Au	Quartz		Ti:sapphire (400 nm, 150 fs)	Metal patterns	Tan (2003)
Polymer composites	Si, Ag		ArF (193 nm, 20 ns), 3ω-Nd:YAG (355 nm)	Fabrication of chemical sensors	Piqué (2003) [31]
DNA material and proteins	Poly-L-lysine, nylon-coated glass	40–65	3ω-Nd:YAG (355 nm, 10 ns)	Biosensors	Fernández-Pradas, Colina (2004–2005) [49,50]
Fungi (Trichoderma) conidia	Glass		KrF (248 nm, 30 ns)	Controlled transfer of organisms	Hopp (2004) [51]
Au, Al	Mica, quartz	50	KrF (248 nm, 28 ns)	Contact masks	Landström (2004)
Blend of LEP and inert polymers (polysterene)	PEDOT: MPS		Nd:YAG (CW)	Organic LED displays	Lee (2004) [44]
Carbon/binder, LiCoO$_2$/carbon/binder	Metal foils	40–60	3ω-Nd:YAG (355 nm)	Fabrication of Li-ion microbatteries electrodes	Wartena (2004) [33]

7.1.2. Laser Transfer Methods

The laser-direct write (LDW) [2] is a laser-induced forward transfer method for fabrication of metal-oxide conformal thin films on flexible and rigid substrates. Mask steps and additional patterning are not necessary, while *in situ* laser annealing is combined.

The hydrogen-assisted laser-induced forward transfer uses the focused laser beam to heat a porous sacrificial layer which contains gases. Such an example is the hydrogenated amorphous silicon (a-Si:H) [3,4]. By heating the layer, the temperature of the confined gases increases, and therefore expands explosively destroying locally the layer while applying a propelling force to the underlying material which is then released and deposited on the receiving substrate. It has to be noted that, in this implementation, lower energy densities than in the case of ablative evaporation are needed.

The matrix-assisted pulsed laser evaporation/direct-write (MAPLE/DW) is a novel laser-assisted direct-write process (originates from the Naval Research Laboratory–NRL) and is capable of producing patterns of a broad range of materials [5]. The materials which can be transferred are in the form of rheological fluid, polymer-based composite or fine powder. The technique is a combined and enhanced implementation of LIFT with the matrix-assisted pulsed laser evaporation (MAPLE) process, a sophisticated extension of pulsed-laser deposition (PLD). The key feature that separates MAPLE/DW from other direct-write techniques is that it is a pyrolytic method with respect to the matrix material but not with respect to the soluble material of interest. The method can be utilized *in situ* micromachining processes. This technique is described in Chapter 6 in detail.

The thermal imaging utilizes the pixelized transfer of organic or other thin film materials [6]. An infrared diode laser, is split into a number of individually addressable spots, and is focused through a donor base at a thin metal layer onto which the transfer material is coated. The conversion of light energy to heat at this interface results in the release of gaseous products and shockwave which transfer the material from the substrate to a flexible or rigid receiver. This technique enables the printing of multiple, successive layers via a dry additive process.

The laser-induced thermal imaging (LITI) [7] is a patterning method for the selective transfer of light emitting polymer composites. In the LITI transfer process, a laser operating in a continuous wave (cw) mode is focused on the surface of a donor film, which is kept in intimate contact with a receiving substrate. The donor consists of a transparent substrate, a light-to-heat conversion layer (LTHC), an interlayer and a transfer light-emitting polymer (LEP) layer. The laser beam is focused through the donor's transparent carrier on the surface of the LTHC and the desired pattern is created by scanning with the laser beam. The conversion of light into heat in the LTHC layer results in the transfer of the polymer layer adjacent to it, under the exposed regions.

7.2. Laser Microprinting for Electronics and Optoelectronics

7.2.1. Conventional Pattern Transfer Processes

Technologies for the patterning of materials are diverse and could be divided in many categories, one of which involves the transfer of patterned materials. The process of simultaneous pattern and material transfer might be either subtractive, where the material is selectively removed from the work piece, or additive, where a pattern of the material is transferred on the work piece. All the techniques that will be discussed in this chapter are additive techniques. In this section, some of the most important conventional pattern transfer processes will be described prior to the presentation of the applications of LIFT.

The term 'pattern transfer' was often taken to be synonymous with lithography and photolithography, in micro and nanotechnologies. The limitations that photolithography now faces are based on the physics of diffraction. Additionally, this method involves a large number of processing steps and is not suitable for all types of materials, since the developing, etching and stripping processes use chemicals that can damage sensitive materials. As a result, it is now worth considering non-photolithographic methods for pattern transfer, with a reduced number of processing steps. Some promising non-photolithographic methods include soft lithography, screen printing and inkjet printing.

Approaches based on physical contact, such as the soft lithographic techniques, are particularly attractive because they circumvent limitations due to diffraction. They allow patterning of a wide range of materials even on curved surfaces. Soft lithography [8,9] is a general term for several techniques that use a patterned elastomeric stamp to shape soft materials. Compared to conventional fabrication methods, soft lithography shows promise by not only achieving the requirement of precise fabrication, but also of decreasing the fabrication cost through low material costs and high fabrication throughout. The strength of this method lies in the patterning of materials (e.g. polymers, biological molecules) that are incompatible with classical resist processes. Alignment issues of this process though remain a challenge.

Screen printing [10] is a simple and environment-friendly method to produce electronic circuitry and interconnections. In this method, patterns are generated by using a 'blade' to squeeze ink through predefined screen masks. It is a purely additive method in which an ink is deposited and patterned in a single step. The ink consists of fine particles and commonly has the viscosity of a paste. The material is wiped over the screen with a straight edge, forcing some of the paste through the open areas of the pattern and the mesh carrier onto the nearby work piece. Low viscosity inks are not appropriate for this method, since they may flow prematurely or uncontrollably through the screen and cause poor resolution. Therefore, screen printing may be most suited for high viscosity solutions and pastes of high molecular weight. Although it was believed at first that screen printing is not suitable for deposition of thin films with less than 100 nm thickness, deposition of a fully organic active layer, having a thickness of several tens of nanometers with the use of screen printing, has also been demonstrated [11]. One of the primary advantages of screen printing is the ability to transfer both the pattern and material of interest onto a substrate at a high rate, compared with photolithography.

An alternate approach to pattern transfer is to shoot the ink onto the work piece at the proper position using the inkjet technology [12,13]. Inkjet based deposition requires no tooling, is non-contact, and is data-driven: no masks or screens are required, since the printing information is stored digitally. Since the original observation by Lord Rayleigh in 1878 that a liquid stream is unstable and tends to break up into individual droplets [14], a large number of inkjet technologies have been developed. Most of these methods fall into two general categories: continuous mode and drop on demand mode. Continuous inkjet (CIJ) technology is based on inducing an electrical charge to the liquid. After the jet breaks up into isolated droplets, the charge remains on the droplets and can be used to deflect them toward the substrate or into an ink collection and recirculation system. In drop-on-demand (DOD) inkjet technology, ink droplets are formed only when required. The two dominant techniques in this area are thermal and piezoelectric DOD printing. In thermal DOD printing, droplets are generated by heating the wall of the ink chamber, causing the formation of vapor bubbles and the ejection of droplets through a nozzle orifice. In piezoelectric DOD printing, a pressure wave in the ink chamber is generated by applying a voltage pulse to a piezoelectric stack or plate, resulting in the formation of droplets at the nozzles. The resolution limits for inkjet are controlled by the minimum drop volume that can be deposited. Inkjet printing technology can dispense spheres of fluid with diameters of 15–200 μm (2 picoliters to 5 nanoliters) at rates of 0–25 KHz per second for single droplets on-demand, and up to 1 MHz for continuous droplets.

This technology can be applied in a broad range of materials and substrates to form a variety of structures, from micro-optic components to biological microarrays. As a result, it may have extensive applications in sensing and detectors.

7.2.2. Laser Printing for Electronics and Power Devices

The laser-induced forward transfer (LIFT) process was firstly demonstrated by Bohandy and colleagues [15]. Copper and silver lines, 50 and 15 μm wide, were produced by single laser pulses of a nanosecond ArF excimer laser, 193 nm, under high-vacuum conditions (10^{-6} mbar). This was the beginning of a systematic study on applications of LIFT in forming of conductive lines such as interconnects and, further, in microelectronics. Transfer of tungsten, titanium and chromium was later demonstrated by Kántor et al. [16,17] and Tóth et al. [18] and while a lot more works have been published since then on LIFT of metals. The dimensions and quality of the transferred patterns in these works have experimental as well as theoretical interest. Let us now see in details the contemporary applications of laser-induced transfer methods to electronics and optoelectronics.

In 1996, LIFT was successfully applied by Sameshima and colleagues [19] to top-gate thin-film transistors (TFTs) fabrication. An aluminum film was firstly deposited on poly-crystalline silicon using a thin a-Si:H sacrificial layer. Aluminum was removed and transferred by heating the a-Si:H film and evaporating the confined hydrogen. An XeCl excimer laser, 308 nm, 30 ns, with energy fluence of 0.2 J/cm^2 was used as laser source. The method was proven to have the potential of simple and low cost application to top-gate TFT fabrication.

Thin films of high-temperature superconductors YBaCuO and BiSrCaCuO have been formed by Fogarassy et al. [20–22] using LIFT. The films were irradiated by an ArF excimer laser, 193 nm, 20 ns, at fluences between 0.05 and 0.5 J/cm^2, an Nd:YAG, 1064 nm, 5 ns, at fluences between 0.1 and 1 J/cm^2 and a frequency-doubled (2ω) Nd:YAG, 532 nm, 100 ns, at fluences from 0.3 to 2.5 J/cm^2. The transferred films were deposited on MgO and silicon substrates. The films were observed to maintain their properties after deposition and were able to obtain superconducting phases. Fabrication of diffractive micro-optical patterns such as binary-amplitude computer-generated holograms (CGH) has been demonstrated by Zergioti and colleagues [23] using either pixel-by-pixel direct writing of the pattern or by projection of a master hologram mask. Single and multilevel structures of metals and oxides [24] have been created. The used laser was an ultra-short KrF excimer laser, 248 nm, 500 fs, with typical energy fluences in the order of 0.35 J/cm^2. Pixels of 3×3 μm^2 were achieved. Such patterns can be used as diffractive microgratings for several applications including security and authentication labels, electronic pricing, optical fiber telecommunications, to name a few.

A KrF excimer laser, 248 nm, 500 fs, has also been used by Mailis et al. [25] in order to print grating chromium structures, with sub-micron resolution on etched optical fibers. The energy fluence was 0.1 J/cm^2. The quality of the deposited structures showed the potential of using LIFT in order to fabricate microstructures onto non-planar and high-curvature surfaces, such as Bragg fiber gratings an important application for the further development of optical fiber sensors and optical telecommunications.

Toet and colleagues [26,27] demonstrated an application of laser-induced forward transfer based on the use of an a-Si:H sacrificial layer to top-gate TFT fabrication in printed silicon. An XeCl excimer laser, 308 nm, 35 ns, with energy fluence of

0.6 J/cm^2 was used. A poly-Si film, firstly deposited on an a-Si:H coated quartz carrier, was transferred onto a glass substrate. This printed silicon was used for top-gate TFT fabrication by conventional photolithography process. The produced TFTs were tested to have well on/off ratios, mobilities comparable to those of a-Si devices (0.1 cm^2/Vs) and typical TFT characteristics.

In 1999, Piqué *et al.* [28,29] applied LIFT and matrix-assisted pulsed laser evaporation/direct-write (MAPLE/DW) to the fabrication of electronics and sensors. A KrF excimer laser, 248 nm, 20 ns, was used to transfer various materials on silicon and commercial PCBs. The energy fluence was in the order of 0.5 J/cm^2. Ag, Au and NiCr metallic lines with thickness of 10 μm were deposited. Micro capacitors BaTiO$_3$ (BTO) and SrTiO$_3$ (STO) have been also fabricated and their thickness was 20–25 μm and 5–20 μm respectively. Further, four-spiral Y$_3$Fe$_5$O$_{12}$ core flat inductors were produced by MAPLE/DW. Finally, chemoresistors made by printing of PECH/ graphite were demonstrated. Tests showed that the material maintained its properties after transfer. Despite the fact that the results were preliminary, it was shown that MAPLE/DW is a viable process for non-destructive printing of polymer materials and composites. Later, the same group [30] improved the conditions and applied MAPLE/DW to deposit metals and composite dielectrics. The obtained electrical properties of the transferred materials were comparable to those produced by conventional methods. An ArF excimer laser, 193 nm, 20 ns, 0.1–0.2 J/cm^2, and a frequency-tripled Nd:YAG, 355 nm, 0.1 J/cm^2, were used respectively by the group of Piqué and colleagues [31] to MAPLE deposition of chemoselective (SXFA) and MAPLE/DW of chemoresistive polymer composites (PECH/carbon) for applications in chemical sensor devices. The performance of the fabricated sensors was found to be equivalent to that of conventionally produced similar sensor.

MAPLE/DW was used by Fitz-Gerald and colleagues [32] in order to create high-resolution display pixels. Dense oxide phosphor powders of Y$_2$O$_3$:Eu (red) and Zn$_2$SiO$_4$:Mn (green) were transferred onto alumina and polymer substrates. A KrF excimer laser at 248 nm with pulse duration 25 ns, at energy fluences of 0.1–2.5 J/cm^2, was used. The emission spectra before and after transfer were identical, indicating that the transfer process did not affect the properties of the materials. Therefore, the MAPLE/DW can be developed to be a viable method for fabrication of high-definition displays.

Wartena *et al.* [33] have published in 2004, the fabrication of Li-ion microbatteries by means of laser-induced transfer. The battery electrodes were produced by the laser-induced transfer of inks of charge-storage materials onto micromachined metal-foil current collectors by a frequency-tripled (3ω) Nd:YAG laser at 355 nm. Such materials are composites of carbon/binder and LiCoO$_2$/carbon/binder. The thickness of the printed electrodes was 40–60 μm with 4 × 4 μm^2 footprints. The cells of 0.16 cm^2 surface area were measured to exhibit similar capacity to 1 cm^2 area sputtered thin-film Li-ion cells. Although the laser-produced cells have less volumetric capacity than the sputtered ones, they can be made thicker and, thereby, use of a smaller footprint can provide the same capacity. Piqué *et al.* [34] concluded to comparative results applying the laser direct-write method to fabricate rechargeable Li-ion microbatteries using a frequency tripled Nd:YVO$_4$ laser, 355 nm (Fig. 2). LIFT was proven to be a viable method for Li-ion microbattery electrodes fabrication, the first step to integrate a microbattery on a device.

The laser direct-write technique has been applied by Kim *et al.* [35] in order to transfer porous, 5–20 μm thick, nano-crystalline titanium oxide (nc-TiO$_2$) layers

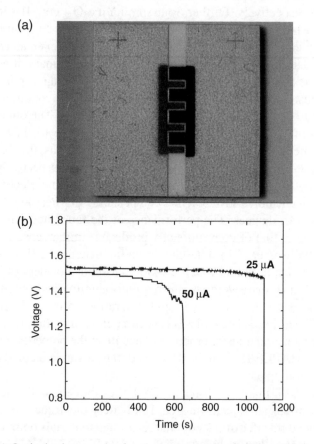

Figure 2: (a) Optical micrograph of a planar Ag$_2$O/Zn alkaline microbattery made by laser direct-write. (b) Discharge behavior from two planar alkaline microbatteries at 25 and 50 μA, respectively, as a function of time. Total electrode mass is ~250 μg in both cases. (Reprinted from [34] with permission from Springer.)

onto TiO_2/F-doped SnO_2 coated glass. A frequency-tripled (3ω) Nd:YVO_4, 355 nm, with energy fluence of 0.1 J/cm^2 was the laser source. The layers are incorporated in dye-sensitized solar cells. The white light power conversion efficiency was measured to be approximately 4.3%, comparable to those fabricated with TiO_2 by standard methods. The avoidance of additional patterning and the ability to control the layer thickness provided by the laser transfer method, showed that it is a promising technique for fabrication of solar cells on flexible and rigid substrates.

7.2.3. Laser Printing for Organic Optoelectronics

In this paragraph, we emphasize on laser microprinting methods and their use for generating simple patterns of organic/polymeric materials as well as, devices, such as organic thin film transistors (TFTs) and organic light emitting devices (OLEDs).

In the early studies of laser induced transfer experiments of polymeric materials, poly(methyl methacrylate) (PMMA) was the most commonly used. Blanchet *et al.* [37] have investigated the potential of UV laser ablation for the deposition of PMMA, while Mito *et al.* [36] have utilized the explosive gas pressure produced by an interlayer for the transfer of a doped PMMA film.

In their work on PMMA transfer, Blanchet *et al.* [37] have used the fourth harmonic of an Nd–YAG laser (266 nm), focused to a 2 mm^2 spot size onto the surface of a solid pellet of the material. The laser-ablated PMMA films were deposited in an argon atmosphere of 200 mTorr, at fluences ranging from 0.01 to 1 J/cm^2. The pulse width was 10 ns with a repetition rate of 10 Hz. The analysis of Blanchet *et al.* has shown that the photon energy initially absorbed is rapidly converted to vibrational heating of the solid. This, results in the pyrolytic decomposition of the target, forming an ablation plume primarily composed of methyl methacrylate (MMA). The explosively ejected monomers arrive at the substrate (KBr, Si and CaF_2) where re-polymerization occurs. The morphology of films deposited at ambient temperature does not depend on the particular substrate while depends on the deposition parameters, especially the temperature and the background pressure.

Karnakis *et al.* [38] have used a KrF excimer laser (248 nm) of 30 ns pulse duration, to irradiate a pyrene-doped PMMA host film for the transfer of the pyrene molecules below the surface of a receiving polybutyl methacrylate (PBMA) substrate. The laser was passed through a beam homogenizer pulse and was focused to 1.2×1.2 cm^2 spot size. The two films, which were both spin coated on quartz substrates, were brought into contact. The host film had a thickness ranging from 0.1 to 4 μm, while the receiving polymer's thickness was maintained at 2 mm. The dopant molecules are expected to absorb photons. The analysis of the fluorescence spectra of the receiver polymer surface indicated the implantation of the transferred pyrene molecules in the subsurface of the receiver polymer, since their fluorescence spectra would be different if the molecules decomposed or just adhered on the polymer surface. The receiver polymer surfaces retained their original flatness in most of the cases.

Nakata et al. [39] have investigated the laser-induced forward transfer of Rhodamine 610 laser dye on a SiO_2 substrate using the second harmonic of a Nd:YAG laser (532 nm, 8 ns). Droplets of an ethanol solution of the donor dye film were applied on a gold base film, which was placed about 15 μm from the SiO_2 substrate. The fluorescence of the transferred films was investigated after optical excitation and the fluorescence emission was optimal at 49 mJ/cm^2 laser fluence in vacuum condition.

The laser microprinting of semiconducting and conducting polymers has also great interest, since they are the core materials for constructing organic thin-film transistors (TFTs). Pentacene is considered to be one of the best candidates for replacing amorphous silicon in organic TFTs. Experimental results of thin pentacene film deposition via UV pulsed laser evaporation were published by Blanchet et al. [40] using the fourth harmonic of an Nd–YAG laser (266 nm, 10 ns). The laser beam was expanded and collimated onto a 3 cm^2 area on the target surface, and the experiments performed under ultrahigh vacuum of 10^{-7} Torr. The mass analysis has shown the absence of lower mass pentacene fragments in the mass spectra indicating that the thin-film transfer occurs via evaporation rather than an ablation process. The semiconducting performance of the laser-evaporated pentacene films was analyzed in a TFT configuration and was found to be comparable to state of the art pentacene thin films grown by thermal evaporation.

The laser transfer of conducting polymers has also been investigated. Polyaniline (PANI) is considered to be unique among conducting polymers not only because it is environmentally stable, but its conductivity can be reversibly controlled. The pixelized transfer of polyaniline (PANI)/nanotube structures was illustrated, by Blanchet et al. [41] using the thermal imaging method. A 40 W infrared diode laser, 780 nm, was splitted to 250 individually addressable spot of 2.7×5 μm^2 area, was focused through the donor base at a thin metal layer onto which the PANI was coated. The conversion of light to heat at this interface results in the release of gaseous products and shockwave which transfer the material from the substrate to the flexible receiver. A 500 μm wide line of composite was printed with this method with a 7 μm gap, which could potentially be used as a source drain of a printed field effect transistor, with a conductivity of 2 S/cm. Using this method, Blanchet et al. [42] have printed a TFT backplane, containing 5000 transistors with 20 μm channel, onto a 50×80 cm flexible substrate. Sources and drains were slightly shifted from the center of the gate due to the lack of built-in registration once the receiver is removed. The maximum misregistration over an area of 4000 cm^2 was less than 200 μm. The electronic systems printed with this method offer attractive characteristics like low-cost processing, mechanical flexibility, large area coverage, etc. which are not easily achieved with established silicon technologies.

Laser-induced thermal imaging (LITI) has been applied for the selective transfer of light-emitting polymer (LEP) composites and the production of full-color organic electroluminescent devices [7]. Most commercial track LEPs are not directly suitable for laser thermal transfer because of their high film strength or cohesion that hampers the precise pattern formation. To overcome this problem, Chin *et al.* [43] have developed small molecule host/LEP blend hybrid systems, to reduce cohesion of the transfer layers and improve the pattern accuracy as well as the efficiency and lifetime of the devices in comparison to conventional PLED technology. In the LITI transfer process a Nd:YAG laser of total power of 8 W operating in continuous wave (cw) mode is focused to a Gaussian spot, with a measured diameter of $300 \times 40 \ \mu m^2$, on the surface of a donor film, which is kept in intimate contact with a receiving substrate. The donor consists of a transparent substrate polyethylene terapthalate (PET), a light-to-heat conversion layer (LTHC), an interlayer and an 80 nm thick LEP layer. The receptor consists of a transparent substrate (glass), an ozone treated ITO film (anode) and a hole conducting layer spin-coated on top of it. The laser beam is focused through the donor's transparent carrier on the surface of the LTHC layer and a galvanometer is used for precision scanning of the laser beam, which creates the desired pattern. The conversion of light into heat in the LTHC layer results in the transfer of the LEP as uniform, high-resolution stripes with good thickness uniformity and electroluminescent performance. Three donor films (red, green and blue) are used sequentially to create a full color display. The hole conducting layer on the receptor surface was either an 80 nm-thick PEDOT:PSS layer or a poly-(9,9-dioctylfluorene) (PFO) hole injection layer with N,N′-diphenyl-N, N′-bis (3-methylphenyl)-(4,4′-diamine) (TPD). The overall position accuracy of the 80 μm transferred LEP stripes is claimed to be less than 2.5 μm. Chin *et al.* have successfully used this method for the production of organic light-emitting diodes (OLEDs) and a full color 2.2″ QCIF AMOLED display, illustrated in Fig. 3.

More recently, Lee *et al.* [44] have also used the LITI method for the transfer of light-emitting polymer (LEP) material. To improve the transfer quality they have modified the PEDOT hole transport layer by adding methacryloxy propyl-trimethoxysilane (MPS), which changes it from hydrophilic to hydrophobic. This makes the modified PEDOT:MPS receiving layer compatible with the transferred hydrophobic LEP and improves the light-emitting efficiency and brightness of the PLED. The efficiency of the PLED with modified PEDOT was 4.8 cd A^{-1} while the efficiency of the PEDOT device was only 3.6 cd A^{-1}. The performance of these devices was also proven to be improved for higher energy doses of the laser and was claimed to be comparable to those of spin-coated devices.

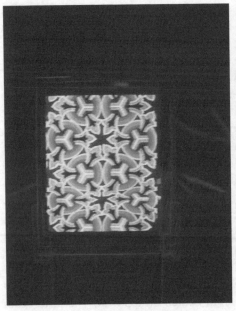

Figure 3: Images of 2.2″ QCIF active matrix full color hybrid-OLED device patterned by laser transfer. (Reprint from [43] with permission from the authors.)

Advantages of the LITI process are high-resolution patterning, good film thickness uniformity, multilayer stack ability and scalability to large-size mother glass. Since it is a dry process, LITI transfer is not affected by the solubility properties of the transfer layer, in contrast with most conventional techniques where the wet processing steps included may affect integrity or properties of the previously deposited layers.

In summary, laser assisted transfer techniques can provide, high resolution methods for manufacturing complex multilayer organic circuits by the sequential printing of solid films. These laser printing techniques circumvent the serious solvent compatibility issues associated with most conventional printing methods. Thin, flexible and ultra-light organic electronic devices can be fabricated at speeds and resolution adequate for the early commercialization of these devices. Although the early results of these techniques are promising, further study is still required before either the materials or the process are refined to adequate performance standards.

7.3. Laser Printing of Biomaterials

The fabrication of microarrays of biological molecules for the development of biosensors is a fast growing field. A common mission unites pharmaceutical companies, academia, biotech and diagnostic firms to employ massive analysis in order to develop products that improve quality of life. The fabrication of advanced microdevices (biosensors, lab-on-chip, medical sensors and actuators) will provide better health studies and fast detection of spread and rare illnesses. Previous techniques that have been developed for deposition of biomaterials are based on spotting [45], inkjet [46] and photolithography [47]. A variety of methods are currently available for depositing biopolymers in biosensor systems. According to most of the published works [48], biopolymer materials are applied onto their binding or adsorbing substrates in solution by simple dispensing or soaking techniques. For achieving localized deposition a variety of microfluidic, gel microcasting and localized activation methods have been described for recent examples. The vast majority of methods for depositing a plurality of biopolymer samples onto solid substrates are liquid deposition processes.

Fernandez-Pradas *et al.* [49], have shown that regular microarrays of double stranded DNA of salmon sperm onto poly-L-lysine coated slides through the LIFT technique using the third harmonic (355 nm) of the Nd:YAG laser beam. To accomplish this, a laser-absorbing layer (titanium) between the laser-transparent support and the solution film was used. Recently, they have compared [50] the fabrication of cDNAs microarrays onto poly-L-lysine treated glass slides by LIFT and pin microspotting. The produced microarrays were submitted to hybridization with the complementary strands of the spotted cDNAs, each one tagged with a different fluorochrome. Comparative fluorescence scanner analysis has revealed that the microarrays transferred through LIFT are equivalent to those transferred through pin microspotting in terms of signal intensity and gene discrimination capacity. Figure 4 shows the fluorescent images of the same microarrays obtained through LIFT (a,b) and pin microspotting (c,d).

Hopp *et al.* [51] have recently presented an investigation on absorbing film assisted laser induced forward transfer of fungus (Trichoderma) conidia, by a KrF excimer laser beam. The laser fluence was varied in the range of 0–2600 mJ/cm^2. Silver thin film of 50 nm, was used to act as absorbing layer and during the process this layer absorbed the laser, transformed the electronic excitation into kinetic which gently desorbed the donor film.

Another direct write technique which combines the laser-induced forward transfer (LIFT) method with the matrix assisted pulsed laser evaporation (MAPLE) has been demonstrated by Ringeisen *et al.* [5,52], who deposited patterns of active proteins, viable *E. coli* and mammalian Chinese hamster ovary cells and also generated patterns of viable *Escherichia coli* by using nanosecond laser pulses. The laser printing of biomaterials by means of MAPLE/DW method requires the assistance of transferring matrix materials. Karaiskou *et al.* [53,54] have presented the deposition of biomaterials by laser direct transfer avoiding the use of any matrix material. The method achieves high spatial resolution by the means of ultrafast laser pulses that avoid melting and vaporization and allow material transfer with narrow angular divergence.

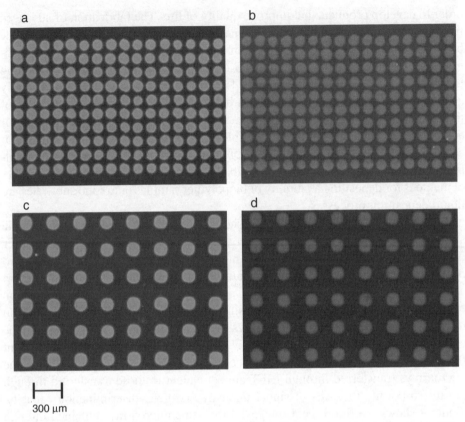

Figure 4: Fluorescence scanner images obtained after hybridization of the microarrays spotted through (a,b) LIFT and (c,d) pin microspotting. The spots in (a) and (c) contain the MAPK3 gene cDNA and the spots in (b) and (d) contain EST2 human gene cDNA. In the hybridization target solution MAPK3 cDNA is tagged with Cy3 (a,c) and the ETS2 cDNA with Cy5 (b,d). (Reprint from [50] with permission from Elsevier.)

Therefore, labile biomaterials are deposited without significant damage since the use of ultrafast laser pulses minimizes the adverse thermal effects of the process and lowers the required energy threshold for transfer. Recently, they presented [55] the microprinting of enzymes on glass slides by means of a hybrid distributed feedback dye laser/KrF excimer laser, delivering ~0.5 ps duration pulses of 10 mJ/pulse, at $\lambda = 248$ nm. Figure 5 shows the enzymatic staining reaction (sodium acetate 0.095M, H_2O_2 0.04%, citric acid 0.065M, Manitol 0.027M, 3,3,5,5-Tetramethyl-benzidine 1.2 mM) of the deposited horseradish peroxidase, demonstrating that the enzyme

Figure 5: Active enzyme staining reaction. The Enzyme deposited on the glass surface is incubated in a solution containing: sodium acetate 0.095M, H_2O_2 0.04%, citric acid 0.065M, D-Manitol 0.027M, 3,3,5,5-Tetramethyl-benzidine 1.2 mM and horseradish peroxidase activity results the precipitation of the chromogenic compound. The developed staining is visualized by Confocal Laser Fluorescent Scanning. (Reprint from [55] with permission from Elsevier.)

remains active after the printing process. This example shows the versatility of the methods for a variety of biomaterials (nucleic acids, proteins, enzymes) that could be easily extended to different other substances such as carbohydrates, lipids, etc. This enzyme is relatively sensitive and the stain developed over the deposited enzyme demonstrates that the enzyme remains active after the Direct Laser printing process. This is very important for the development of applications where both printing resolution and biological activity are required, such as antibody or antigen microarrays. The technique is performed over a wide range of biomaterials. The retained enzyme mainly depends on the binding capacity of this surface. These printing results demonstrate that the LIFT process is wide applicable and can be used to directly write micro-patterns of biological molecules, such as DNA, proteins and enzymes without the assistance of any transferring matrix material.

7.4. Physics of the LIFT Method – Time-Resolved Imaging Diagnostics

The complexity of the physical processes involved in the laser induced forward transfer and the dependence on the irradiating laser fluence as well as the laser pulse duration can be studied by applying time resolved diagnostics.

Several groups have studied the temporal evolution of the absorption of a laser pulse in a confined target layer. Most of the early works in laser transfer were focused on the study physics of the process under irradiation of long infrared pulses. Lee *et al.* [56] have studied the dynamics of the laser ablation transfer of films irradiated by 250 ns, 1064 nm pulses. The films consisted of dye suspensions in polymer resins optimized for absorption in the IR wavelength range. The transfer was imaged by a stroboscopic microscope illuminated by a 25 ps, 532 nm pulsed Nd:YAG frequency doubled laser. They observed subsonic transfer velocities in the order of 233 m/s (0.75 Mach) while the film was detached at about 35 ns after the irradiation started. Tolbert *et al.* [57] have studied the laser ablation transfer of films by stroboscopic shadowgraphic imaging. They used a 150 ns, 1064 nm Nd:YAG laser beam to transfer films consisting of color coatings on transparent polymer substrates or polyester substrates with dynamic release layer (metallic aluminum). Their measurements show that the coating material leaves the film's surface at $t \approx 50$ ns at a velocity near 1 Mach (340 m/s). The coating ablated from films prepared with a dynamic release layer, consisted of much larger particles than those from films with out a dynamic release layer. Furthermore, Hare *et al.* [58] have studied the fundamental mechanisms of exposure by near-infrared pulses of multilayer films exposed to even longer 10 µs, 1064 nm YAG pulses. They used top view time resolved optical microscopy and observed microscopic images of the sample's surface using a home build dye laser of (500 ps, 500 nm). They observed debonding of the surface coating at about 1.5 µs after the initiation of irradiation. By following a combined approach, Bullock *et al.* [59] have studied the laser induce back ablation of 1 µm thick Al targets after irradiation by 10 ns, 1064 nm Nd:YAG pulses. They measured the integrated reflectivity of the Al surface and the transverse propagation of the resulting vapor plume from shadowgraphs using 140 fs, 800 nm probe laser pulses synchronized to the irradiating laser beam. Under vacuum the velocity of the Al vapor was measured to be in the order of 60 km/s. Furthermore the reflectivity measurements revealed that there seems to be negligible contribution of the possible Al plasma formation to the surface reflectivity for plasma densities of less than 30%, the critical density. The dynamic behavior of the emissive species during LIFT was studied by Nakata *et al.* [60] by observing two dimensional laser induced fluorescence. They irradiated Au films (20 to 500 nm thickness) deposited on glass substrates by 9 ns, 400 nm pulses and observed the dynamic behavior of Au atoms and emissive particles both in vacuum and in atmospheric air. They measured velocities of 2 km/s for Au atoms in vacuum and 100 m/s for emissive particles. They also observed reflection of the ejected species when a receiver substrate was placed in the vicinity (Fig. 6).

On the other hand, the study of the dynamics of the laser induced transfer in the short pulse regime revealed clear differences to the long pulse regime. One of the first attempts in the short pulse duration regime was that of Shoen *et al.* [61] who have studied the compressional shocks generated after the irradiation of Cu foils by infrared 1054 nm, 10 ps pulses. The Cu foils were sandwiched between two fused silica windows of 40 to 250 µm thickness. By using an interferometric method the

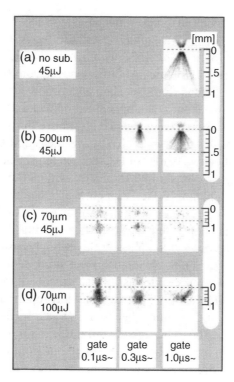

Figure 6: Interaction of ablated atoms and particles with the substrate. The film thickness was 100 nm. (a) Observed without substrate; the distances between the donor film and the substrate were (b) 500 μm and (c), (d) 70 μm. The energy was 45 μJ except for (d), with 100 μJ. (Reprinted from [60] with permission from Springer.)

magnitude and the temporal profile of the shock pressure were deduced. The pressure dependence on fluence was observed to be linear and the temporal profile of the wave was a rapid rise and tail witch followed a $t^{1/2}$ dependence. Later on, Bullock *et al.* [62] have studied the laser induced back ablation of aluminum thin film targets, of 250 nm to 500 nm thickness, after irradiation by 2–3.6 ps, 1053 nm laser pulses. The ejected Al plume was probed by a two color probe pulse ($\lambda_1 = 1053$ nm, $\lambda_2 = 527$ nm) producing shadowgraphs or two color time delayed interferograms when an interferometer was used. Their measurements show that in a low fluence range the ejected plume follows the irradiating pulse Gaussian spatial profile and the edge velocity scales reasonably well with the constant amount of absorbed energy over the ablated mass (2 km/s for 250 nm Al, 5.2 J/cm²). For high fluences the edge velocity saturates and the ejected plume does not follow a Gaussian shape. This is attributed to the increasing absorption of laser energy in the transparent glass substrate due to non-linear absorption mechanisms such as avalanche ionization leading to a saturation of the transmitted energy to the Al target. This effect was further clarified by Moore *et al.* [63] whom have studied shock compression phenomena by irradiating 1 μm thick Al films deposited on BK7 glass and sapphire substrates with 120 fs, 800 nm pulses. They also observed that non-linear effects such as surface optical breakdown and optical Kerr effect flatten the initial Gaussian intensity profile of the laser beam at fluence just above threshold. Furthermore, by an elegant experiment Gahagan *et al.* [64] have measured the rise time and the particle and average shock velocities of laser-generated shock wave in vapor plated Al films using frequency domain interferometry with sub-ps time resolution. The irradiating laser was a 800 nm, 130 fs Ti:sapphire and the measured shock velocities were in the order of 5 km/s corresponding to pressures of 30–50 kbar and to a shock front that extends only a few tens of lattice spacing. The 10–90% rise times were found to be in the order of 6.25 ps for target thickness ranging from 250 nm to 2 μm. Following that, Funk *et al.* [65] have used high resolution frequency domain interferometry to make ultrafast measurements of the induced changes in the optical properties of thin aluminum targets. They reported particle velocities of 300 m/s during the first 5 ps after the laser irradiation. By following a similar approach Gahagan *et al.* [66] have applied ultrafast time-resolved 2D interferometric microscopy to measure the shock wave breakout from thin metal films. They estimated the 10–90% rise time of the pressure profile at ~53.7 ps after the irradiating pulse for a 250 nm Al sample. The peak surface velocity was measured in this case to be ~600 m/s which correspond to a peak stress of 4.7 GPa. These series of experiments although were not performed for the direct purpose of studying LIFT in particular, they provide quite interesting indirect information since the experimental conditions under which the Al targets were irradiated were exactly the same as those of LIFT.

Almost all the above-mentioned studies were focused on the dynamics of infrared or visible laser induced transfer. The spectral range of the most commonly used laser sources are either Nd:YAG for the nanosecond or picasecond pulses and Ti:sapphire for the fs pulses. On the other hand in the UV spectral range, excimer lasers are mostly used. Papazoglou *et al.* [67] studied the dynamics of LIFT by using 248 nm, 0.5 ps UV laser pulses. The target material was InO_x thin films, deposited on fused silica substrates. The transfer process was investigated by means of time-resolved shadowgraphic imaging. The velocity of the ejected material was 400 ± 10 m/s at 300 ns after irradiation, with a slight dependence on the laser fluence. The fraction of the material ejected by the sub-picosecond LIFT process, which is visible to the shadowgraphic technique, was measured to be highly directional with narrow angular divergence (3°). A comparative study [68] of the dynamics of laser forward transfer of Cr for 0.5 ps and 15 ns, 248 nm laser pulses has been performed by the same group. They observed that the directionality of the ejected material was very high (3°) in the case of the sub-ps laser used as a pump source. Figure 7 shows the time resolved stroboscopic schlieren imaging of the dynamics of the transfer process of Cr thin film material which is irradiated by 0.5 ps, 248 nm pulses. The ejected plume is imaged by 30 ns, 308 nm low intensity pulses. The material ejection (dark area) and shock wave formation in air are clearly visible.

On the other hand in the case of nanosecond laser pulses, the ejected material front was hemispherical in shape corresponding to a high divergence (~50°). The laser transfer of biological materials was also observed [69] to be highly directional when sub-ps UV laser pulses was used for the transfer of the material. Furthermore, Yamada *et al.* [70] have studied the optimization of the LIFT process for Ni thin film deposited on fused silica substrates by varying laser fluence and substrate receiver distance. They irradiated the target films by 30 ns, 248 nm laser pulses and monitored the transfer process by imaging the ejected plume with an ICCD camera. They observed reflection of the ejected plume when a receiver substrate was placed near the target. Sano *et al.* [71] have studied the dynamics of the laser induced forward transfer of Au and Ni thin films irradiated by 30 ns, 248 nm laser pulses. They observed that the film is removed during the laser pulse and that the ejected material velocity is increased as the laser fluence is increased.

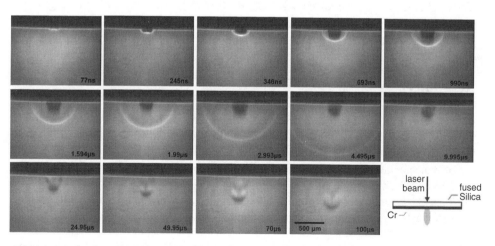

Figure 7: Time resolved stroboscopic schlieren imaging of the dynamics of the UV laser transfer process. The target material (Cr of 20 nm thickness, deposited on fused silica) is irradiated by 0.5 ps, 248 nm pulses. The ejected plume is imaged by 30 ns, 308 nm low intensity pulses. The material ejection (dark area) and shock wave formation in air are clearly visible.

7.5. Summary and Future Aspects

The laser-induced forward transfer (LIFT) is a mature fabrication technique utilizing lasers to remove thin film material from a transparent support and deposit it onto a suitable substrate. It has been demonstrated that LIFT is capable of fabricating mesoscopic patterns of organic, biological and inorganic materials. This chapter has covered the transfer of materials for optoelectronic, electronic and biotechnological applications.

The authors of this chapter believe that LIFT will find further potential applications, as a rapid prototyping tool, especially in the area of polymers for OLEDs and optoelectronic devices and in the area of biomaterials for biosensors. Further improvement of the LIFT includes improvement on the instrumentation and on the basic investigations of the laser-transfer interactions.

Acknowledgments

Experiments were carried out at the Ultraviolet Laser Facility operating at IESL-FORTH and supported by the EU through the Research Infrastructures activity of FP6 (Project: Laserlab-Europe; Contract No: RII3-CT-2003-506350). The Project is co-funded by the European Social Fund (75%) and National Resources (25%) through the project **PYTHAGORAS 68/829**. Last the authors are grateful to A. Pique, P. Serra, Min Chul Suh, Y. Nakata and their publishers for kindly providing permission for use of published materials.

References

[1] Bohandy, J., Kim, B.F. and Andrian, F.J., Metal deposition from a supported metal film using an excimer laser, *Journal of Applied Physics*, **1986**, *60*(4), 1538–1539.
[2] Kim, H., Kushto, G.P., Arnold, C.B., Kafafi, Z.H. and Piqué, A., Laser processing of nanocrystalline TiO_2 films for dye-sensitized solar cells, *Applied Physics Letters*, **2004**, *85*(3), 464–466.
[3] Toet, D., Thompson, M.O., Smith, P.M., Carey, P.G. and Sigmon, Th.W., Thin film transistors fabricated in printed silicon, *Japanese Journal of Applied Physics*, **1999**, *38*(10A), Part 2, L1149–L1152.
[4] Toet, D., Smith, P.M., Sigmon, T.W. and Thompson, M.O., Experimental and numerical investigations of a hydrogen-assisted laser-induced materials transfer procedure, *Journal of Applied Physics*, **2000**, *87*(7), 3537–3546.
[5] Piqué, A. and Chrisey, D.B. (Eds.), *Direct-write Technologies for Rapid Prototyping Applications*, Academic Press, **2002**.

[6] Blanchet, G.B., US Patent 5, 523, 192, **1995**.

[7] Wolk, M.B., Baude, P.F., Florezak, J.F., McCormick, F.B. and Hsu, Y., *Thermal transfer element and process for forming organic electroluminescent devices*, US Patent 6, 582, 876, **2003**.

[8] Xia, Y., Rogers, J.A., Paul, K.E. and Whitesides, G.M., Unconventional methods for fabricating and patterning nanostructures, *Chemical Review*, **1999**, *99*, 1823–1848.

[9] Michel, B., Bernard, A., Bietsch, A., Delamarche, E., Geissler, M., Juncker, D., Kind, H., Renault, J.-P., Rothuizen, H., Schmid, H., Schmidt-Winkel, P., Stutz, R. and Wolf, H., Printing meets lithography: Soft approaches to high-resolution patterning, *IBM Journal of Research and Development*, **2001**, *45*(5), 697–720.

[10] Bao, Z., Rogers, J.A. and Katz, H.E., Printable organic and polymeric semiconducting materials and devices, *Journal of Materials Chemistry*, **1999**, *9*, 1895–1904.

[11] Pardo, D.A., Jabbour, G.E. and Peyghambarian, N., Application of screen printing in the fabrication of organic light-emitting devices, *Advanced Materials*, **2000**, *12*(17), 1249–1252.

[12] Wallace, D.B., Cox, W.R. and Hayes, D.J., *Direct Write Technologies for Rapid Prototyping Applications*, Pique, A. and Chrisey, D.B. (Eds.), Academic Press, Boston, **2002**.

[13] Sirringhaus, H. and Shimoda, T., Inkjet printing of functional materials, *MRS Bulletin*, **2003**, 802–806.

[14] Rayleigh, J.W.S., On the instability of jets, (pp. 4–13), *Proceedings of the London Mathematical Society*, **1879**, *10*(4), 4–13.

[15] Bohandy, J., Kim, B.F., Adrian, F.J. and Jette, A.N., Metal deposition at 532 nm using a laser transfer technique, *Journal of Applied Physics*, **1988**, *63*(4), 1158–1162.

[16] Kántor, Z., Tóth, Z. and Szörényi, T., Laser-induced forward transfer: The effect of support-film interface and film-to-substrate distance on transfer, *Applied Physics A*, **1992**, *54*, 170–175.

[17] Kántor, Z., Tóth, Z. and Szörényi, T., Metal pattern deposition by laser-induced forward transfer, *Applied Surface Science*, **1995**, *86*, 196–201.

[18] Tóth, Z., Szörényi, T. and Tóth, A.L., Ar+ laser-induced forward transfer (LIFT): A novel method for micrometer-size surface patterning, *Applied Surface Science*, **1993**, *69*, 317–320.

[19] Sameshima, T., Laser beam application to thin film transistors, *Applied Surface Science*, **1996**, *96–98*, 352–358.

[20] Fogarassy, E., Fuchs, C., Kerherve, F., de Unamuno, S. and Perriere, J., Laser-induced forward transfer: A new approach for the deposition of high Tc superconducting thin films, *Journal of Materials Research*, **1989**, *4*(5), 1082–1086.

[21] Fogarassy, E., Fuchs, C., Kerherve, F., Hauchecorne, S. and Perriere, J., Laser-induced forward transfer of high-Tc YBaCuO and BiSrCaCuO super-conducting thin films, *Journal of Applied Physics*, **1989**, *66*(1), 457–459.

[22] Fogarassy, E., Fuchs, C., de Unamuno, S., Perriere, J. and Kerherve, F., High-Tc super-conducting thin-film deposition by laser-induced forward transfer, *Materials and Manufacturing Processes*, **1992**, *7*(1), 31–51.

[23] Zergioti, I., Mailis, S., Vainos, N.A., Papakonstantinou, P., Kalpouzos, C., Grigoropoulos, C.P. and Fotakis, C., Microdeposition of metal oxide structures using ultra-short laser pulses, *Applied Physics A*, **1998**, *66*, 579–582.

[24] Koundourakis, G., Rockstuhl, C., Papazoglou, D., Klini, A., Zergioti, I., Vainos, N.A. and Fotakis, C., Laser printing of active optical microstructures, *Applied Physics Letters*, **2001**, *78*(7), 868–870.

[25] Mailis, S., Zergioti, I., Koundourakis, G., Iliadis, A., Grigoropoulos, C.P., Papakonstantinou, P., Vainos, N.A. and Fotakis, C., Etching and printing of diffractive optical microstructures by a femtosecond excimer laser, *Applied Optics*, **1999**, *35*(11), 2301–2308.

[26] Toet, D., Thompson, M.O., Smith, P.M., Carey, P.G. and Sigmon, Th.W., Thin film transistors fabricated in printed silicon, *Japanese Journal of Applied Physics*, **1999**, *38*(10A), Part 2, L1149–L1152.

[27] Toet, D., Smith, P.M., Sigmon, T.W. and Thompson, M.O., Experimental and numerical investigations of a hydrogen-assisted laser-induced materials transfer procedure, *Journal of Applied Physics*, **2000**, *87*(7), 3537–3546.

[28] Piqué, A., Chrisey, D.B., Auyeung, R.C.Y., Fitz-Gerald, J.M., Wu, H.D., McGill, R.A., Lakeou, S., Wu, P.K., Nguyen, V. and Duignan, M., A novel laser transfer process for direct writing of electronic and sensor materials, *Applied Physics A*, **1999**, *69*, S279–S284.

[29] Chrisey, D.B., Piqué, A., Fitz-Gerald, J.M., Auyeung, R.C.Y., McGill, R.A., Wu, H.D. and Duignan, M., New approach to laser direct writing active and passive mesoscopic circuit elements, *Applied Surface Science*, **2000**, *154–155*, 593–600.

[30] Chrisey, D.B., Piqué, A., Modi, R., Wu, H.D., Auyeung, R.C.Y. and Young, H.D., Direct-writing of conformal mesoscopic electronic devices by MAPLE/DW, *Applied Surface Science*, **2000**, *168*, 345–352.

[31] Piqué, A., Auyeung, R.C.Y., Stepnowski, J.L., Weir, D.W., Arnold, C.B., McGill, R.A. and Chrisey, D.B., Laser processing of polymer thin films for chemical sensor applications, *Surface Coating Technology*, **2003**, *163–164*, 293–299.

[32] Fitz-Gerald, J.M., Piqué, A., Chrisey, D.B., Rack, P.D., Zelenik, M., Auyeung, R.C.Y. and Lakeou, S., Laser direct writing of phosphor screens for high-definition displays, *Applied Physics Letters*, **2000**, *76*(11), 1386–1388.

[33] Wartena, R., Curtright, A.E., Arnold, C.B., Piqué, A. and Swider-Lyons, K.E., Li-ion microbatteries generated by a laser direct-write method, *Journal of Power Sources*, **2004**, *126*, 193–202.

[34] Piqué, A., Arnold, C.B., Kim, H., Ollinger, M. and Sutto, T.E., Rapid prototyping of micropower sources by laser direct-write, *Applied Physics A*, **2004**, *79*, 783–786.

[35] Kim, H., Kushto, G.P., Arnold, C.B., Kafafi, Z.H. and Piqué, A., Laser processing of nanocrystalline TiO_2 films for dye-sensitized solar cells, *Applied Physics Letters*, **2004**, *85*(3), 464–466.

[36] Mito, T., Tsujita, T., Masuhara, H., Hayashi, N. and Suzuki, K., Hollowing and transfer of poly-methyl-methacrylate film propelled by laser ablation of Triazeno polymer film, *Japanese Journal of Applied Physics*, **2001**, *40*, L805–L806.

[37] Blanchet, G.B., Deposition of poly(methyl-methacrylate) films by UV laser ablation, *Macromolecules*, **1995**, *28*, 4603–4607.

[38] Karnakis, D.M., Goto, M., Ichinose, N., Kawanishi, S. and Fukumura, H., Forward-transfer laser implantation of pyrene molecules in a solid polymer, *Applied Physics Letters*, **1998**, *73*(10), 1439–1441.

[39] Nakata, Y., Okada, T. and Maeda, M., Transfer of laser dye by laser-induced forward transfer, *Japanese Journal of Applied Physics*, **2002**, *41*(7B), Part 2, L839–L841.

[40] Blanchet, G.B. Fincher, C.R. and Malajovich, I., Laser evaporation and the production of pentacene films, *Journal of Applied Physics*, **2003**, *94*(9), 6181–6184.

[41] Blanchet, G.B., Fincher, C.R. and Gao, F., Polyaniline nanotube composites: A high-resolution printable conductor, *Applied Physics Letters*, **2003**, *82*(8), 1290–1292.

[42] Blanchet, G.B., Loo, Y., Rogers, J.A., Gao, F. and Fincher, C.R., Large area, high resolution, dry printing of conducting polymers for organic electronics, *Applied Physics Letters*, **2003**, *82*(3), 463–465.

[43] Chin, B.D., Suh, M.C., Kim, M.H., Kang, T.M., Yang, N.C., Song, M.W., Lee, S.T., Kwon, J.H. and Chung, H.K., High efficiency AMOLED using hybrid of small molecule and polymer materials patterned by laser transfer, *Journal of Information Display*, **2003**, *4*(3), 1–5.

[44] Lee, J.Y. and Lee, S.T., Laser-induced thermal imaging of polymer light-emitting materials on poly-(3,4-ethylenedioxythiophene): Silane hole-transport layer, *Advanced Materials*, **2004**, *16*(1), 51–54.

[45] Schena, M., Shalon, D., Davis, D.W. and Brown, P.O., Quantitative monitoring of gene expression patterns with a complementary DNA microarray, *Science*, **1995**, *270*, 467–470.

[46] Hughes, T.R., Mao, M., Jones, A.R., Burchard, J., Marton, M.J., Shannon, K.W., Lefkowitz, S.M., Ziman, M., Schelter, J.M., Meyer, M.R., Kobayashi, S., Davis, C., Dai, H. Y., He, Y.D.D., Stephaniants, S.B., Cavet, G., Walker, W.L., West, A., Coffey, E., Shoemaker, D.D., Stoughton, R., Blanchard, A.P., Friend, S.H. and Linsley, P.S., Expression profiling using microarrays fabricated by an ink-jet oligonucleotide synthesizer, *Nature Biotechnology*, **2001**, *19*, 342–347.

[47] Fodor, S.P.A., DNA sequencing: Massively parallel genomics, *Science*, **1997**, *277*, 393–395.

[48] Cullen, D.C., Brown, R.G. and Lowe, C.R., Detection of immuno-complex formation via surface plasmon resonance on gold-coated diffraction gratings, *Biosensors*, **1987–8**, *3*(4), 211–225.

[49] Fernández-Pradas, J.M., Colina, M., Serra, P., Dominguez, J. and Morenza, J.L., Laser-induced forward transfer of biomolecules, *Thin Solid Films*, **2004**, *453–4*, 27–30.

[50] Colina, M., Serra, P., Fernández-Pradas, J.M., Sevilla, L. and Morenza, J.L., DNA deposition through laser induced forward transfer, *Biosensors and Bioelectronics*, **2005**, *20*, 1638–1642.

[51] Hopp, B., Smausz, T., Antal, Zs., Kresz, N., Bor, Zs. and Chrisey, D., Absorbing film assisted laser induced forward transfer of fungi (Trichoderma conidia), *Journal of Applied Physics*, **2004**, *96*(6), 3478–3481.

[52] Ringeisen, B.R., Chrisey, D.B., Pique, A., Young, H.D., Modi, R., Bucaro, M., Jones-Meehan, J. and Spargo, B.J., Generation of mesoscopic patterns of viable Escherichia coli by ambient laser transfer, *Biomaterials*, **2002**, *23*, 161–166.

[53] Karaiskou, A., Zergioti, I., Fotakis, C., Kapsetaki, M. and Kafetzopoulos, D., Microfabrication of biomaterials by the sub-ps laser-induced forward transfer process, *Applied Surface Science*, **2003**, *208–209*, 245–249.

[54] Kafetzopoulos, D., Zergioti, I., Fotakis, C. and Thiraios, G., *European patent pending*, **2003**.

[55] Zergioti, I., Karaiskou, A., Papazoglou, D.G., Fotakis, C., Kapsetaki, M. and Kafetzopoulos, D., Time resolved schlieren study of sub-picosecond and nanosecond laser transfer of biomaterials, *Applied Surface Science*, **2005**, *247*(1–4), 584–589.

[56] Lee, I.Y.S., Tolbert, W.A. Dlott, D.D., Doxtader, M.M., Foley, P.M., Arnold, D.R. and Ellis, E.W., Dynamics of laser ablation transfer imaging investigated by ultrafast microscopy, *J. Imag. Sci. Tech.*, **1992**, *36*(2), 180–187.

[57] Tolbert, W.A., Lee, I.Y.S., Doxtader, M.M., Ellis, E.W. and Dlott, D.D., High–speed color imaging by laser ablation transfer with a dynamic release layer: fundamental mechanisms, *J. Imag. Sci. Tech.*, **1993**, *37*(4), 411–421.

[58] Hare, D.E., Rhea, S.T., Dlott, D.D., D'Amato, R.J. and Lewis, T.E., Fundamental mechanisms of lithographic printing plate imaging by near-infrared lasers, *J. Imag. Sci. Tech.*, **1997**, *41*(3), 291–300.

[59] Bullock, A.B., Bolton, P.R. and Mayer, F.J., Time-integrated reflectivity of laser induced back ablated aluminum thin film targets, *Journal of Applied Physics*, **1997**, *82*(4), 1828–1831.

[60] Nakata, Y. and Okada, T., Time resolved microscopic imaging of the laser induced forward transfer process, *Applied Physics A*, **1999**, *59*, S275–S278.

[61] Schoen, P.E. and Campilo, A.J., Characteristics of compressional shocks resulting from picosecond heating of confined foils, *Applied Physics Letters*, **1984**, *45*, 1049–1051.

[62] Bullock, A.B. and Bolton, P.R., Laser induced back ablation of aluminum thin films using picosecond laser pulses, *Journal of Applied Physics*, **1999**, *85*(1), 460–465.

[63] Moore, D.S., Gahagan, K.T., Reho, J.H., Funk, D.J., Buelow, S.J., Rabie, R.L. and Lippert, T., Ultrafast nonlinear optical method for generation of planar shocks, *Applied Physics Letters*, **2001**, *78*(1), 40–42.

[64] Gahagan, K.T., Moore, D.S., Funk, D.J., Rabie, R. L., Buelow, S.J. and Nicholson, J.W., Measurement of shock wave rise times in metal thin films, *Physics Review Letters*, **2000**, *85*(15), 3205–3208.

[65] Funk, D.J., Moore, D.S., Gahagan, K.T., Buelow, S.J., Reho, J.H., Fisher, G.L. and Rabie, R.L., Ultrafast measurement of the optical properties of aluminum during shock-wave breakout, *Physics Review B*, **2001**, *64*, 115114.

[66] Gahagan, K.T., Moore, D.S., Funk, D.J., Reho, J.H. and Rabie, R.L., Ultrafast interferometric microscopy for laser driven shock wave characterization, *Journal of Applied Physics*, **2002**, *92*(7), 3679–3682.

[67] Papazoglou, D.G., Karaiskou, A., Zergioti, I. and Fotakis, C., Shadowgraphic imaging of the sub-ps laser induced forward transfer process, *Applied Physics Letters*, **2002**, *81*, 1594–1596.

[68] Zergioti, I., Papazoglou, D.G., Karaiskou, A., Fotakis, C., Gamaly, E. and Rode, A., A comparative schlieren imaging study between ns and sub-ps laser forward transfer of Cr, *Applied Surface Science*, **2003**, *208–209*, 177.

[69] Karaiskou, A., Zergioti, I., Fotakis, C., Kapsetaki, E. and Kafetzopoulos, D., Microfabrication of biomaterials by the sub-ps laser induced forward transfer process, *Applied Surface Science*, **2003**, *208*, 245.

[70] Yamada, H., Sano, T., Nakayama, T. and Miyamoto, I., Optimization of laser-induced forward transfer process of metal thin films, *Applied Surface Science*, **2002**, *197–198*, 411.

[71] Sano, T., Yamada, H., Nakayama, T. and Miyamoto, I., Experimental investigation of laser-induced forward transfer process of metal thin films, *Applied Surface Science*, **2002**, *186*, 221.

[72] Baseman, R.J., Froberg, N.M., Andreshak, J.C. and Schlesinger, Z., Minimum fluence for laser blow-off thin gold films at 248 and 532 nm, *Applied Physics Letters*, **1990**, *56*(15), 1412–1414.

[73] Schultze, V. and Wagner, M., Laser-induced forward transfer of aluminium, *Applied Surface Science*, **1991**, *52*, 303–309.

[74] Lätsch, S., Hiraoka, H., Nieveen, W. and Bargon, J., Interface study on laser-induced material transfer from polymer and quartz surfaces, *Applied Surface Science*, **1994**, *81*, 183–194.

[75] Esrom, H., Zhang, J.-Y., Kogelschatz, U. and Pedraza, A.J., New approach of laser-induced forward transfer for deposition of patterned thin metal films, *Applied Surface Science*, **1995**, *86*, 202–207.

[76] Pimenov, S.M., Shafeev, G.A., Smolin, A.A., Konov, V.I. and Vodolaga, B.K., Laser-induced forward transfer of ultra-fine diamond particles for selective deposition of diamond films, *Applied Surface Science*, **1995**, *86*, 208–212.

[77] Poon, C.C. and Tam, A.C., Laser-induced forward transfer of carbon and chromium films in gases of one atmosphere pressure, *CLEO Digest*, **1996**, 377–378.

[78] Zergioti, I., Mailis, S., Vainos, N.A., Ikiades, A., Grigoropoulos, C.P. and Fotakis, C., Microprinting and microetching of diffractive structures using ultrashort laser pulses, *Applied Surface Science*, **1999**, *138–139*, 82–86.

[79] Papakonstantinou, P., Vainos, N.A. and Fotakis, C., Microfabrication by UV femtosecond laser ablation of Pt, Cr and indium oxide thin films, *Applied Surface Science*, **1999**, *151*, 159–170.

[80] Bähnisch, R., Gross, W. and Menschig, A., Single-shot, high repetition rate pattern transfer, *Microelectronics Engineering*, **2000**, *50*, 541–546.

[81] Yamada, H., Sano, T., Nakayama, T. and Miyamoto, I., Optimization of laser-induced forward transfer process of thin metal films, *Applied Surface Science*, **2002**, *197–198*, 411–415.

Chapter 8

Recent Progress in Direct Write 2D and 3D Photofabrication Techniques with Femtosecond Laser Pulses

J. Koch, T. Bauer, C. Reinhardt and B.N. Chichkov

Laser Zentrum Hannover e.V., Hannover, D-30419 Germany

Abstract

Some of the rapidly advancing femtosecond laser technologies like ablative micro- and nanostructuring, two-photon polymerization, multiphoton activated processing, nanotexturing of metals, deep drilling and cutting, 3D photofabrication, including application examples, are reviewed in this chapter.

Keywords: Critical dimension (CD); Galvano-scanner; Photofabrication; Sub-micrometer; Surface Plasmon polaritons (SPP); Two-photon polymerization (2PP); Ultrashort laser pulses.

8.1. Laser Photofabrication Technique/Introduction

The invention of femtosecond lasers has opened many exciting possibilities for high-resolution material processing. For the first time optical energy and laser pulses controllable in time and space with unprecedented precision became available. It is not surprising that these high-quality optical pulses have revolutionized photofabrication technologies. Due to concentrated efforts of many groups in the world, direct ablative microstructuring of all kind of solid materials has been studied and many related advantages have been established. The outcome was a novel (contact-, post-processing-, and damage-free) laser micromachining technology. Femtosecond lasers enabled processing of a wide range of materials (including

Recent Advances in Laser Processing of Materials
J. Perrière, E. Millon and E. Fogarassy (Editors)

heat sensitive and thermo-reactive) with a sub-micrometer resolution. At present, nearly arbitrary shaped 2D and 3D structures can be produced by direct write photofabrication techniques using femtosecond laser pulses. In this chapter we present a brief review of recent progress in using these techniques for drilling, cutting, surface texturing, and 3D shaping.

8.1.1. Ablative Micro- and Nanostructuring with Femtosecond Laser Pulses

Applications of ultrashort laser systems for high-quality processing of different materials are very rapidly growing. The overwhelming majority of the research groups working in this field are using Ti:Sapphire femtosecond laser systems with 1 kHz repetition rate. Ti:Sapphire laser systems are pumped by visible laser radiation, which is usually provided by frequency-doubled diode-pumped solid-state lasers. Novel femtosecond lasers are based on ytterbium tungstate (Yb:KGW) or other materials and can be directly pumped by laser diodes. These systems are more attractive for industrial applications.

During the last few years, femtosecond lasers have established themselves as excellent and universal tools for direct ablative microstructuring of solid materials [1–8]. The main features of material processing with femtosecond laser pulses are: (i) efficient, fast, and localized energy deposition, (ii) well-defined deformation and ablation thresholds, and (iii) minimal thermal and mechanical damage of the substrate material. When, for example, one compares results of microstructuring of a thin chromium layer with femtosecond and nanosecond laser pulses, advantages of femtosecond lasers become evident (see Fig. 1).

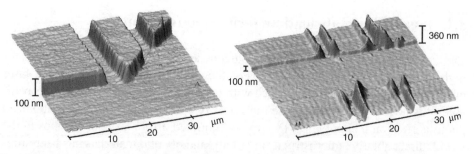

Figure 1: Atomic force microscope images of a chromium layer microstructured with 225 nJ, 150 fs laser pulses (left) and 270 nJ, 10 ns laser pulses (right) using 780 nm Ti:Sapphire laser radiation. Note the different height scalings in these pictures.

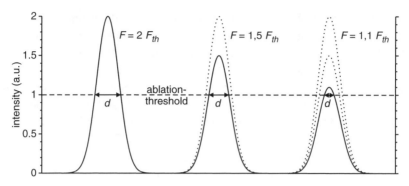

Figure 2: Schematic illustration of how one can overcome the diffraction limit by taking advantage of the well-defined ablation threshold.

Choosing femtosecond laser parameters close to the ablation threshold one can produce perfect microstructures without any visible thermal or mechanical damage. It is well known that the minimum achievable structure size is determined by the diffraction limit of the optical system and is of the order of the radiation wavelength. However, this is different for ultrashort laser pulses. By taking advantage of the well-defined ablation (in general, modification) threshold one can beat the diffraction limit by choosing the peak laser fluence slightly above the threshold value [1] (see the schematic illustration in Fig. 2). In this case, only the central part of the beam can modify the material and it becomes possible to produce subwavelength structures. The minimum structure size is determined by the equation

$$ d = d_0 \left(\ln \frac{F}{F_{\text{th}}} \right)^{1/2} \tag{1} $$

where F is the laser fluence, F_{th} is the threshold fluence for ablation, and d_0 is the beam diameter.

In principle, this technique enables arbitrarily small structure sizes. But in practice, there is a limit due to intensity fluctuations of the laser pulses and beam-pointing instabilities. Therefore, the minimum achievable structure size is limited by the beam profile quality and stability of the femtosecond laser system.

There are a lot of applications of ablative micro- and nanostructuring with femtosecond laser pulses working close to the ablation threshold. However, when a high processing speed is required (in case of drilling or cutting) one is forced to work at much higher laser fluences, well above the ablation threshold. At these laser fluences, the energy introduced into a workpiece and the corresponding thermal load are high. The ablation depth per pulse is determined by the energy transfer into the target due to electron thermal conduction and/or due to the generated shock wave. This is similar to ablation with nanosecond and longer laser pulses. Therefore, in the high fluence regime, one usually does not expect any remarkable advantages for material processing with femtosecond lasers. But they still exist. For example, one can drill more efficiently and produce deeper holes due to the lower thermal losses and negligible hydrodynamic expansion of the ablated material during the femtosecond laser pulse. The latter allows 'plasma shielding' of laser radiation to be avoided. Moreover, for deep drilling applications the most important criteria are the hole geometry and quality, which are much better when femtosecond lasers are applied. The explanation of this effect is quite simple. High laser intensities are required to rapidly drill a through-hole with femtosecond pulses. After the through-hole is drilled, the high intensity part of the laser pulse propagates directly through the hole without absorption. At the edge of the laser pulse, at laser intensities close to the ablation threshold, only a small amount of material is removed. Starting from this moment, the interaction of femtosecond laser pulses with the workpiece occurs in the low-fluence regime, in which all advantages of femtosecond lasers can be realized. One can consider this as an 'integrated' low-fluence femtosecond laser postprocessing, which is responsible for the excellent hole quality, or as a low-fluence finishing [9].

The effect of low-fluence finishing can be visualized by making a cut through a piece of metal foil with femtosecond lasers. By moving a 125 μm piece of magnesium foil perpendicular to the laser beam with a constant velocity of 20 mm/min,

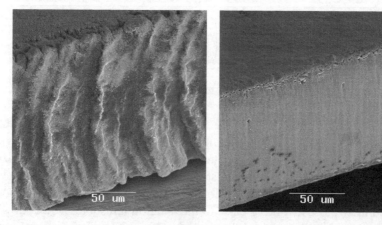

Figure 3: Scanning electron microscope images of a cut through 125 μm magnesium foil produced by femtosecond laser pulses at the cutting speed of 20 mm/min after a single (*left*) and five (*right*) laser passes.

we produced different cuts with 0.6 mJ laser pulses at a 1 kHz repetition rate. In Fig. 3, scanning electron microscope (SEM) images of the cut cross sections produced after a single (left) and five (right) laser passes are shown. One can see that after the first pass, the laser cut through the foil leaves a fairly rough surface. During the subsequent laser passes, the most intense central part of the laser pulse propagates through the cut without absorption. The interaction with the cut walls occurs only at the pulse edge, where the local laser intensity is low. In this low-fluence regime, only a small amount of material can be removed from the walls, which is equivalent to a 'polishing' effect that remarkably improves the surface quality.

Thus, not only at laser fluences close to the ablation threshold, but also at high laser fluences and high processing speed, ablative microstructuring with femto-second lasers allows to produce excellent results.

8.1.2. Two-Photon Polymerization (2PP) Technique

One of the rapidly advancing femtosecond laser technologies is 3D microstructuring of photosensitive materials by two-photon polymerization (2PP) technique [10,11]. 2PP is very attractive technique for the fabrication of 3D structures with a resolution down to 100 nm. For 2PP and 3D material processing, we usually

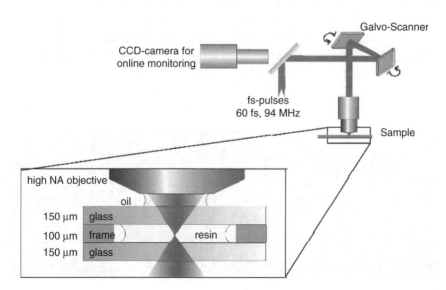

Figure 4: Principle experimental setup for the fabrication of 3D structures by 2PP.

apply near-infrared Ti:Sapphire femtosecond laser oscillators (at 800 nm). When femtosecond laser pulses are tightly focused into the volume of a photosensitive material (or photoresist), they initiate 2PP process by, for example, transferring liquid into the solid state. This allows the fabrication of any computer-generated 3D structure by direct laser 'recording' into the volume of a photosensitive material [12,13].

The principle experimental setup used for 2PP microstructuring is shown in Fig. 4. The expanded laser beam is guided through the x–y galvo-scanner into the high NA immersion oil objective, focusing femtosecond laser pulses into the volume of the resin. The CCD camera mounted behind the last mirror enables online process monitoring. The sample is mounted on a 3D piezo positioning stage. By moving the laser focus three-dimensionally inside the resin one is able to write complex 3D structures with resolution down to 100 nm. Photosensitive materials used have to be transparent at the wavelength of the laser radiation.

Very promising 'negative' photosensitive materials applied in our work are called Ormocers (organic modified ceramics) [14]. Ormocers are organic–inorganic hybrid polymers containing a highly crosslinkable organic network as well as inorganic components. These materials have high optical quality, as well as chemical and thermal stability. The polymerization process in Ormocers are initiated by radically reacting photo initiators (Irgacure369 Ciba). In addition, Ormocers are biocompatible, which opens a broad range of biomedical applications [15]. A few microstructuring examples are demonstrated below.

Figure 5 shows microcapsules fabricated by 2PP which can be used for direct drug delivery. High flexibility of 2PP allows to fabricate these structures directly on the stent surface. Another negative photosensitive material used in our work is a commercially available SU8, which is an epoxy-based resin. In this case, the crosslinking of molecules is initialized by photoacid generators (PAGs). Under light

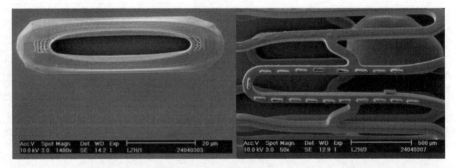

Figure 5: Microcapsules for direct drug delivery fabricated on metallic stents by 2PP.

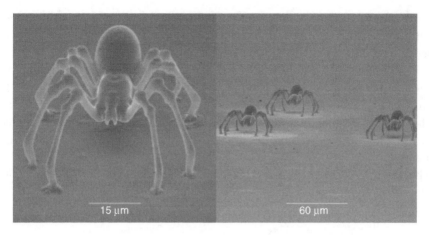

Figure 6: Microspider array fabricated by 2PP of SU8.

exposure, the PAGs form strong acids, which by the application of heat (post baking at 95°C) lead to irreversible cross-linking of the polymer. SU8 is solid, while Ormocers are highly viscous liquids. This results in somewhat different irradiation and processing strategy of these materials [13]. In solid SU8 one can fabricate more complex structures than that is possible with liquid Ormocers. In Fig. 6, a SEM image of microspiders fabricated by 2PP in SU8 is shown. The main body of the spider is located 6 µm above the substrate and is supported by eight 2 µm diameter legs.

8.1.3. Multiphoton Activated Processing (MAP)

Multiphoton activated material processing has very important advantages over processes allowing absorption of single photons – an increased spatial resolution and the possibility of photofabrication inside transparent materials, like it has been demonstrated above with 2PP technique. Here we discuss further examples of multiphoton activated processing (MAP) of transparent materials.

Due to the non-linear nature of the interaction of femtosecond laser pulses with transparent materials, simultaneous absorption of several photons is required to initiate ablation from the surface or any structural modifications in the volume. Multiphoton absorption produces initial free electrons that are further accelerated by the femtosecond laser electric field. These electrons induce avalanche ionization and optical breakdown, and generate a microplasma. The subsequent expansion of the microplasma results in the fabrication of a small structure at the target surface. The diameter of such a femtosecond laser drilled hole not only depends on the energy distribution in the laser-matter interaction area but also on the ablation threshold. There is also a dependence on the energy band gap of the material. To overcome a wider band gap more photons are needed. This results in a stronger localization of the ionization process to the peak intensity region of the laser pulse. The smaller absorption volume leads to a reduced hole diameter. This provides further possibilities for the fabrication of sub-diffraction limited structures.

The minimum structure size (for constant fluence ratio F/F_{th}) that can be produced by femtosecond laser pulses in materials with an energy band gap is determined by the equation

$$d = \frac{k\lambda}{\sqrt{q}NA} \tag{2}$$

where λ is the radiation wavelength, q is the number of photons required to overcome the energy band gap, NA is the numerical aperture of the focusing optics, and k is the material-dependent proportionality constant ($k < 1$). To illustrate this equation, Fig.7 shows schematically the effective beam profile and the ablation hole diameter for materials where one, two, and four photons are needed for each ionization process.

The reproducibility of the MAP technique depends on two factors. First, material defects can locally change the band-gap energy (and the ablation threshold) resulting in a change of the produced structure size. Second, avalanche ionization gains importance when longer laser pulses are applied. This results in a growing influence on the number of free electrons existing in the laser-matter interaction

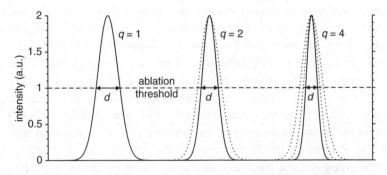

Figure 7: Effective beam profile and ablation hole diameter d for materials where $q = 1$, 2, and 4 photons are needed for each ionization process.

area before the onset of ionization. For tightly focused laser pulses one cannot speak of a homogeneous distribution of free electrons (coming from material defects or doping). Therefore, high reproducible laser structuring of transparent materials with small structure sizes by MAP technique can be performed only by ultrashort laser pulses (e.g. for SiO2: <100 fs, or even better: <50 fs [16]) when the role of avalanche ionization is not critical.

Besides laser ablation, writing of optical waveguides in transparent materials is an important example of the MAP technique. Here, the small density and refractive index change produced near the threshold are suitable for direct writing of wave-guides and other photonic devices, as illustrated in Fig. 8 [17,18]. Voids which can be formed at higher laser energy are ideal for binary data storage because of their high optical contrast [19].

Figure 8: Principle setup for optical waveguide microfabrication in transparent materials.

8.2. High-Resolution 2D Photofabrication Technique with Femtosecond Laser Pulses

Photolithography allows parallel processing of a large number of devices. Therefore, it is a relatively cheap and fast technology for mass production. In many cases where this technology is not practical or too expensive, femtosecond laser photofabrication technique can be applied. The time needed for a femtosecond laser based photo-fabrication process depends on the speed of sample and/or beam positioning systems, the required number of laser pulses, and the laser pulse repetition rate.

When surface structuring with single laser pulses or a small laser pulse overlap is possible, one can reduce processing time by using a so-called 'on-the-fly' technique. In this case, the desired surface pattern is subdivided into a number of rows. Each row is scanned with a constant translational velocity. According to the pattern of each row the laser is switched on and off 'on-the-fly'. The advantage of this technique (over a technique using single positioning procedures for each pattern position) becomes evident by comparing the required processing times in the following example. Let us assume that the task is to fabricate a regular hole array consisting of 500×500 holes with a hole-to-hole separation distance of 1 μm. For the laser technique based on single positioning procedure this results in a number of 2,50,000 positioning steps. Assuming a time needed for positioning of one hole of approximately 1 s (including the time required for the relaxation of the positioning stage vibration), the total processing time would be about 70 h, which is not practical. In contrast to this, using the 'on-the-fly' technique in combination

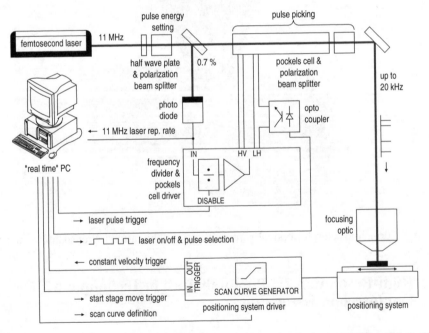

Figure 9: Schematic setup and trigger plan for the on-the-fly laser 2D photofabrication technology.

with a 1 kHz laser system one row of 500 holes can be fabricated in approximately 3 s (including the time needed for the move to the start position of the next row). This results in a total processing time of 25 min, which can be reduced further by using a laser system with higher repetition rate.

The realization of the 'on-the-fly' technology needs a careful setting of trigger signals to the positioning system and the laser pulse picker device. A sketch of a realized setup together with the trigger connections is shown in Fig. 9. The switching laser signal according to the desired pattern is generated by a real time computer and is adjustable up to 20 kHz. This rate and the distance between the dots of the structure determine the translational velocity of the positioning system. The scan curve definition together with the start trigger signal for the translation move is given to the positioning system. Only when a constant velocity has been reached, the laser on/off signal is applied to the Pockels cell.

An example for the high flexibility and positioning accuracy of the presented 'on-the-fly' 2D photofabrication technique is illustrated in Fig. 10. This figure shows a SEM microscopy of the famous Pablo Picasso's 'Don Quixote' painting, made on an area of $600 \times 800 \ \mu m$. The femtosecond laser produced 'Don Quixote' is fabricated from a regular array of sub-micrometer holes with 600 nm diameter and a hole-to-hole separation distance of 1.5 μm. This example is given here to demonstrate that 2D microstructuring of metal films can be performed with a very high quality and reproducibility using femtosecond laser pulses.

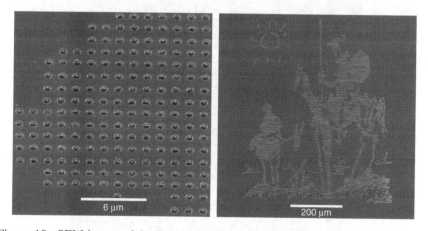

Figure 10: SEM image of the Pablo Picasso's – 'Don Quixote' microscopy (*right*), fabricated in 100 nm gold on glass using on-the-fly technology, consisting of regular holes (see a detail on the *left* side).

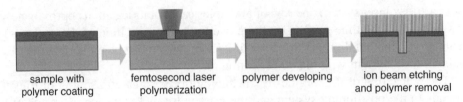

sample with   femtosecond laser   polymer developing   ion beam etching
polymer coating   polymerization   and polymer removal

Figure 11: Transfer of structures produced in positive photoresist into gold surface by ion etching.

Further important example for high-resolution patterning of different solids is a three-step procedure illustrated in Fig. 11, which can be referred as femtosecond lithography (FL). Using positive (or negative) photoresist as a coating material, high quality microstructuring of metals, semiconductors, and dielectrics can be produced by two-photon illumination of photoresist followed by subsequent development and pattern transfer. The critical dimension (CD), defined as a minimum structure size, which can be realized with two-photon illumination technique using femtosecond laser pulses (at 800 nm wavelength) depends on the photosensitive material and can be lower than 100 nm.

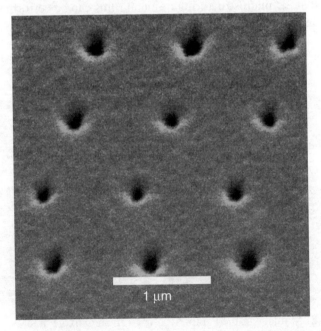

Figure 12: Holes produced in the gold surface by ion etching of structured samples coated with positive photoresist, minimum hole diameter is 150 nm.

Using three-step procedure and positive photosensitive resist AZ1506 from *Clariant* as a coating material on metal surface, structural dimensions down to 150 nm in gold have been demonstrated in first experiments (see Fig. 12). This photoresist is structured by two-photon bond breaking induced by femtosecond laser illumination. The resist layer had a thickness of 500 nm. Its spectral sensitivity is in the range of 310–440 nm, giving a large two-photon absorption cross section and thus a good response to the laser wavelength around 800 nm. Some irregularity in the size of the fabricated holes appear due to slightly different number of pulses used for the illumination of photoresist or due to slightly different photoresist thickness.

The 'on-the-fly' technique combined with the three-step procedure shown in Fig. 11 is very promising approach for the fabrication of high-quality 2D structures.

8.3. Nanotexturing of Metals by Laser-Induced Melt Dynamics

Femtosecond laser material processing with structure sizes in the micrometer range has a universal character. Nearly arbitrary patterns and structures can be fabricated in almost all solid materials. But the well-defined ablation threshold and the non-linearity of laser-matter interaction allows even further decreasing of structural dimensions. As mentioned before, nanostructures can be fabricated by direct ablation with tightly focused femtosecond laser pulses [20]. High quaity holes and structures with a resolution (structure size) down to 100 nm are possible.

But with structure sizes below the micrometer range, femtosecond laser material processing forfeits its universal character. The dependence of shape and size on material properties gains in importance. In some metals, femtosecond laser fabricated sub-micrometer structures are dominated by solidified flow dynamics in molten material. In this section we concentrate on this phenomenon and present a technique for laser texturing of metals without ablation.

In metals, femtosecond laser pulses induce a strong thermal non-equilibrium between electron and lattice subsystems. Relaxation of this non-equilibrium (energy transfer to the lattice) occurs after the laser pulse and is determined by the rate of electron–phonon coupling, electron thermal diffusivity, and other material properties. A slower increase of the lattice temperature results in a longer existence of the molten phase. This in turn, allows flow dynamics in the molten material to affect ablation results. Since the effect of heat flow away from the focal region is much more critical for smaller dimensions, tight focusing should intensify this process. Using thin metal layers on the top of low heat conducting substrates like glass instead of a bulk metal, heat flow can be considered as only 2D (nearly no heat flow into the substrate). This results in a further increase of the time the molten phase

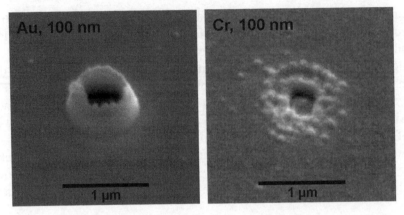

Figure 13: Scanning electron microscope (SEM) images of single pulse ablation results
of 100 nm thick chromium (left) and gold layers (right).

can exist, and therefore, makes flow dynamic processes in the molten material
more important.

Figure 13 shows ablation results of 100 nm thick gold and chromium layers
processed by single 30 fs laser pulses at 800 nm and a fluence of 150% of the ablation threshold fluence. The strong melt dynamic observable in gold is absent in
chromium. Whereas the electron–phonon coupling strength in gold is weak, it is
strong in chromium. Considering this, it can be understood that in gold, ablation
occurs out of the molten phase, whereas in chromium, it occurs almost without
melting [21].

Metals with weak electron–phonon coupling seem to be less suitable for
femtosecond laser sub-micrometer structuring. As can be seen in Fig. 13, the hole
diameter in gold is bigger than in chromium. Furthermore, there is a strong burr
formation in gold which is absent in chromium.

But one can also make use of laser-induced melt dynamics. By adjusting the laser
fluence slightly below the ablation threshold (but above the melting threshold),
laser surface structuring/texturing without ablation becomes possible. In a certain
range of femtosecond laser parameters such a process can be used for the fabrication of nanojets on thin gold films on glass. This technology is very reproducible
(see Fig. 14) and allows one to perform large-area patterning. Nanojets always
appear on a bump-like structure. Both processes, the formation of a bump-like
corrugated surface and nanojets, have well-defined thresholds. When the laser
pulse energy is increased, the bump-like structure first appears on the target
surface. Then the nanojets are formed. Figure 15 shows SEM images of structures
fabricated with different laser pulse energies. In Fig. 16, dependencies of the

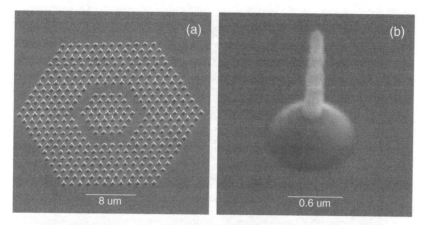

Figure 14: SEM images of an array of nanojets fabricated in a 60 nm thick gold film with femtosecond laser pulses (a) and a single nanojet in detail (b).

height (measured from the target surface) of the bump and nanojet structures on the laser pulse energy are shown. Broken arrows illustrate that on starting from a 16 nJ laser pulse energy (2.5 J/cm^2) the process becomes unstable and the microbump and/or the nanojet structures are destroyed. The microbumps are hollow inside. Their shell thickness is determined by the thickness of the molten layer. The length of the nanojets can exceed 1 μm and depends on the laser parameters and the gold layer thickness. We found the longest nanojets on a layer thickness around 60 μm.

Since the adhesive strength of gold on a glass substrate is very weak compared to the adhesive strength e.g. chromium on glass, one could hold this responsible for the observed differences in the ablation results (see Fig. 13). The formation of microbumps could be a result of local stripping of the gold layer from the glass substrate. But then, these structures should not appear on gold layers much thicker than the penetration depth of the laser pulse energy (about 100 nm for gold [22]). Figure 17 shows a similar structure to that observed on gold layers up to 100 nm, but here the gold layer thickness is 2500 nm. So, the weak adhesion strength of gold to the glass substrate can be excluded from being the reason for the formation of these structures. However, it is possible that it boosts their formation. The microbump and nanojet in Fig. 17 are smaller than that produced with the same laser parameters on gold layers with a thickness below than 100 nm. The shorter existence of the molten phase in thick gold layers due to the better heat conduction of the underlying solid gold (instead of an underlying glass substrate) could also be responsible for this effect.

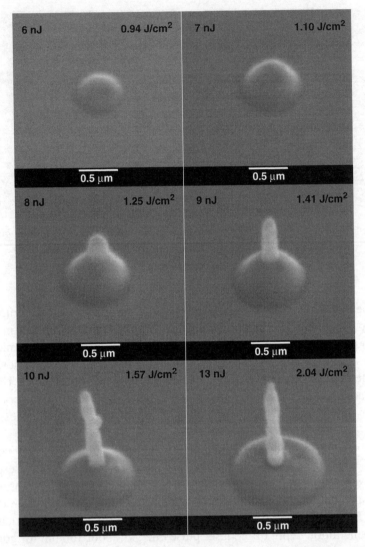

Figure 15: SEM images of sub-wavelength surface structures produced in a 60 nm gold layer by tightly focused femtosecond laser pulses with different pulse energy.

The dynamic picture of the processes induced by a femtosecond laser pulse in a molten gold layer is not yet fully understood. One can try to first explain it by already known effects.

Qualitatively, the jet formation is analogous to that induced by a droplet fall into a glass filled with water. The surface waves initiated by the droplet fall collide at

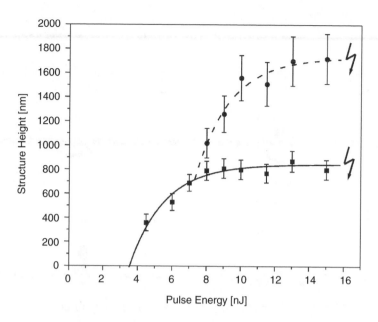

Figure 16: Dependencies of the bump-like structure (soild curve) and nanojet height (dashed curve) on the laser pulse energy (60 nm gold layer thichness). Broken arrow illustrate that starting from a 16 nJ laser pulse energy the process becomes unstable and bump and/or nanojet structures are destroyed.

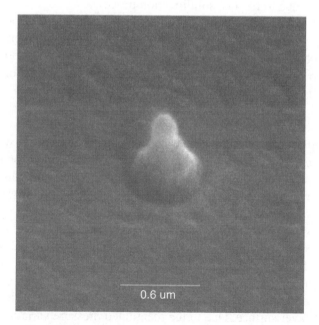

Figure 17: SEM image of a nanojet in a 2500 nm thick gold layer.

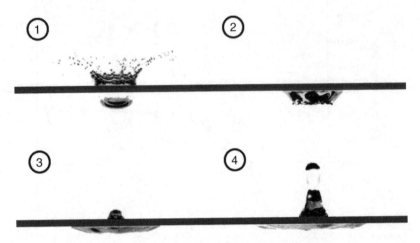

Figure 18: Formation of a splash after a droplet fall in water.

the center and form a liquid jet, i.e. a splash (see Fig. 18). Approximately the same can happen in a molten metal film heated by a single femtosecond laser pulse. The waves initiated in melt collide at the center and produce, at first, a molten, and then, a solid jet due to the fast solidification process.

Maybe not the molten but the solidification dynamics is responsible for the formation of microbumps and nanojets. Solidification of the molten metal starts at the melt periphery and very rapidly proceeds to the center. This process can also induce surface modifications and can be responsible for the formation of the hollow bump structures and nanojets.

Ivanov and Zhigilei investigated the kinetics and microscopic mechanisms of laser melting and disintegration of thin free-standing gold films irradiated by a short laser pulse in a computational model [23]. They found strong pressure variations of a few cycles during the existence of the molten phase leading to disintegration, i.e. the formation of a bubble inside the gold film. Whereas in their simulations the free-standing gold film thickness melts completely, allowing dilatation in two directions, in our experiments a fixed boundary (glass substrate or solid gold) suppresses dilatation in one direction and possibly boosts it in the other direction (perpendicular outwards from the free gold surface). This could explain the huge bump-like structure. The nanojet could be a result of this dynamic during bubble expansion. It appears at the lateral position with highest temperature and highest expansion velocity of the bubble in a direction perpendicular to the layer surface. Due to the mentioned pressure variations, the process of nanojet formation could proceed stepwise, like a displacement pump, meaning

0.6μm

Figure 19: Swellings along a nanojet, threshold-filtered cut-out of Fig. 3.2(b).

that the nanojet becomes longer after every bubble pressure increase. An indication for this can be found in the SEM images of nanojets. There are several swellings along the nanojets (see Fig. 19), where the jet growth could have been slowed or interrupted. But one has to take into account, that all the SEM images represent structures after solidification, which is a non-instantaneous process with its own dynamics, which can also be responsible for the formation of nanojets.

The demonstrated technology gives access to the fabrication of high-resolution high-aspect ratio, direct-write textures in some metallic surfaces without ablation, and therefore, without debris contamination. One possible application of this technology could be the fabrication of metallic band-gap nanostructures utilizing surface plasmon polaritons (SPP) as information carriers (see below).

8.4. Deep Drilling and Cutting with Ultrashort Laser Pulses

Besides applications related to the generation of nano- and submicron structures, ultrashort pulsed lasers are very versatile tools for the efficient structuring of almost all kinds of solid materials. Concerning industrial applications, cutting, drilling, and structuring of metals and other construction materials like technical ceramics or glasses are the most developed so far. Especially for applications, where highly precise structures without noticeable thermal influences are required, ultrashort pulsed lasers offers a great potential. Although this kind of lasers, at present, cannot offer productivity in terms required for mass production, there are exciting applications where the unique properties of ultrashort laser pulses lead to machining quality being not accessible with other methods. Some of the applications will be presented in the following.

Diseases of coronary vessels, like the obstruction of the vessels due to plaque are typically treated with catheter operations and by the implantation of vessel support structures, so-called stents. Stents are being used for a long time, but still show significant rates of post-operative complications like restenosis. One approach for the reduction of such complications is the use of temporary implants, which necessitates the use of special materials, which can dissolve or degrade within the body after the operation. Such materials could be polymers on a Polylactide basis (e.g. PLLA), but they are typically very sensitive to thermal load during manufacturing processes. Cutting of such polymers using CO_2 lasers comes along with undesired modification of the polymer properties. Application of UV radiation for cutting the polymer results in a strong modification of the polymer molecular weight distribution and therefore, in the lack of biocompatibility. The manufacturing of such implants has successfully been carried out using ultrashort laser sources (see Fig. 20). The same technology has also been successfully used for the structuring of X-ray opaque materials, such as tantalum (Fig. 21).

High precision structuring capabilities are also needed for the fabrication of shielding masks for ion-implantation processes in semiconductor industries. As demonstrated in Fig. 22, the shielding masks can be structured with femtosecond lasers.

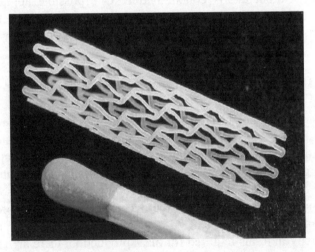

Figure 20: Biodegradable PLLA stent.

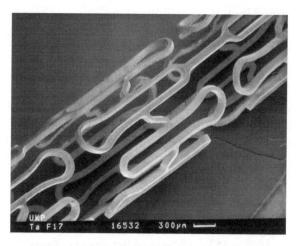

Figure 21: X-ray opaque stent from tantalum.

Figure 22: Shielding mask for ion-implantation process.

Applications of ultrashort pulsed lasers are not limited to metallic or polymeric materials, also different optical materials, for example, glass or quartz can be microstructured. Figure 23 shows the surface of a photocathode window structured by femtosecond laser pulses (followed by a simple thermal post-processing). The introduction of such windows in optical detectors increases the detection and conversion efficiency from photons to electrons especially in the near infrared spectrum, significantly extending the range of operation of such detectors [24,25].

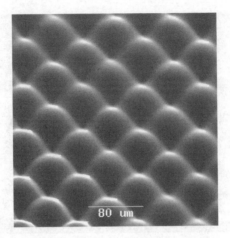

Figure 23: fs-laser structured cathode surface.

The huge range of applications in femtosecond laser technology is obvious. This field also offers niches, where the structures can be made exclusively by ultrashort pulsed lasers. Those niches are typically not accessible by other manufacturing techniques, e.g. EDM techniques or wire erosion processes due to heat conduction or stability problems. Here, minimal achievable structural dimensions are typically around 80–100 µm, which can easily be matched by ultrashort ablation processes. Especially for the fuel injection technology in the automotive sectors, reduction of nozzle diameters are of high interest. This allows to scale down fuel consumption and emissions. Such nozzles can be realized with highest precision without thermal damage to the remaining material by ablation with ultrashort laser pulses [9] (see Fig. 24).

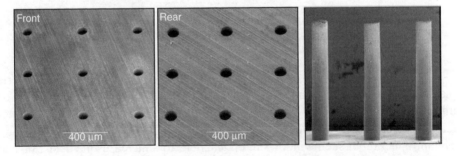

Figure 24: Drilling in 1 mm stainless steel; details of nozzle in- and outlet as well as replica of the channel geometry.

8.5. 3D Photofabrication and Microstructuring

Femtosecond laser 'rapid prototyping' technology based on 2PP technique, originally developed for the fabrication of photonic devices, has proven to be an important technology for a flexible fabrication of 3D structures with micro- and nanoscaled dimensions [10–14]. One important example is the photofabrication of 3D microstructures by curing photosensitive organic–inorganic hybrid polymers (ORMOCER®) by femtosecond laser pulses (see Section 8.1.2). These polymers offer advantageous biological properties, coming close to the bioactivity of Extracellular Matrices (ECMs) [15].

In this chapter, experiments which have been carried out to compare laser-induced single- and 2PP using UV and infrared laser systems are described. The lateral dimensions of the structures produced by 2PP are significantly smaller compared to the UV-polymerized structures. A further advantage, compared to the UV-polymerized samples, is the low sensitivity of the 2PP against cross-linking of densely packaged structures. In Fig. 25, grid structures produced by IR and UV light pulses are compared. In case of IR femtosecond laser pulses, there are no

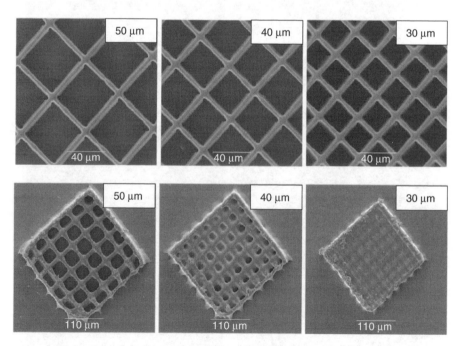

Figure 25: SEM pictures of the grid structures generated by 2PP, showing no sign of undesired cross-linking effects (top), UV-generated structures with an excessive cross-linking (bottom).

signs of cross-linking of densely packaged structures. Whereas with UV light, structures with a spacing less than 30 μm cannot be fabricated, because of the excessive cross-linking, transforming the grid structure into a solid body.

2PP by infrared laser pulses offers additional advantages for the fabrication within the volume of the photosensitive resins, as described above. Choosing a sufficiently long Rayleigh length of the optical focusing system, polymerization through the complete height of the resin layer can be carried out with a single IR laser pulse. In addition to the low tendency of cross-linking, very high aspect ratio structures can be produced without the necessity of vertical movement of the

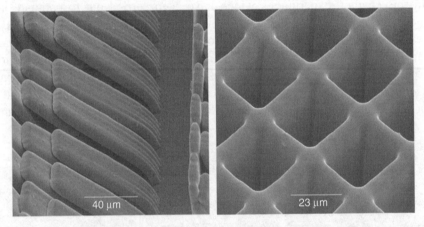

Figure 26: SEM-images of the test structures fabricated by 2PP with the aspect ratio of 30 (left), High-aspect ratio grid structure (right).

Figure 27: SEM-Pictures of 3D structures generates by 2PP using single pulses.

laser focus position. By this method, test structures with the aspect ratio better than 30 have been fabricated (see Fig. 26).

Truly 3D structures can be realized by introducing a suitable processing strategy. Synchronization of laser pulses and scanner movement as well as multiple irradiation steps give access to a fast, efficient manufacturing process of strongly undercut structures. The following pictures (see Fig. 27) demonstrate the flexibility of this method for the generation of different types of structures. The creation of structures with a 1 mm^2 dimension is just a matter of seconds.

For many applications, e.g. for the fabrication of scaffolds for tissue engineering or medical implants, porous, truly 3D structures are required. As demonstrated above, this can be realized in an efficient manner by using 2PP technology.

8.6. Application Examples

The femtosecond laser processing of different materials with the various techniques described above finds many applications in science and technology. Among the scientific applications, the fabrication of sub-micrometer scale structures is of special interest for the realization of photonic crystals and for guiding and manipulation of SPP. The last field, known as plasmonics, is rapidly developing. The potential of femtosecond laser 2D- and 3D structuring for this particular application will be discussed in the following Section (8.6.1).

3D structuring of positive and negative photoresists using femtosecond laser pulses offers interesting possibilities for the fabrication of sub-micrometer channels and cavities. These structures may serve as microfluidic components, with interesting applications for the realization of dye-based microlasers, as will be presented in the second Section (8.6.2).

8.6.1. Plasmonics

Miniaturization of optical components is of great interest for information technology. However, for transportation of optical signals within glass fibers, structural dimensions of the order of the light wavelength are usually required. Resolution of optical data storage devices is determined in the same way by the diffraction limit of light. For radiation in the near infrared spectral range this is one order of magnitude larger than the electronic tracks on a microchip. Further miniaturization of optical elements down to the sub-wavelength size is very important and requires new concepts. The use of SPP propagating on metal surfaces as information carriers in a kind of 'optical circuit' is one of such concepts.

Surface plasmons, in general, are coherent oscillations of the electron gas at the interface of a metallic conductor with a dielectric. Together with the electromagnetic field associated with these oscillations, the quantized oscillations are called surface plasmon polaritons (SPP). The electromagnetic field of the SPPs is restricted to the interface and its amplitude decays exponentially to both sides from the metal surface (so called evanescent field). The SPPs can propagate along the conductor-dielectric-boundary with a wave vector k_{SPP} determined by $k_{SPP}^2 = k_L^2(\varepsilon_m\varepsilon_d/(\varepsilon_m + \varepsilon_d))$, where ε_m and ε_d are the complex dielectric functions of the metal and the dielectric, respectively. Figure 28 shows schematically the charge and field distribution in the SPPs. For each frequency ω, the wave vector of the surface plasmon polariton is larger than that one of the corresponding free light wave k_L [26].

To excite SPP's on a metallic surface, the wave vector mismatch has to be compensated, which can be accomplished in the so-called Kretschmann- or Otto-configurations (see Fig. 29) or by using a grating structure.

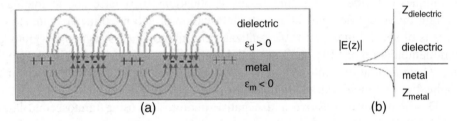

Figure 28: Charge and field configuration of a surface plasmon polariton (a) and the corresponding exponentially decaying electric field amplitudes in both media.

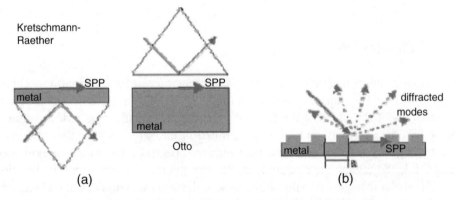

Figure 29: Excitation of SPP's on flat metal surfaces using Kretschmann- and Otto-configurations (a) or a grating structure (b).

In the Kretschmann-configuration the SPP's are launched, when light impinges on the prism–metal interface under an angle θ greater than the angle for total internal reflection [27]. If the metal film is thin enough, the evanescent field associated with the reflection extends through the metal and its wave vector is matched with that of the SPP's when $k_{SPP} = k_L\, n_{prism} \sin\theta$ is fulfilled. Another possibility is to use a grating-coupling method [28], where the missing momentum between the SPP and the exciting light wave is supplied by the reciprocal lattice vectors $k_{grating}$ of the periodic grating structure, given by $k_{grating} = 2\pi/a$, where a is the grating period.

Due to the larger momentum vector, SPP's can propagate on structures much smaller than the wavelength of the corresponding light wave. In this way it is possible to circumvent the diffraction limit. However, the distance that SPPs can travel is limited by the absorption in metal and is therefore restricted to a few tens or hundreds of micrometers [29], which determines the size of an 'optical circuit'.

For the guiding of the SPP's on metal surfaces, there exist different approaches. First, the metal layer on a dielectric substrate can be shaped in a way that it directs the SPP's to a desired location. Second, the metal layer can be structured to provide the guiding properties. That can be done by the fabrication of band-gap structures or by dielectric plasmon waveguides. Guiding of SPP's has been demonstrated along a 200 nm wide and 50 nm broad metal stripe on a glass substrate with a mode width of 115 nm. This is 1/7 of the wavelength used for excitation, demonstrating the huge potential of SPP's for the miniaturization of optical components [30]. Other schemes apply aligned nanoparticles, where each particle can be considered as a resonator for certain plasmon frequencies (coupled resonant oscillator waveguide, CROW) [31]. Switching of plasmons in analogy to the switching of electrical currents on integrated circuits (transistors) has also been demonstrated. Below the potential of femtosecond laser technologies for the fabrication of SPP waveguide structures is discussed.

The different SPP waveguide schemes can be realized with femtosecond laser material processing using the methods described in the previous chapters. Metallic waveguides on dielectric substrates can be produced by structuring of a thin layer of positive photoresist, subsequent gold-sputtering, and lift-off procedure. After lift-off of the non-irradiated photoresist (and the gold layer on top of it), arbitrary shaped 2D gold structures can be produced. In Fig. 30 two bend waveguide structures after illumination and development of the positive photoresist (a) and after gold-sputtering and lift-off procedure (b) are shown.

2D photonic band-gap structures can be fabricated by direct laser ablation or laser-induced melt dynamics. The first process results in the creation of holes, whereas the latter leads to the formation of bump- and jet-like structures. A hexagonal surface patterning will inhibit the propagation of SPPs at certain frequencies. For these SPPs waveguiding can be realized in the defect (non-structured) lines.

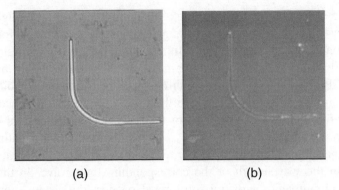

(a)   (b)

Figure 30: Fabrication of metallic waveguides on glass substrates. (a) is the microscope image of the structured positive resist layer, (b) shows the metallic waveguide after the lift-off.

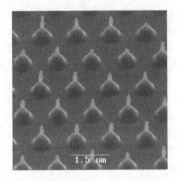

Figure 31: Hexagonal pattern fabricated by laser induced melt dynamic in a 50 nm gold layer.

Figure 31 gives an example of a hexagonal pattern of bump- and jet-like structures produced by femtosecond laser pulses in a 50 nm thin gold layer on a glass substrate.

Two-photon polymerization can be used to build arbitrary shaped dielectric plasmon waveguides on the gold surface. For the fabrication of these dielectric waveguides a negative photoresist is used. After the polymerization process the non-irradiated resist is dissolved, leaving the desired structures on the metal surface. Dielectric waveguides made from ORMOCER® on a gold surface are shown in Fig. 32.

The non-linear character of the structuring processes allow in all the cases the fabrication of structures with a resolution in the sub-wavelength-range, e.g. smaller

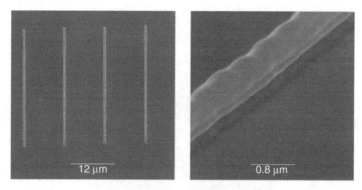

Figure 32: Dielectric plasmon waveguides made from ORMOCER on a gold surface. The right picture is a close-up showing the waveguide surface quality.

than 800 nm. Microstructuring examples demonstrated in this section show that femtosecond laser technologies can play an important role in the fabrication of sub-micrometer structures for guiding and manipulation of SPP and for the development of plasmonic devices.

8.6.2. Microfluidics

Nowadays, microfluidics finds many applications in chemical, biochemical, and medical analysis [32]. The major advantage of this technology is in the miniaturization of the fluid volume and amount of reagents which are required. Due to the dimensions of the microfluidic channels, in the range of several micrometers, the flow of the liquid is completely laminar, with Reynolds numbers in the order of 1. Microfluidics is of special interest for the development of microlasers and microcavities for light amplification in dye solutions. An integration of these components on a microchip is attractive for chemical and biological applications. Microlasers can serve as coherent light sources for an optical analysis of reactants, where also the reaction takes place on the same chip. This is important for the realization of micro-total analytic systems (μTAS) and complete 'lab-on-a-chip' devices [33].

For the fabrication of imbedded microfluidic components such as cavities and channels, the structuring of positive photoresist with femtosecond laser pulses is very promising technique. Figure 33 shows microchannels fabricated in S1813-photoresist.

In [33] the fabrication of microfluidic lasers in photoetchable glasses has been demonstrated. This fabrication procedure includes several steps: writing of the

Figure 33: SEM picture of microchannels with different diameters fabricated by structuring of S1813 positive photoresist.

structure, baking, etching, and post-baking of the sample. More simple procedure can be realized with positive photoresists. An example of a simple ring resonator fabricated with femtosecond lasers is shown in Fig. 34. It consists of a microcavity surrounded by four 90° mirrors. The emitted light from a dye solution inside the cavity can be directed around the resonator by total internal reflection [33]. The picture represents a horizontal cut through the device, which also can be fabricated completely buried in the resist and the dye can be injected via microchannels.

This technique provides a simple method for the fabrication of various imbedded microstructures and microfluidic devices. It allows to produce arbitrary shaped and very complex systems of channels and cavities.

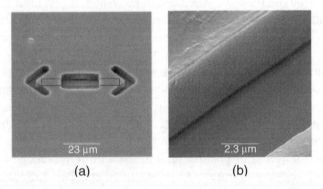

(a)   (b)

Figure 34: A microring laser resonator fabricated in S1813 photoresist. (a) shows a cut through the cavity and the mirrors together with one possible round trip inside the resonator, (b) is a close-up of a mirror, showing the good surface quality.

8.7. Summary and Outlook

Recent progress in direct write 2D and 3D photofabrication technologies has been discussed. Due to non-linear nature of many basic laser-matter interaction processes, application of ultrashort laser systems allows one to overcome the diffraction limit and to produce 2D and 3D sub-wavelength microstructures. This is very powerful technology which is still in a rapidly developing stage.

We hope that the results demonstrated in this book will stimulate further research and developments in this very exciting field of laser physics and material processing technologies.

Acknowledgments

The authors gratefully acknowledge very important contribution from our colleagues who have been involved into different parts of this work: G. Kamlage, F. Korte, A. Ostendorf, A. Ovsianikov, S. Passinger, and J. Serbin. This work has been supported by the German research association (DFG) and the European Network of Excellence 'Plasmo-Nano-Devices'.

References

[1] Pronko, P.P., Dutta, S.K., Squier, J., Rudd, J.V., Du, D. and Mourou, G., *Opt. Commun.,* **1995**, *114,* 106.

[2] Momma, C., Chichkov, B.N., Nolte, S., von Alvensleben, F., Tünnermann, A. and Welling, H., *Opt. Commun.,* **1996**, *129,* 134.

[3] Chichkov, B.N., Momma, C., Nolte, S., von Alvensleben, F. and Tünnermann, A., *Appl. Phys. A,* **1996**, *63,* 109.

[4] Stuart, B.C., Feit, M.D., Herman, S., Rubenchik, A.M., Shore, B.W. and Perry, M.D., *J. Opt. Soc. Am. B,* **1996**, *13,* 459.

[5] Simon, P. and Ihlemann, J., *Appl. Phys. A,* **1996**, *63,* 505.

[6] Kautek, W., Krüger, J., Lenzner, M., Sartania, S., Spielmann, C. and Krausz, F., *Appl. Phys. Lett.,* **1996**, *69,* 3146.

[7] Nolte, S., Momma, C., Jacobs, H., Tünnermann, A., Chichkov, B.N., Wellegehausen, B. and Welling, H., *J. Opt. Soc. Am. B,* **1997**, *14,* 2716.

[8] Horwitz, J.S., Krebs, H.-U., Murakami, K. and Stuke, M., (Eds.), in *Laser Ablation, Proceedings of the 5th International Conference,* **1999**.

[9] Kamlage, G., Bauer, T., Ostendorf, A. and Chichkov, B.N., *Appl. Phys. A,* **2003**, *77,* 307–310.

[10] Cumpston, B.H., Ananthavel, S.P., Barlow, S., Dyer, D.L., Ehrlich, J.E., Erskine, L.L., Heikal, A.A., Kuebler, S.M., Lee, I.-Y.S., McCord-Maughon, D., Qin, J., Röckel, H., Rumi, M., Wu, X.-L., Marder, S.R. and Perry, J.W., *Nature,* **1999**, *398,* 51

[11] Kawata, S., Sun, H.-B., Tanaka, T. and Takada, K., *Nature,* **2001**, *412,* 697.

[12] Serbin, J., Egbert, A., Ostendorf, A., Chichkov, B.N., Houbertz, R., Domann, G., Schulz, J., Cronauer, J., Fröhlich, L. and Popall, M., *Opt. Lett.* **2003**, *28,* 301

[13] Serbin, J., Ovsianikov, A. and Chichkov, B.N., *Opt. Express,* **2004**, *12,* 5221.

[14] Houbertz, R., Fröhlich, L., Popall, M., Streppel, U., Dannberg, P., Bräuer, A., Serbin, J. and Chichkov, B.N., *Adv. Eng. Mater.,* **2003**, *5,* 551

[15] Doraiswamy, A., Pats, T., *et al., Mater. Res. Soc. Proc.,* **2005**, *845.*

[16] Kaiser, A., Rethfeld, B., Vicanek, M. and Simon, G., *Phys. Rev. B,* **2000**, *61,* 11437.

[17] Korte, F., Adams, S., Egbert, A., Fallnich, C., Ostendorf, A., Nolte, S., Will, M., Ruske, J.-P., Chichkov, B.N. and Tünnermann, A., *Optics Express,* **2000**, *7,* 41.

[18] Will, M., Nolte, S., Chichkov, B.N. and Tünnermann, A., *Appl. Opt.,* **2002**, *41,* 4360.

[19] Watanabe, W., Toma, T., Yamada, K., Nishii, J., Hayashi, K. and Itoh, K., *Opt. Lett.,* **2000**, *25,* 1669.

[20] Korte, F., Serbin, J., Koch, J., Egbert, A., Fallnich, C., Ostendorf, A. and Chichkov, B.N., *Appl. Phys. A,* **2003**, *77,* 229.

[21] Korte, F., Koch, J. and Chichkov, B.N., *Appl. Phys. A,* **2004**, *79,* 879.

[22] Bonn, M., Denzler, D.N., Funk, S., Wolf, M., Wellershoff, S.-S. and Hohlfeld, J., *Phys. Rev. B,* **2000**, *61,* 1101.

[23] Ivanov, D.S. and Zhigilei, L.V., *Appl. Phys. B,* **2003**, *68,* 64114.

[24] Harmer, S.W. and Townsend, P.D., *J. Phys. D,* **2003**, *36,* 1–7.

[25] Vazquez, G.V., Townsend, P.D., Magarahbi, M., Bauer, T. and Gonzales, M., *Phys. Stat. Sol. (a),* **2003**, *198,* 465–477.

[26] Raether, H., Surface plasmons on smooth and rough surfaces and on gratings, *Springer Tracts in Modern Physics,* **1988**, *111.*

[27] Kretschmann, E., Raether, H. and Naturforch. Z., *Zeitschrift für Naturforschung,* **1968**, *A 23,* 2135.

[28] Hutley, M. and Meystre, P., *Opt. Comm.,* **1976**, *19,* 431.

[29] Lamprecht, B., Krenn, J.R., Schider, G., Leitner, A., Aussenegg, F.R. and Weeber, J.C., *Appl. Phys. Lett.,* **2001**, *79,* 51.

[30] Krenn, J.R. and Aussenegg, F.R., *Phys. Journal,* **2002**, *1,* 39.

[31] Krenn, J.R., Dereux, A., Weeber, J.C., Bourillot, E., Lacroute, Y., Goudonnet, J.P., Schider, B., Gotschy, W., Leitner, A., Aussenegg, F.R. and Girard, C., *Phys. Rev. Lett.,* **1999**, *82,* 2590.

[32] Weil, B.H. and Yager, P., *Science,* **1999**, *283,* 346.

[33] Cheng, Y., Sugioka, K. and Midorikawa, K., *Opt. Lett.,* **2004**, *29,* 2007.

Chapter 9

Self-Organized Surface Nanostructuring by Femtosecond Laser Processing

J. Reif[a,b], *F. Costache*[a,b] *and M. Bestehorn*[a]

[a]*Brandenburgische Technische Universität Cottbus, Konrad-Wachsmann-Allee 1, 03046 Cottbus, Germany;* [b]*IHP/BTU JointLab, Cottbus, Germany*

Abstract

At the bottom of ablation craters produced by femtosecond laser ablation, regular periodic patterns develop with a typical feature size of 100–200 nm. This chapter reviews the present and previous scenarios of pattern formation upon laser ablation and gives an overview over laser nanostructuring of solid surfaces. Finally, possible applications are discussed.

Keywords: Femtosecond laser ablation; Laser Induced Periodic Surface Structures (LIPSS); Nanostructuring of solid; Ripples structure; Surfaces; Surface instability.

9.1. Introduction

The tradition of laser materials processing has been almost as long as that of the laser itself. In present technology, it has gained its undisputed position in many fields such as, e.g. photo-lithography in the semiconductor industry, or cutting, drilling, and welding in the automobile industry. Most machining procedures are based on laser ablation, i.e. the removal of material upon the impact of intense laser pulses. In recent years, particularly ultrashort-pulse lasers have found specific application when it comes to ultra-high precision machining [1–6]. However, closer inspection of the processed material often reveals (for certain combinations of laser fluence and number of pulses) periodic structures on the treated areas, in the literature referred to as 'Laser Induced Periodic Surface Structures' (LIPSS) or 'Ripples' [7] (cf. Fig. 1). Sometimes, these structures may be detrimental to the desired 'clean' machining [8–11].

Recent Advances in Laser Processing of Materials
J. Perrière, E. Millon and E. Fogarassy (Editors)

Figure 1: Ablation crater with ripples structure at the bottom. The enlarged lower panel
exhibits even more complicated nanostructures.

On the other hand, regular nanostructures bear the potential for interesting applications in several emerging fields [12]. In micro- and nanoelectronics, the coupling
of biologically active materials to silicon technology, e.g. for sensor devices, can
be achieved by self-assembly of thin films on nanostructured silicon substrate
[13–17]. Regular two-dimensional arrays of nanotips on Si substrate can serve as
a field emitter matrix for integrated display technology [18,19]. Also, the application of regular nanostructures for magnetic applications is an emerging field [20].
It appears obvious to use LIPSS for such purposes.

In this contribution we, therefore, discuss the generation of LIPSS in more detail,
in particular the fundamental mechanisms. We start, in Section 9.2, with reviewing
the classical picture, developed in the 1980s, assuming interference effects as the

origin for ripples. However, as shown in Section 9.3, recent research indicates that this explanation cannot account for many peculiar features observed for the generated structures. Consequently, we present a new model for the fundamental mechanisms in Section 9.5, after reviewing, in Section 9.4, the fundamentals of laser ablation dynamics. Finally, in Section 9.6, possible applications will be considered in some detail.

9.2. Classical Picture of Laser Induced Periodic Surface Structure (LIPSS) Formation

Laser induced periodic surface structures (LIPSS) or ripples have been known ever since the early days of the laser and of laser/matter interaction [21,22], followed by a manifold of investigations on many target materials and under different irradiation conditions [7,18,23–38]. Early models assumed that these ripples might be the result of (frozen) acoustic surface waves [23]. Very often, however, a periodicity at the order of the wavelength was observed. Further, the ripples orientation strongly depended on the polarization of the incident laser light. These observations made evolve a model, explaining the ripples as a result of an inhomogeneous energy input, due to an interference between the incident laser and a surface scattered wave [24–27].

In 1982/1983, John Sipe and coworkers refined this idea [7,28], assuming, in more detail, that the surface projection of the incident light interferes with random periodicities in the target selvedge region, correlated to surface roughness, such that $|\mathbf{k}_{proj} \pm \mathbf{k}_{rough}| = |\mathbf{k}|$. This is similar to the interference, in k-space, of electron waves with the periodicity of the crystalline lattice, and corresponds to respective resonances in the selvedge polarization, resulting in a modulated intensity and absorption pattern at the target. The inhomogeneous energy input is directly translated into a conformal variation of ablation depth, and the ripples structure is an image of the interference pattern. The situation is sketched in Fig. 2.

Then, the periodicity in the modulation pattern, i.e. the ripples wavelength Λ, is controlled by the incident wavelength λ and by the incident angle θ, and is given by:

$$\Lambda = \frac{\lambda}{n(1 \pm \sin\theta)} \tag{1}$$

where the target index of refraction n enters for the propagation of the scattered wave in the selvedge, and the $\pm$ accounts for the scattered wave co-propagating and counter-propagating to the surface projection of the incident wave. As experimentally

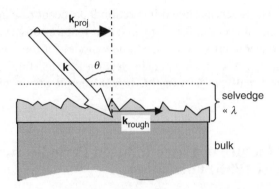

Figure 2: Typical geometry for the 'interference model'.

observed, this model also predicts these ripples to be perpendicular to the polarization, excluding longitudinal effects. However, for light with a polarization component perpendicular to the surface (*p*-polarization), also ripples parallel to the polarization should be allowed, with a periodicity of

$$\Lambda = \frac{\lambda}{n(\cos\theta)} \tag{2}$$

For many years, this model was applied very successfully to explain the observation of many kinds of LIPSS though, often, no attention was paid to features in the generated structures that were more or less contradictory, such as bifurcations in the ripples lines, meandering structures, or two dimensional arrays of cones. Also patterns of very deep trenches between very flat regular plateau lines do not really fit into the model. Already in 1989, Sipe *et al.* considered additional processes like self-organization during cooling for pattern formation in laser produced melts [39] under cw irradiation.

9.3. Typical Laser Induced Surface Nanostructures

Recently, the research on LIPSS found new attention in connection with ultra-precise materials processing with ultra-short laser pulses. Though the pulse durations there are, generally, shorter than the assumed ablation times, still ripples structures are observed [30–37]. As in earlier observations, two different ripples systems perpendicular to each other can occur (Fig. 3), with the feature size in one of them substantially smaller than the wavelength.

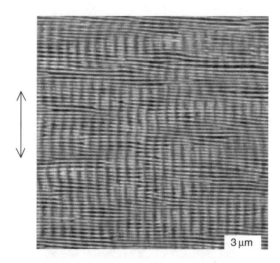

Figure 3: Two perpendicular ripples systems, produced with a Ti:Sapphire laser (800 nm) on BaF_2. The laser polarization is indicated by the double arrow.

Further even more peculiar features are, e.g. broad flat crests with very narrow trenches in between, with a meandering overall structure [11,33–35] (Fig. 4), or two-dimensionally ordered conical structures [36–37] (Fig. 5).

Most experiments have been performed using linearly polarized light. Then, the orientation of the generated structures is strongly dependent on the light polarization, the 'fine' ripples always being perpendicular to the light electric field (Fig. 6).

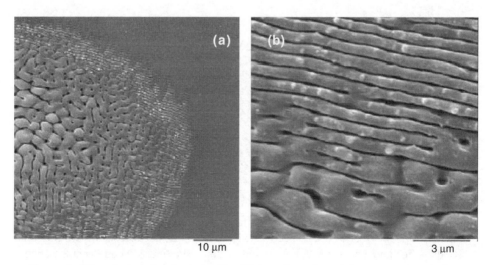

Figure 4: Meandering structures on Si with broad crests and deep, narrow trenches.

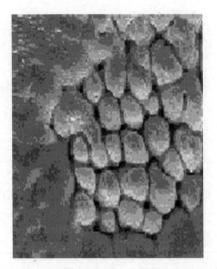

Figure 5: Two-dimensional array of cones.

Recent experiments with circularly polarized light show, however, similar aligned structures, independent of the sense of polarization (Fig. 7).

Another interesting feature can be seen in Fig. 6(c): the nanostructures exhibit very many bifurcations: lines are split into two and then recombined or joined with other lines further on. Closer inspection reveals that this appears to be a very general feature of these structures. In Fig. 6(a), the startling result is shown that the ripples

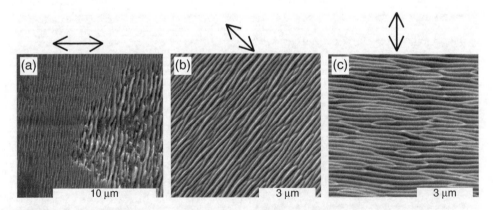

Figure 6: Periodic structure-dependence on laser beam polarization (indicated by arrow), rotated: 0° (a), 40° (b), 90° (c), respectively; Note in (a) the coexistence of two different periods: a fine and a coarse one, both oriented alike. The finer structure is at the crater edge, the coarser one in the center.

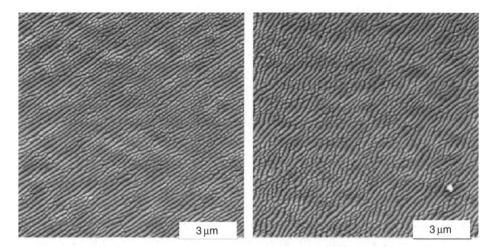

Figure 7: Periodic structures obtained with circular polarization (left: left circular; right: right circular).

period depends on the intensity: in the center of the spot, it is about 10 times larger than at the edge, with a discontinuous transition. Anyway, in many of the experiments presented here, no distinct influence of the wavelength or of the angle of incidence on the ripples spacing could be observed. Instead other factors seem to be important such as input fluence, number of pulses, absorptivity. In Fig. 8, it is shown that the irradiation dose, i.e. the product of incident fluence and number

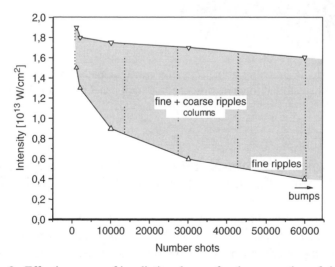

Figure 8: Effective range of irradiation dosage for the generation of ripples.

of pulses, should play an important role for the ripples formation. There are strong indications, that also a 'history' can be found, starting with individual nanoparticles (e.g. spheres) which then align to chains (Fig. 9) and finally kind of coagulate to structures as in Fig. 4. Generally, it has been shown that the patterns are fully developed only after a series of incident pulses [31,35,38], indicating a certain positive feedback.

Finally, the nanostructures are very similar to those observed under, in principal, completely different conditions: when solid surfaces are sputtered with an ion beam, also periodic structures of about similar feature size are generated [40], as is shown in Fig. 10.

These results boosted the idea to propose self-organization from instabilities, which already has been very popular to explain the ion sputtering results, as the responsible mechanism [30], which has gained more and more attraction, recently, fostered by the present understanding of the fundamental mechanisms for laser ablation.

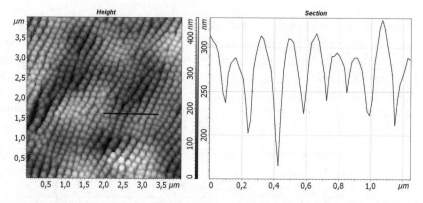

Figure 9: Periodic structures at the edge of a crater, obtained with circular polarization (imaged with an atomic force microscope; the right panel shows the height variations along the indicated line).

9.4. Dynamics of Femtosecond Laser Ablation from Dielectrics and Semiconductors: Generation of Surface Instabilities

The present picture of the ablation process indicates, indeed, the generation of a considerable instability at the target surface. In particular, ultra-fast experiments, separating laser–material interaction from subsequent laser-plume and plume-target interactions, have contributed substantially to a better understanding.

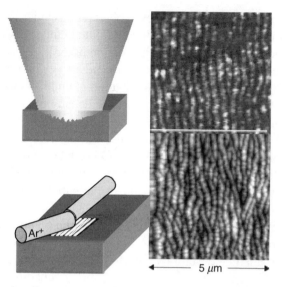

Figure 10: Comparison between patterns from laser ablation (upper) and ion sputtering (lower [40]).

At laser intensities well below the single shot damage threshold, Coulomb explosion upon multiphoton surface ionization has been established as the principal origin of positive ion emission from dielectrics [41–44]. In this picture, the intense laser pulses induce multiphoton ionization. Electrons which are close to the surface can gain sufficient energy to immediately leave the target, leaving behind holes of positively charged ions. Since this process is assumed to occur almost instantaneously, because of the low electron mass, the holes cannot be filled at the same speed (this holds for dielectrics and for semiconductors [45,46] where electrons are sufficiently slow). Consequently, if the density of holes is sufficiently high, the surface becomes electrostatically unstable and some of the positive ions will be ejected. This can occur with rather high kinetic energy of several 10 eV up to about 100 eV, and not only atomic ions but also larger clusters are ejected.

This process can be significantly enhanced by the presence of resonant defect states in the band gap, which are, usually, created during a first irradiation phase of several laser shots. The quantity of ejected material under these conditions is relatively small and amounts to an ablation of only a monolayer or even less per laser pulse [47,48]. Therefore, this is not really the regime of interest for effective material patterning. In this regime, the density of electrons that do *not* leave the surface is not sufficient to transfer enough energy via hot-electron–phonon collisions to the lattice to provoke substantial target heating [49,50–53], nor to provide sufficiently fast conduction electrons to generate avalanche ionization.

At slightly higher laser intensity, however, observations indicate that a new process sets in and turns to become dominant: the additional emission of neutral particles as a result of rapid hot-electron thermalization [43,52,53]. In this case, the ablation rate can be at the order of a few hundred nanometers per pulse [43]. This regime is characterized by a bimodal velocity distribution of emitted species, with an additional slower ion component. In this regime, the density of conduction band electrons is greatly larger. Additionally, the surface ionization tends to saturate because of space charge effects. Thus the conduction band electrons can be heated by free carrier absorption and impact processes [50] to produce a hot electron gas which, subsequently, relaxes to the lattice by electron–phonon collisions [51,52]. Since typical time constants for this process are in the sub-picosecond to picosecond range [54], the surface region can be very rapidly heated to very high temperatures far above the melting temperature and even the critical temperature without any significant volume expansion, however, immediately (on a sub-picosecond time scale) resulting in a lattice destabilization [55]. To release the corresponding high pressure, the material expands adiabatically as a homogeneous, superheated liquid to reach the Van der Waals two-phase regime, bounded by the binodal curve. Crossing the spinodal to the metastable regime will result in the formation of large gas bubbles surrounded by a liquid skin. Since the gas expansion is faster than that of the liquid skin, the pressure will rise again until the skin ruptures and the gas is released. This is the typical situation responsible for a phase explosion [52,53,56].

Both scenarios are characterized by a high degree of instability, indicating that models of nonlinear dynamics should be appropriate to understand the following relaxation processes of self-organized structure formation.

M. Garcia and coworkers demonstrated that, already at a very early stage of the laser-target interaction, the absorption of an ultra-short laser pulse and the associated excitation of electrons, leads to a destabilization of the crystal lattice within less than a picosecond for strongly coupled materials like dielectrics or semiconductors [55]. This is shown in Fig. 11.

9.5. Self-Organized Pattern Formation from Instabilities: Femtosecond Laser Ablation and Ion Beam Erosion

Structure formation by self-organization has been a topic of theoretical and experimental studies for more than thirty years [57], and has been applied successfully to explain similar structures to LIPSS, obtained in ion sputtering [40,58–61]. The corresponding model is based on the assumption that incident ions can ionize surface atoms of the target. The resulting surface charge leads to an erosion of a surface layer due to Coulomb repulsion. Taking into account further processes like

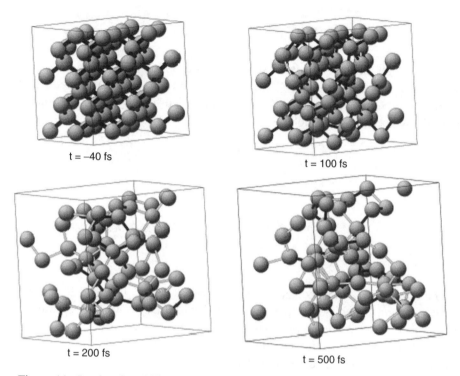

Figure 11: Lattice destabilization (diamond) upon excitation of binding electrons (from [55]).

surface diffusion, a stability analysis shows that – depending on the incident ion energy and their penetration depth – a plane surface can become unstable and a new, modulated spatially periodic structure develops – the ripples [62].

The surface instability is due to the appearance of a negative surface tension. Corresponding theoretical models are known from hydrodynamics of thin films and have been studied extensively during the last ten years [63–66]. The idea is now to consider the region of instability similar to a thin film with a modulated thickness. From a comparable model describing the results of structure formation upon ion sputtering, a rate equation can be developed which takes into account the competition between surface roughening due to particle emission and surface smoothing because of self diffusion. This rate equation is of the KPZ type (Kardar, Parisi, Zhang) [67]:

$$\frac{\partial}{\partial t} h = -V[h]\sqrt{1 + (\nabla h)^2} - D\Delta^2 h \tag{3}$$

where h denotes the film modulation height and the potential

$$V[h] = \int_F dx'dy'f\left(h[x,y] - h[x+x',y+y'],x',y',\vartheta\right)$$

accounts for all relevant forces. Assuming a finite action of the smoothing parameter f, this is a nonlocal integro-differential equation, which can be reduced to a partial differential equation of the Kuramoto–Sivashinsky-, Cahn–Hilliard-, or Swift–Hohenberg-Type [57,68,69]:

$$\frac{\partial}{\partial t}h = -V_0 + \gamma\frac{\partial h}{\partial x} + v_x\frac{\partial^2 h}{\partial x^2} + v_y\frac{\partial^2 h}{\partial y^2} - D\Delta^2 h$$
$$+ \frac{\lambda_x}{2}\left(\frac{\partial h}{\partial x}\right)^2 + \frac{\lambda_y}{2}\left(\frac{\partial h}{\partial y}\right)^2 + higher\ orders$$

(4)

The general idea is to consider the situation as shown in Fig. 12. A film (the instable region) of thickness a is modulated with height $h(x, y)$, representing a series of hills and valleys. For a particle at a hill site, the surface tension then is higher than at the valley site, tending to pull the particle downhill to minimize the surface. On the other hand, assuming Coulomb explosion as the ablation driving force, we find that for a uniform surface charge density, the valley-ion has more neighboring holes than the hill-ion and feels, therefore, a stronger repulsion. As a consequence, the ablation process tends to deepen the valleys and the diffusion process tends to fill them.

Equation (1) describes this competition and yields the following result: first (i.e. during the first numerical iterations, corresponding to the first pulses), periodic structures develop and grow linearly. However, this growth saturates at a certain modulation height, and typical 'coarsening' sets in: the ripples merge into larger and larger scaled structures (cf. Fig. 4). During the linear phase, the ripples structures are very similar to those observed experimentally (Fig. 13). Also, the

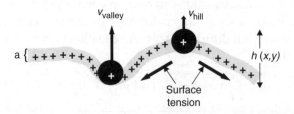

Figure 12: Theoretical model of the surface instability.

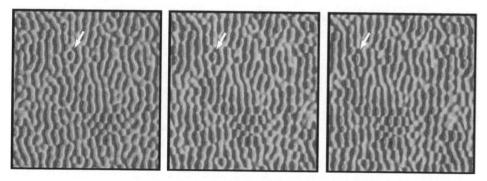

Figure 13: Theoretical ripples patterns. The panels show several subsequent iterations. Note the traveling of the pattern (the arrow points to one point-like structure which approaches the left margin).

influence of irradiation dose is compatible with the theoretical scenario. An interesting result of the modulations is also shown in Fig. 13: the ripples seem to travel from iteration to iteration, similar to a sand dune in the desert.

So far, however, the theoretical model is still in its initial stadium. It is yet far from yielding quantitative predictions, and it cannot yet identify the relevant control parameter for the ripples orientation, which for linearly polarized light appears to be the polarization and for circular polarization is still completely unknown.

9.6. Summary and Outlook

We showed that laser ablation, in particular with ultra-short laser pulses can result in regular periodic nanostructures at the interaction site. Discussing the classical model of inhomogeneous energy coupling as the origin of these structures showed that, in fact it cannot account for many of the observed features. More likely, the nanostructures seem to be the result of a self-assembly from a surface instability created by the ablation process.

Further, we demonstrated, that the typical feature size of 100–250 nm under our conditions does not depend on wavelength or on angle of incidence, but may be sensitively controlled by an appropriate choice of irradiation parameters like intensity, fluence, and irradiation dose. As already mentioned, a large number of applications has been proposed, such as the formation of tip arrays [19,37,70,71], maskless production of nanospaced line arrays for future close packed microelectronic structures [12], MEMS [72], nanostructuring soft polymers [9–11,36,73], and hard diamond [32]. Future applications, certainly, depend on a better understanding of the relevant control parameters.

References

[1] Nolte, S., Momma, C., Jacobs, H., Tünnerman, A., Chichkov, B.N., Wellegehausen, B. and Welling, H., *J. Opt. Soc. America B,* **1997**, *14*, 2716.

[2] Krüger, J. and Kautek, W., *Laser Physics,* **1999**, *9*, 30.

[3] Haigh, R., Hyden, D., Longo, P., Neary, T. and Wagner, A., *Proc. SPIE,* **1998**, *3546*, 477.

[4] Hertel, I.V., Stoian, R., Rosenfeld, A., Ashkenasi, D. and Campbell, E.E.B., *Riken Review,* **2001**, *32*, 23–30.

[5] Génin, F. (Ed.), Special Issue '*Femtosecond Techniques for Materials Processing*', *Appl. Phys. A,* **2003**, *76*, 299.

[6] Krüger, J. and Kautek, W., *Proc. SPIE,* **1996**, *2403*, 436.

[7] van Driel, H.M., Sipe, J.E. and Young, J.F., *Phys. Rev. Lett.,* **1982**, *49*, 1955.

[8] Krüger, J. and Kautek, W., *Appl. Surf. Sci.,* **1996**, *96–98*, 430.

[9] Bonse, J., Sturm, H., Schmidt, D. and Kautek, W., *Appl. Phys. A,* **2000**, *71*, 657.

[10] Bonse, J., Wrobel, J.M., Krüger, J. and Kautek, W., *Appl. Phys. A,* **2001**, *72*, 89.

[11] Bonse, J., Baudach, S., Krüger, J., Kautek, W. and Lenzner, M., *Appl. Phys. A,* **2002**, *74*, 19.

[12] Himpsel, F.J., Kirakosian, A., Crain, J.N., Lin, J.-L. and Petrovykh, D.Y., *Sol. State Commun.,* **2001**, *117*, 149.

[13] Wang, J., *Nuc. Acids Res.,* **2000**, *28*, 3011.

[14] Schaeferling, M., Schiller, S., Paul, H., Kruschina, M., Pavlickova, P., Meerkamp, M., Giammasi, C. and Kambhampati, D., *Electrophoresis,* **2002**, *23*, 3097.

[15] Richard, C., Balavoine, F., Schultz, P., Ebbesen, T.W. and Mioskowski, C., *Science,* **2003**, *300*, 775.

[16] Yan, H., Park, S.H., Finkelnstein, G., Reif, J.H. and La Bean, T.H., *Science,* **2003**, *301*, 1882.

[17] Li, Q., Gao, H., Wang, Y., Luo, G. and Ma, J., *Electroanalysis,* **2001**, *13*, 1342.

[18] Bai, X.D., Zhi, C.Y., Liu, S., Wang, E.G. and Wang, Z.L., *Sol. State Commun.,* **2003**, *125*, 185.

[19] Pedraza, A.J., Fowlkes, J.D. and Lowndes, D.H., *Appl. Phys. A,* **1999**, *69*, S731.

[20] Skomski, R., *J. Phys.: Condens. Matter,* **2003**, *15*, R841.

[21] Birnbaum, M., *J. Appl. Phys.,* **1965**, *36*, 3688.

[22] Siegrist, M., Kaech, G. and Kneubühl, F.K., *Appl. Phys.,* **1973**, *2*, 45.

[23] Maracas, G.N., Harris, G.L., Lee, C.A. and McFarlane, R.A., *Appl. Phys. Lett.,* **1978**, *33*, 453.

[24] Emmony, D.C., Howson, R.P. and Willis, L.J., *Appl. Phys. Lett.,* **1973**, *23*, 598.

[25] Brueck, S.R.J. and Ehrlich, D.J., *Phys. Rev. Lett.,* **1982**, *48*, 1678.

[26] Zhou, G., Fauchet, P.M. and Siegman, A.E., *Phys. Rev. B,* **1982**, *26*, 5366.

[27] Keilmann, F., *Phys. Rev. Lett.*, **1983**, *51*, 2097.

[28] Sipe, J.E., Young, J.F., Preston, J.S. and van Driel, H.M., *Phys. Rev. B*, **1983**, *27*, 1141; Young, J.F., Preston, J.S., van Driel, H.M. and Sipe, J.E., *Phys. Rev. B*, **1983**, *27*, 1155; Young, J.F., Sipe, J.E. and van Driel, H.M., *Phys. Rev. B*, **1984**, *30*, 2002.

[29] Yabe, A. and Niino, H., in *Laser Ablation of Electronic Materials*, Fogarassy, E. and Lazare, S. (Eds.), (p. 199), North-Holland, **1992**.

[30] Henyk, M., Vogel, N., Wolfframm, D., Tempel, A. and Reif, J., *Appl. Phys. A*, **1999**, *69*, S355.

[31] Costache, F., Henyk, M. and Reif, J., *Appl. Surf. Sci.*, **2002**, *186*, 352–357; Reif, J., Costache, F., Henyk, M. and Pandelov, S., *Proc. SPIE*, **2002**, *4760*, 980; Reif, J., Costache, F., Henyk, M. and Pandelov, S.V., *Appl. Surf. Sci.*, **2002**, *197–198*, 891; Costache, F., Henyk, M. and Reif, J., *Appl. Surf. Sci.*, **2003**, *208–209*, 486; Reif, J., Costache, F., Eckert, S. and Henyk, M., *Appl. Phys. A*, **2004**, *79*, 1229.

[32] Wu, Q., Ma, Y., Fang, R., Liao, Y. and Yu, Q., *Appl. Phys. Lett.*, **2003**, *82*, 1703.

[33] Coyne, F., Magee, J.P., Mannion, P., O'Connor, G.M. and Glynn, T.J., *Appl. Phys. A*, (online first) DOI: 10.1007/ s00339-004-2605-2, **2004**.

[34] Borowiec, A. and Haugen, H.K., *Appl. Phys. Lett.*, **2003**, *82*, 4462.

[35] Costache, F., Kouteva-Arguirova, S. and Reif, J., *Solid State Phenomena*, **2004**, *95–96*, 635–640; Costache, F., Kouteva-Arguirova, S. and Reif, J., *Appl. Phys. A*, **2004**, *79*, 1429; Reif, J., Costache, F.A., Eckert, S., Kouteva-Arguirova, S., Bestehorn, M., Georgescu, I., Semerok, A.F., Martin, P., Gobert, O. and Seifert, W., *Proc. SPIE*, **2004**, *5662*, 737; Reif, J., Costache, F. and Kouteva-Arguirova, S., *Proc. SPIE*, **2004**, *5448*, 756.

[36] Baudach, S., Bonse, J. and Kautek, W., *Appl. Phys. A*, **1999**, *69*, S395.

[37] Her, T.-H., Finlay, R.J., Wu, C. and Mazur, E., *Appl. Phys. A*, **2000**, *70*, 383.

[38] Lam, Y.C., Tran, D.V., Zheng, H.Y., Murukesha, V.M., Chai, J.C. and Hardt, D.E., *Surf. Rev. and Lett.*, **2004**, *11*, 217.

[39] Preston, J.S., van Driel, H.M. and Sipe, J.E., *Phys. Revl. B*, **1989**, *40*, 3942.

[40] Erlebacher, J., Aziz, M.J., Chason, E., Sinclair, M.B. and Floro, J.A., *Phys. Rev. Lett.*, **1999**, *82*, 2330.

[41] Reif, J., *Opt. Eng.*, **1989**, *28*, 1122.

[42] Reif, J., Henyk, M. and Wolfframm, D., *Proc. SPIE*, **2000**, *3933*, 26.

[43] Stoian, R., Ashkenasi, D., Rosenfeld, A. and Campbell, E.E.B., *Phys. Rev. B*, **2000**, *62*, 13167; Stoian, R., Rosenfeld, A., Ashkenasi, D., Hertel, I.V., Bulgakova, N.M. and Campbell, E.E.B., *Phys. Rev. Lett.*, **2002**, *88*, 097603.

[44] Henyk, M., Wolfframm, D. and Reif, J., *Appl Surf. Sci.*, **2000**, *168*, 263.

[45] Roeterdink, W.G., Juurlink, L.B.F., Vaughan, O.P.H., Dura Diez, J., Bonn, M. and Kleyn, A.W., *Appl. Phys. Lett.*, **2003**, *82*, 4190.

[46] Cheng, H.-P. and Gillaspy, J.D., *Phys. Rev. B*, **1997**, *55*, 2628.

[47] Reif, J., Fallgren, H., Nielsen, H.B. and Matthias, E., *Appl. Phys. Lett.*, **1986**, *49*, 930.

[48] Reif, J., Costache, F. and Henyk, M., *SPIE Proc. Series*, **2002**, *4426*, 82.

[49] Lenzner, M., Krüger, J., Sartania, S., Cheng, Z., Spielmann, C., Mourou, G., Kautek, W. and Krausz, F., *Phys. Rev. Lett.*, **1988**, *80*, 4076.

[50] Kaiser, A., Rethfeld, B., Vicanek, M. and Simon, G., *Phys. Rev. B*, **2000**, *61*, 11437.

[51] Stuart, B.C., Feit, M.D., Herman, S., Rubenchik, A.M., Shore, B.W. and Perry, M.D., *Phys. Rev. B*, **1995**, *53*, 1749.

[52] Kelly, R. and Miotello, A., *Appl. Surf. Sci.*, **1996**, *96–98*, 205.

[53] Zhigilei, L.V., Kodali, P.B.S. and Garrison, B.J., *Chem. Phys. Lett.*, **1997**, *276*, 269.

[54] Guizard, S., D'Oliveira, P., Daguzan, P., Martin, P., Meynardier, P. and Petite, G., *Nucl. Instr. Meth. Phys. B*, **1966**, *116*, 43; Quéré, F., Guizard, S., Martin, P., Petite, G., Gobert, O., Meynardier, P. and Perdrix, M., *Appl. Phys. B*, **1999**, *68*, 459.

[55] Jeschke, H.O., Garcia, M.E., Lenzner, M., Bonse, J., Krüger, J. and Kautek, W., *Appl. Surf. Sci.*, **2002**, *197–198*, 839.

[56] Cavalleri, A., Sokolowski-Tinten, K., Bialkowski, J., Schreiner, M. and von der Linde, D., *J. Appl. Phys.*, **1999**, *85*, 3301.

[57] Cross, M.C. and Hohenberg, P.C., *Rev. Mod. Phys.*, **1993**, *65*, 851.

[58] Mayer, T.M., Chason, E. and Howard, A.J., *J. Appl. Phys.*, **1993**, *76*, 1633.

[59] Chason, E., Mayer, T.M., Kellermann, B.K., McIlroy, D.T. and Howard, A.J., *Phys. Rev. Lett.*, **1994**, *72*, 3040.

[60] Habenicht, S., Lieb, K.P., Koch, J. and Wieck, A.D., *Phys. Rev. B.*, **2002**, *65*, 115327.

[61] Sigmund, P., *Phys. Rev.*, **1969**, *184*, 383; Sigmund, P., *J. Mater. Sci.*, **1973**, *8*, 1545.

[62] Bradley, R.M. and Harper, J.M.E., *J. Vac. Sci. Technol.*, **1988**, A6, 2390.

[63] Oron, A., *Phys. Rev. Lett.*, **2000**, *85*, 2108.

[64] Oron, A., Davis, S.H. and Bankoff, S.G., *Rev. Mod. Phys.*, **1997**, *69*, 931.

[65] Bestehorn, M. and Neuffer, K., *Phys. Rev. Lett.*, **2001**, *87*, 046101.

[66] Bestehorn, M., Pototsky, A. and Thiele, U., *Eur. Phys. J.*, **2003**, *B33*, 457.

[67] Kardar, M., Parisi, G. and Zhang, Y.C., *Phys. Rev. Lett.*, **1986**, *56*, 889.

[68] Kuramoto, Y., *Chemical Oscillations, Waves, and Turbulence*, Springer, Berlin, **1984**.

[69] Cahn, J.W. and Hilliard, J.E., *J. Chem. Phys.*, **1958**, *28*, 258.

[70] Kawakami, Y. and Ozawa, E., *Appl. Phys. A*, **2000**, *71*, 453.

[71] Younkin, R.J., PhD Thesis; Harvard University, Cambridge, **2001**.

[72] Ozkan, A.M., Malshe, A.P., Railkar, T.A., Brown, W.D., Shirk, M.D. and Molian, P.A., *Appl. Phys. Lett.*, **1999**, *75*, 3716.

[73] Bolle, M. and Lazare, S., *J. Appl. Phys.*, **1993**, *73*, 3516.

Chapter 10

Three-Dimensional Micromachining with Femtosecond Laser Pulses

W. Watanabe and K. Itoh

Department of Material and Life Science, Graduate School of Engineering, Osaka University, 2-1 Yamadaoka, Suita, Osaka 565-0871, Japan

Abstract

When femtosecond laser pulses are focused inside the bulk of transparent materials, the intensity in the focal volume becomes high enough to produce permanent structural modifications. Femtosecond lasers allow us to accomplish micro-machining in three-dimensional space in transparent materials. In this chapter, we will review the fabrication of optical elements in glasses with femtosecond laser pulses, also including waveguides, couplers, gratings, diffractive lenses, memory, and microchannels.

Keywords: Coupler; Femtosecond laser pulse; Filamentation; Glass; Grating; Photonic device; Refractive-index change; Waveguide.

10.1. Introduction

Optical elements such as waveguides, couplers, and gratings are key in optical communication. Communication technology in practical use is based on optical fibers and integrated optical circuits. Optical fibers utilize one-dimensional space and planar waveguides utilize two-dimensional space. With the increasing data capacity in communications, data storage, signal, and information processing, there is an increasing need for the integration of optical elements and devices. Femtosecond laser pulses can locally induce structural change in the bulk of transparent materials and open the door to three-dimensional micromachining.

Transparent materials cannot linearly absorb visible or near-infrared light, because the energy of a single photon is below the energy required to promote

Recent Advances in Laser Processing of Materials
J. Perrière, E. Millon and E. Fogarassy (Editors)

an electron from the valence to the conduction band. Multiphoton absorption has the ability to promote electrons from the valence to the conduction band by using near-infrared light. Conventional visible and near-infrared light could not excite multiphoton absorption owing to the low photon energy at these wavelengths. The use of high-energy femtosecond lasers, however, can excite multiphoton absorption. The focal point of near-infrared femtosecond lasers can be located in the bulk of the transparent material. When intense femtosecond laser pulses are focused inside the bulk of a transparent material, the intensity in the focal volume becomes high enough to cause nonlinear absorption, which leads to localized modification in the focal volume, while leaving the surface unaffected. Using near-infrared femtosecond lasers, different types of tracks, such as void, color centers, scattering damage, and refractive-index changes have been produced in a wide variety of transparent materials. The type of track produced depends on the laser parameters (i.e. wavelength, pulse duration, energy, repetition rate), the focusing conditions, and the materials (Fig. 1).

The use of femtosecond laser pulses enables one to fabricate three-dimensional photonic structures including waveguides [1–12], couplers [13–17], gratings [18–27], and three-dimensional binary data storage [28–32], lenses [33–35], and channels [36–40]. In this chapter, we present three-dimensional micromachining and the fabrication of photonic devices in the bulk of glass with femtosecond laser pulses.

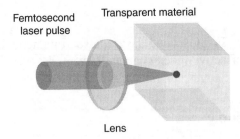

Femtosecond laser pulse   Transparent material

Lens

Figure 1: Schematic of the focusing setup used to produce structures in bulk glass using femtosecond laser pulses.

10.2. Femtosecond Laser-Induced Refractive-Index Change

10.2.1. Induction of Refractive-Index Change

In 1996, it was reported that refractive-index change was induced in the focal spot when near-infrared femtosecond laser pulses were focused inside a wide variety of glasses [1]. The induction of refractive-index change at the focal point has been reported to the order of 10^{-2} to 10^{-4}. Several mechanisms have been proposed for inducing refractive-index changes. In thermal model, energy deposited by the laser melts the material in the focal volume, and the subsequent resolidification leads to density variations in the focal region [4]. Raman spectroscopy of the modified glass indicates an increase in the number of 4- and 3-membered ring structures in the silica network, although 5- and 6-membered ring structures are dominant before laser irradiation [41,42]. Infrared reflectance spectroscopy indicates that slight changes in the Si–O–Si bond angle occur when the glass is irradiated with femtosecond pulses, leading to densification of the glass [24].

In the color center model, color center plays a role. Fluorescence spectroscopy of the modified glass shows a broad fluorescence band at 630 nm [41,42]. This result indicates the formation of non-bridging oxygen hole centers (NBOHC) by femtosecond pulses. These defects and color centers could, in principle, produce refractive-index change [1]. The color centers, however, can be annealed away by heating the glass to about 400°C while the refractive index does not revert to its original value [5]. The color centers are not considered to be responsible for refractive-index change [43]. The physical mechanism responsible for refractive-index change is still under investigation.

10.2.2. *Fabrication of Waveguides*

By translation of the focal point, waveguides can be fabricated inside the bulk glass [1–12]. Techniques for fabricating waveguides can be divided into two categories: side writing and parallel writing. Figures 2(a) and (b) show the schema of

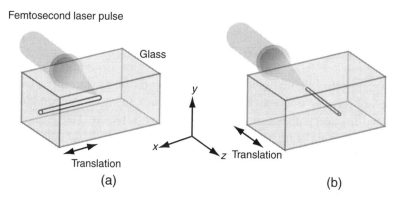

Figure 2: Fabrication of waveguides embedded in glass, (a) Side writing and (b) Parallel writing.

fabrication of waveguides embedded in glass by side writing and parallel writing, respectively. In side writing, the sample is translated perpendicularly to the laser beam. This method can thus fabricate long waveguides, but the symmetry of the core of the waveguide is broken by the intensity distribution. In particular, when the laser pulses are focused by a high-numerical aperture (NA) objective to obtain symmetry in the core, the three-dimensional volume of the sample cannot be fully accessed because of the short working distance. In parallel writing, laser pulses are focused into the sample by a low-NA lens having a long working distance, and the sample is translated parallel to the propagation axis of the laser pulses. The length of the waveguide is then restricted by the working distance of the focusing objective. Parallel writing can, however, fabricate waveguides having symmetrical cores and make full use of the three-dimensional volume. The quality of the waveguide may be degraded by the laser pulses being affected by spherical aberration at different depths. It has been demonstrated that damping losses below 0.8 dB/cm are achieved [5,6].

10.3. Refractive-Index Change by Filamentation

10.3.1. Filamentation

When a laser pulse is focused in a transparent material, nonlinear propagation of the pulse leads to filamentation, a phenomenon whereby the laser beam maintains a near-constant beam waist over many Rayleigh lengths due to a temporal balance between self-focusing and plasma defocusing. Using femtosecond lasers, different types of filamentary tracks, such as color centers, scattering damage, and refractive-index changes have been produced in silica glass [3,44–47].

10.3.2. Induction of Refractive-Index Change

Figure 3 shows the schematic of the optical setup for observation of a filament and the region of refractive-index change in silica glass [3]. Femtosecond laser pulses are focused inside synthesized silica glass. Laser pulses of 130 fs duration were generated by a Ti:sapphire amplifier at a wavelength of 800 nm and a 1 kHz repetition rate. An image of the refractive-index change was observed in the yz-plane perpendicular to the optical axis by use of a transilluminated optical microscope.

Optical images of a filament and the region of refractive-index change observed in the yz-plane are shown in Fig. 4. Figure 4(a) shows an image of a single filament during excitation by laser pulses without illumination by a halogen lamp. The laser

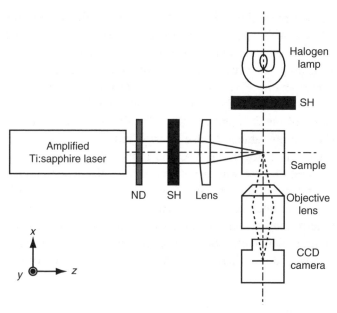

Figure 3: Schematic of the optical setup for observation of a filament and the region of refractive-index change in silica glass (ND: neutral-density filter; SH: shutter).

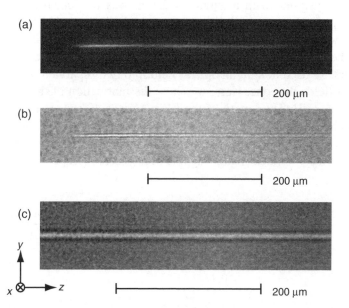

Figure 4: Relationship between a filament and the region of refractive-index change: (a) a single filament, (b) optical image of the region of refractive-index change, (c) expanded view of the region of refractive-index change.

beam was focused by a single lens with a focal length of 100 mm (NA, 0.01). Figure 4(b) shows an image of the refractive-index change induced after 2 min of exposure to focused laser pulses. The image was observed under illumination by the halogen lamp. The region of refractive-index change coincides spatially with that of the filament. Figure 4(c) shows the expanded image of the region of refractive-index change, and the profile is quite smooth.

The formation of the filament and the refractive-index change are summarized in Table 1. The column labeled 'Pulse Energy' indicates the range of formation of a single filament. The lower the NA, the longer the region of refractive-index change. The saturation time was also dependent on NA. The maximum refractive-index change and the diameter were, however, independent of NA. Refractive-index change is estimated to be 0.8×10^{-2}. To investigate thermal stability, the sample with the refractive-index change was treated at 1150°C, which is a higher temperature than the annealing point of silica glass, for 1 h. Observation through a microscope revealed that thermal treatment cured the region of refractive-index change.

10.3.3. Waveguide and Coupler

A straight waveguide with a length of 2 mm was fabricated by translating the sample at 1 µm/s along the optical axis [3]. A He–Ne beam of 632.8 nm was coupled. Figure 5 shows the near-field output image and its cross section. The near-field pattern indicated a single-mode propagation regime at the wavelength of 632.8 nm.

Couplers are used to split an optical field into two separated part and to couple two optical fields into one. Here, we show the fabrication of directional couplers in silica glass [17]. The length of the filament was 40 µm. For two-dimensional translation of the filament, the front of the filament propagates while bending, due to the previously induced refractive-index change region. After exposure, smoothly connected refractive-index change regions were observed. By combining the straight

Table 1: Summary of refractive-index change for various NAs of focusing lens

NA	Pulse energy [µJ/pulse]	Length [µm]
0.05	1.1–2.0	400–500
0.1	0.77–2.41	120–350
0.3	0.55–2.83	30–150
0.44	0.55–1.77	15–70
0.55	0.49–0.62	8–15

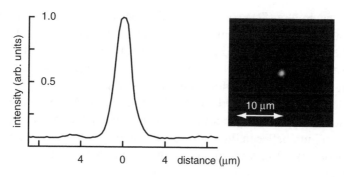

Figure 5: Waveguide output profile at 632.8 nm. A near-field image is shown on the right-hand side.

and curved waveguides, we fabricated directional couplers. Figure 6(a) shows a schema of the fabricated directional coupler. The coupler consists of two wave-guides. One is a 2 mm long, straight waveguide and the other is a curved waveguide that is connected to another straight section. The latter waveguide has 17 mm radius arcs and 0.5 mm length at one end. This straight sector is parallel to the straight waveguide with a 4 μm center-to-center distance. Figure 6(b) shows a side view of a part of the fabricated directional coupler, obtained by using a transmitted microscope.

Coupling properties were investigated by focusing a He–Ne laser beam with a wavelength of 632.8 nm into the straight waveguide and monitoring the near-field

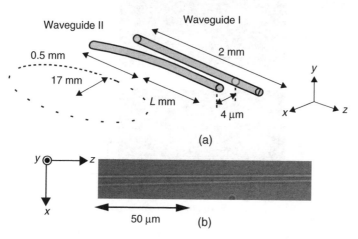

Figure 6: (a) Schematic of a directional coupler. (b) Optical image of the fabricated coupler.

output patterns. Figures 7(a) and (b) show the near-field output patterns of couplers with $L = 1.0$ mm and $L = 0.5$ mm, respectively. The gap between the two waveguides at the end plane was 11.5 μm. The left spot was the output from the straight waveguide and the right spot was from the curved waveguide. Figure 7 shows that the splitting ratios of the directional couplers with $L = 1$ mm and $L = 0.5$ mm are approximately 1:1 and 1:0.5, respectively. The cores of the couplers are close enough that the fundamental modes propagating in each core overlap partially in the cladding region between the cores. Such evanescent wave coupling can lead to the transfer of optical power from one core to another, acting as splitting of the optical power.

We will demonstrate the realization of three-dimensional directional couplers. Figure 8(a) shows the schema of a three-dimensional directional coupler consisting of three waveguides: a 2 mm long straight waveguide (waveguide I) and two curved waveguides that are connected to straight sections (waveguide II and III). The curved waveguides have arc radii of 17 mm. The straight sections of waveguide II and III were parallel to the straight waveguide at a 4 μm center-to-center separation in the x and y directions, respectively. Lengths of the straight sectors in waveguide II and waveguide III were 0.5 and 1.0 mm, respectively. The He–Ne laser beam was coupled to the straight waveguide and the near-field output patterns were monitored. Figure 8(b) shows the near-field output patterns of

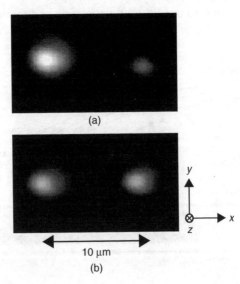

(a)

(b)

10 μm

Figure 7: Near-field pattern of coupler output at 632.8 nm for (a) $L = 0.5$ mm and (b) $L = 1.0$ mm.

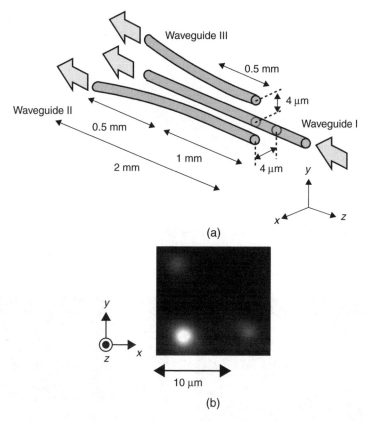

Figure 8: (a) Schematic of a three-dimensional directional coupler. (b) Near-field patterns of coupler output when coupling to a He–Ne laser.

beams from the coupler at a wavelength of 632.8 nm. The beam was split among the three waveguides at different intensities. When broadband light (450 to 700 nm) was coupled to the straight waveguide, spectra were different at the output because coupling properties are dependent on wavelength. The technique for fabricating spectroscopic couplers in glass could thus have potential applications in wavelength division, including in spectral filters.

10.3.4. Grating

The volume grating with a fine period is an important device because of its large diffraction angle, high diffraction efficiency, and high wavelength selectivity.

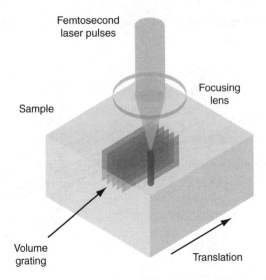

Femtosecond
laser pulses

Focusing
lens

Sample

Volume
grating

Translation

Figure 9: Schematic for fabrication of a volume grating in silica glass.

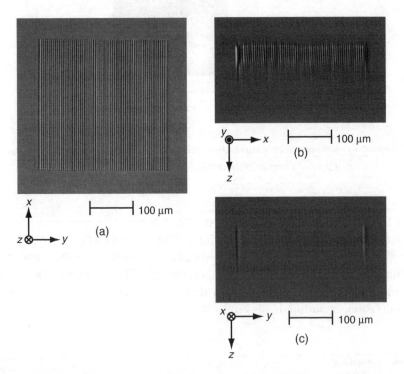

Figure 10: Optical images of the fabricated grating with the period of 5 μm and
thickness of 150 μm. The fabrication energy was 1.0 μJ and the translation speed of the
filament was 1 μm/s. (a) Top view, (b) Side view (xz-plane), (c) Side view (yz-plane).

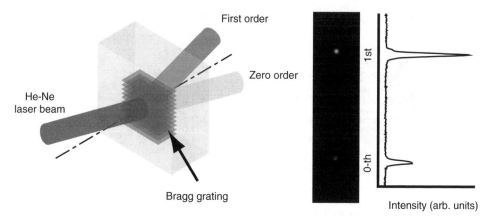

Figure 11: Schematic of the Bragg grating embedded in silica glass and its diffraction.

In this section, we report the fabrication of Bragg gratings in silica glass [20]. Figure 9 shows the schema of the fabrication of the volume grating. When the filament was scanned for 300 μm along the x-axis (perpendicular to the optical axis) at the speed of 1 μm/s, a layer of refractive-index change with a thickness of 2 μm was induced and 60 layers along the y-axis were stacked with a sample displacement. Figure 10 shows optical images of the fabricated grating with the period of 5 μm. Figure 10(a) shows the top view of the grating observed from the z-axis. Figures 10(b) and (c) are the side views that are observed from the x- and the y-axes, respectively. The diffraction efficiency of the grating was measured with a cw He–Ne laser at the wavelength of 632.8 nm as shown in Fig. 11. The angle of the He–Ne beam with respect to the grating vector was adjusted to achieve the maximum diffraction efficiency. The maximum diffraction efficiency of 74.8% was obtained when the grating had a period of 3 μm. The gratings can be used as mirrors or reflectors in silica glass.

10.3.5. Diffractive Lens

The phase-type diffractive lens, called a kinoform, has a focusing property. The lens consists of a series of disks centered at one point, forming alternating annular zones with different refractive indexes to induce phase retardation (Fig. 12(a)). This lens is usually approximated by regions having multiple refractive-index levels.

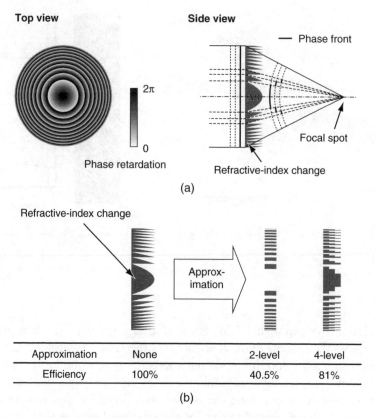

Figure 12: (a) Schematic of phase-type diffractive lens. (b) Two-level and four-level approximation of the phase-type diffractive lens.

For two-level and four-level approximation, the theoretical efficiencies are 40.5 and 81%, respectively, as shown in Fig. 12(b). The fabrication of two-level phase-type diffractive lenses was demonstrated by inducing a refractive-index change in silica glass [34]. The efficiency of this lens reached ~40% at a wavelength of 404 nm. However, the efficiency was only 17% at a wavelength of 642 nm because the length of the refractive-index region was only 30 μm, and therefore, the phase difference between the alternating zones did not reach π at the longer wavelength. To overcome these problems, multilevel phase-type diffractive lenses were fabricated by using filamentation of femtosecond laser pulses [35] (Fig. 13). A filament with a length of 30 μm was formed. Refractive-index change was induced in the phase retardant regions. The other regions were not touched. The lens consisted of 10.5 Fresnel zones. Figure 14 shows the fabricated two-level phase-type diffractive lens consisting of 4 layers. The top and side view of the lens were shown

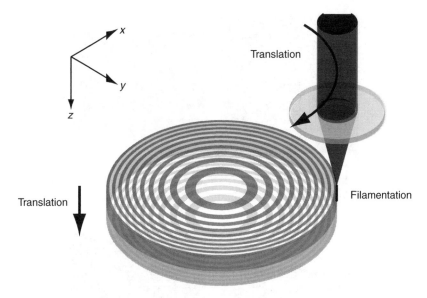

Figure 13: Schematic for fabrication of diffractive lenses in silica glass using filamentation.

in Figs 14(a) and (c). Figure 14(b) shows the magnified view of the part of Fig. 14(a). To measure the resolution limit and the efficiency, a He–Ne laser beam with a wavelength of 632.8 nm was incident upon the lens at the normal incidence. The profile of the focal spot was shown in Fig. 14(d). The resolution limit of the focal spot was 13.1 μm. The efficiency was 37.6% at a wavelength of 632.8 nm. To increase total phase retardation, two-level diffractive lenses with multiple layers were fabricated to increase the overall thickness of the diffractive lenses. The maximum efficiency of the two-level diffractive lenses was 37.6% at a wavelength of 632.8 nm. Furthermore, four-level diffractive lenses were shown to provide a maximum efficiency of 56.9%.

10.4. Void

10.4.1. Formation of Void

Tightly focusing femtosecond laser pulses with a high NA lens produced submicron-damage inside a wide variety of transparent materials. The damage appears as cavities or voids with diameters of only 200 nm to 1 μm, surrounded by densified material [28–30]. An important feature of the void is the large difference of refractive-index

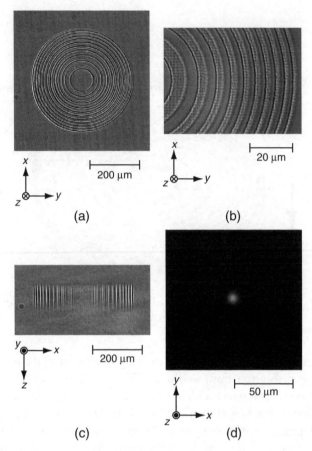

Figure 14: (a) Top view of fabricated phase-type diffractive lens. (b) Magnified image. (c) Side view. (d) Focal spot when He–Ne laser beam (632.8 nm) was incident upon the lens.

change between a cavity and the surrounding region. The dot-by-dot translation of the focal point allows three-dimensional optical data storage [28–32].

10.4.2. Binary Data Storage

The bit status in three-dimensional memory can be rewritten [31,32]. Femtosecond laser pulses were focused inside silica glass using 0.55-NA objective to create voids. Eight successive pulses were launched normally. Figures 15(a–d) show the behavior of a void that was seized and moved [31]. First, for comparison, we created three voids along the y-axis at a depth of 300 μm beneath the surface,

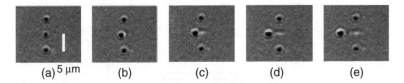

(a) 5 μm   (b)   (c)   (d)   (e)

Figure 15: (a)–(d) Side views of optical seizing and movement of a void. (e) Central void is moved towards input plane by 5 μm.

with a spacing of 5 μm between each void. Then we moved the focal point of the objective forefront of the central void and translated the focal point to the z direction by 0.5 μm with one exposure. By repeating the step, the void was moved by 5 μm, as is shown in Fig. 15(e). Using this technique, the lateral movement of the void along the axis perpendicular to the beam propagation axis was achieved by shifting the focal region of the laser pulses [32].

Let us discuss the mechanism of the motion of the void. When a femtosecond laser pulse is tightly focused, at the focus, the laser intensity becomes high enough to induce nonlinear absorption of the laser energy by the material through multiphoton and avalanche photoionization process, resulting in optical breakdown and the formation of a high-density plasma. This hot plasma explosively expands into the surrounding materials. This microexplosion occurs inside the material and results in a permanently damaged region of the void. After shifting the focal region, optical breakdown is most likely to occur near the front interface of the void. There are probable defects and easily ionized surface states near the interface that enhance the breakdown process. When optical breakdown occurs at this front interface, the debris of the new one fills in the old void. Then we observed that the void moves in an upstream direction.

10.4.3. Fresnel Zone Plate

We will present the fabrication of an amplitude-type diffractive lens by embedding voids inside silica glass. The amplitude-type diffractive lens, also called a Fresnel zone plate, consists of a series of disks centered at one point forming alternating annular zones with different transmittance of light. When either all the even or all the odd zones are blocked, the zone plate has a focusing property. The maximum theoretical diffraction efficiency at the primary focal point under plane-wave illumination is $1/\pi^2$ (about 10.1%). Such an amplitude-type diffractive lens can be realized by embedding voids in silica glass [33]. Figure 16 shows the schematic of the designed Fresnel lens. The designed Fresnel lens has the primary

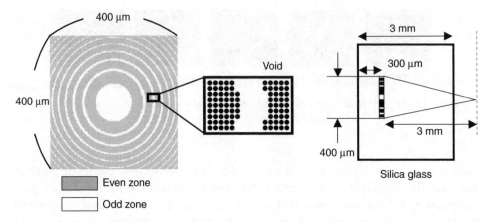

Figure 16: Schematic of fabrication of Fresnel zone plate. We constructed only the zone plate that passes the odd zones and obstructs the even zones. Even zones are fabricated by embedding voids.

focal length of 3 mm at the wavelength of 632.8 nm. The size of the zone plate is 400×400 μm. In this condition, the radius of the outer zone is 200 μm, which corresponds to eleven odd zones.

The Fresnel zone plate was fabricated by embedding voids at the depth of 300 μm beneath the sample of silica glass. The sample was displaced dot-by-dot in the xy-plane perpendicular to the laser propagation axis by steps of 1 μm. Figure 17 shows an optical image of the fabricated Fresnel zone plate by embedding voids. The voids were embedded only in the even zones. When the He–Ne laser beam is incident on the fabricated zone plate, the beam converges in on the primary focal spot on the optical axis. The measured spot size was 7.0 μm and agreed with the theoretical value of 6.1 μm. The measured diffraction efficiency was 2.0%. Compared to diffractive lens using the induction of refractive-index change, this technique is easy to fabricate, however, the diffraction efficiency is limited below the 10.1%.

10.5. Two-Beam Interference

10.5.1. Holographic Grating

A surface-relief grating is a key element in optical communication, especially in dense wavelength division multiplexing. In conventional holographic data storage, an entire page of information is stored at once as an optical interference pattern

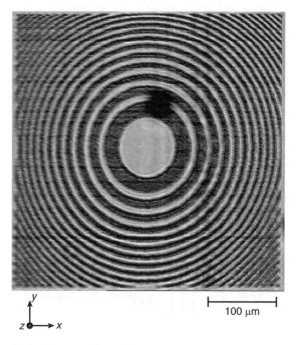

Figure 17: Optical image of the fabricated Fresnel zone plate by embedding
the two-dimensional array of voids. The image was observed under illumination
by halogen lamp.

of two coherent laser beams within a photosensitive optical medium such as photore-
fractive crystals and photopolymers. Recently, it has been shown that grating can be
encoded on the surface of nonphotosensitive glasses by two-beam interference of
a single near-infrared femtosecond laser pulse [23]. This holographic technique
has high productivity because only one femtosecond pulse is needed to form the
grating. Another significant advantage is the ready tunability of the recorded grating period by straightforward adjustment of the angle between the two incident
beams. Figure 18 shows the typical optical setup for two-beam interference. The
laser pulse was split into two equal intensity beams (Beam 1 and Beam 2) that
were redirected at approximately equal angles into the sample. The optical path
lengths were adjusted to give perfect overlap of the two beams both spatially and
temporarily to form an interferogram. The recorded grating can be read out by
either of the two recording beams when the objective was focused on the recording
plane. Using this technique, the gratings can be fabricated inside soda-lime glass
and silica glass [22,25].

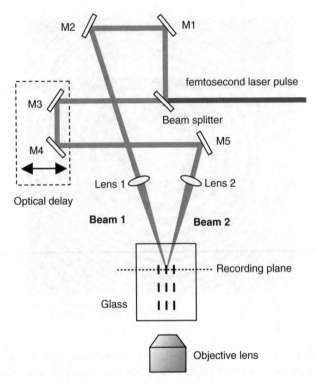

Figure 18: Top view of experimental setup for formation of gratings inside glass. The recording plane is in the *xy*-plane and through the overlapped focal points of the two incident beams (Beam 1 and Beam 2).

10.5.2. Holographic Memory

We will present experimental results of holographic data storage on the surface of fused silica, soda lime, and lead glasses by two-beam interference of a single femtosecond laser pulse [26]. A top view of the experimental schematics for recording of a data image is shown in Fig. 19(a). A laser pulse is split into reference and object beams. The Fourier transform configuration is used to record the information. The reference beam is focused by a lens L_R of 500 mm focal length. The object beam is expanded by a Galileo telescope and transmitted through a data mask. The data bearing object beam is focused or Fourier transformed by the lens L_O of 50 mm focal length. A hologram is written when the reference beam intersects with the object beam and their interference fringes are recorded in the sample. The data image can then be reconstructed by illumination of the hologram with the reference beam (Fig. 19(b)). By use of a second lens behind the sample

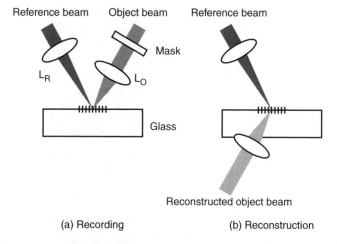

(a) Recording   (b) Reconstruction

Figure 19: Top view of experimental schematics for (a) hologram recording and (b) reconstruction of data.

to perform a second Fourier transformation, the data information can be retrieved by a CCD camera.

The experimental results are shown in Fig. 20. Figures 20(a) and (b) show recorded interferograms on silica glass and the reconstructed image, respectively. When the soda-lime glass with a lower threshold is used, more fringes can be recorded. Figure 20(c) illustrates a hologram on a soda-lime glass plate under the same experimental conditions. The reconstructed image shown in Fig. 20(d) is better than that in Fig. 20(b) on fused silica compared with dimmer spots at the four corners of the reconstructed image. Because the lead glass has a much lower threshold, a hologram is recorded with weaker energy. Figure 20(e) shows a typical microhologram on lead glass. The reconstructed data image is shown in Fig. 20(f). All nine spots are retrieved.

10.6. Microchannel

Microchannels fabricated in glass have the potential for use in microphotonics, microelectronics, microchemistry, and biology, particularly in total analysis systems (μ-TAS). To obtain a three-dimensional channel, some researchers adopted multi-step methods: the processing of glass by both femtosecond laser pulses and subsequent selective chemical etching of the modified region [36–38]. In this section, we show microchannel drilling inside silica glass with femtosecond laser pulses by in-water ablation [39,40].

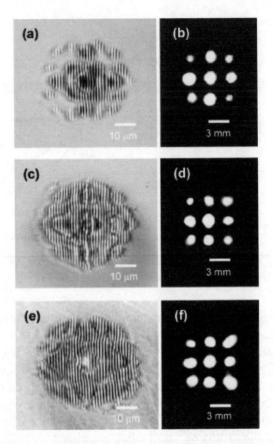

Figure 20: Holograms recorded on fused silica, soda-lime, and lead glass and their corresponding reconstructed images. (a) Hologram on fused silica glass and (b) the reconstruction data; (c) Hologram on soda-lime glass and (d) the reconstruction data; (e) Hologram on lead glass plate and (f) the reconstruction data.

Figure 21 shows a schematic of the fabrication of a straight channel from the rear surface of the sample. The laser beam was first focused on the rear surface of the sample. The ablation occurs at the interface between silica and water. After drilling begins, the focal point is translated toward the front surface step by step. Because dust or ablated particles can disperse in water, the effects of redeposition and blocking are greatly reduced compared with drilling in air. Water can flow into the channel due to the capillary effect, and a high-aspect-ratio channel without tapering can be fabricated by moving the sample. The dead-end straight channel with a diameter of 2 μm and a length of 650 μm was fabricated by focusing 3.2 μJ pulses, which corresponds to the aspect ratio of 325:1. Figure 22 shows the optical

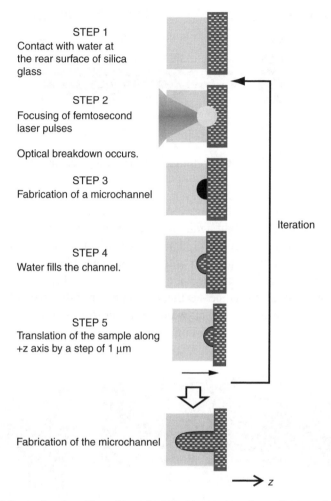

STEP 1
Contact with water at
the rear surface of silica
glass

STEP 2
Focusing of femtosecond
laser pulses

Optical breakdown occurs.

STEP 3
Fabrication of a microchannel

Iteration

STEP 4
Water fills the channel.

STEP 5
Translation of the sample along
+z axis by a step of 1 μm

Fabrication of the microchannel

z

Figure 21: Schematic of making channels from the rear surface of a sample in contact with distilled water.

image of a through-channel fabricated at the energy of 150 μJ/pulse. In this image the channel was filled with water. The channel had a the length of 1 mm and a diameter of 50 μm.

Figure 23 shows aspects of a square-wave shaped hole drilled inside silica by two-dimensional translation of the sample. The diameter of the microchannel was correspondingly reduced from 7 to 4 μm by varying the incident energy from 4 to 1 μJ. Figure 23(a) shows the optical image when water was in the microchannel; the hole was bright. Figure 23(b) shows the image after water went out; the dry

Front surface   Rear surface

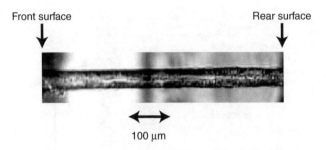

100 µm

Figure 22: Optical image of the through-channel.

channel was dark. To verify the empty structure, the dry microchannel was refilled with water and it became bright again. As water went out, the microchannel turned dark gradually. As shown in Fig. 23(c), the part of the microchannel with water was still bright, while the other part without water was black.

Figure 24 shows the optical image of a three-dimensional spiral microchannel. An important feature of microchannels is that the diameter of the channel is not tapered. This microchannel drilling using in-water ablation from the rear surface provides us with the freedom of fabricating complex three-dimensional microchannels.

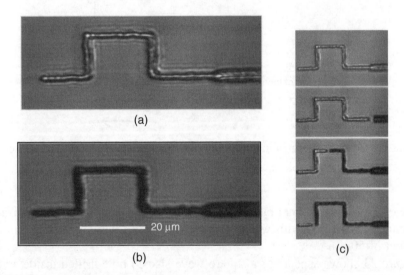

(a)

20 µm

(b)   (c)

Figure 23: A square-wave shaped hole drilled by translating the sample in different directions, (a) when water was in the three-dimensional microchannel, the microchannel was bright, (b) after water went out, the dry microchannel was dark, (c) the video clips of the drying process of the refilled microchannel.

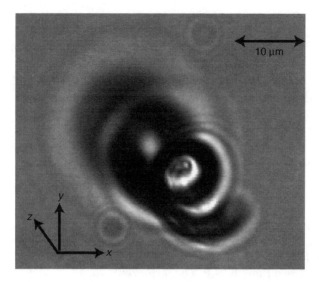

Figure 24: Optical image of a spiral microchannel.

10.7. Summary

We presented three-dimensional micromachining in the bulk of glass with femto-second laser pulses, and demonstrated the fabrication of micro optical component in glasses, including waveguides, couplers, gratings, diffractive lenses, memory, and microchannels.

Acknowledgments

We wish to thank K. Yamada, Y. Li, T. Toma, D. Kuroda, T. Asano, T. Ishizuka, and Y. Iga for assistance with the experiments and J. Nishii (Kansai Center, National Institute of Advanced Industrial Science and Technology) for useful discussions.

References

[1] Davis, K.M., Miura, K., Sugimoto, N. and Hirao, K., *Opt. Lett.*, **1996**, *21*, 1729.
[2] Miura, K., Qiu, J., Inouye, H., Mitsuyu, T. and Hirao, K., *Appl. Phys. Lett.*, **1997**, *71*, 3329.
[3] Yamada, K., Watanabe, W., Toma, T., Nishii, J. and Itoh, K., *Opt. Lett.*, **2001**, *26*, 19.
[4] Schaffer, C.B., Brodeur, A., Garcia, J.F. and Mazur, E., *Opt. Lett.*, **2001**, *26*, 93.

[5] Will, M., Nolte, S., Chichkov, B.N. and Tuennermann, A., *Appl. Opt.*, **2002**, *41*, 4360.
[6] Nolte, S., Will, M., Burghoff, J. and Tuennermann, A., *Appl. Phys. A*, **2003**, *77*, 109.
[7] Cerullo, G., Osellame, R., Taccheo, S., Marangoni, M., Polli, D., Ramponi, R., Laporta, P. and De Silvestri, S., *Opt. Lett.*, **2002**, *27*, 1938.
[8] Salinia, A., Nguyen, N.T., Nadeau, M.-C., Petit, S., Chin, S.L. and Vallée, R., *J. Appl. Phys.*, **2003**, *93*, 3724.
[9] Chan, J.W., Huser, T.R., Risbud, S.H., Hayden, J.S. and Krol, D.M., *Appl. Phys. Lett.*, **2003**, *82*, 2371.
[10] Sikorski, Y., Said, A.A., Bado, P., Maynard, R., Florea, C. and Winick, K.A., *Electron. Lett.*, **2000**, *36*, 228.
[11] Osellame, R., Taccheo, S., Cerullo, G., Marangoni, M., Polli, D., Ramponi, R., Laporta, P. and De Silvestri, S., *Electron. Lett.*, **2002**, *38*, 964.
[12] Tokuda, Y., Saito, M., Takahashi, M., Yamada, K., Watanabe, W., Itoh, K. and Yoko, T., *J. Non-Cryst. Solids*, **2003**, *326–327*, 472.
[13] Homoelle, D., Wielandy, W., Gaeta, A.L., Borrelli, E.F. and Smith, C., *Opt. Lett.*, **1999**, *24*, 1311.
[14] Streltsov, M. and Borrelli, N.F., *Opt. Lett.*, **2001**, *26*, 42.
[15] Minoshima, K., Kowalevicz, A.M., Hartl, I., Ippen, E.P. and Fujimoto, J.G., *Opt. Lett.*, **2001**, *26*, 1516.
[16] Minoshima, K., Kowalevicz, A.M., Ippen, E.P. and Fujimoto, J.G., *Opt. Express*, **2002**, *10*, 645.
[17] Watanabe, W., Asano, T., Yamada, K., Itoh, K. and Nishii, J., *Opt. Lett.*, **2003**, *28*, 2491.
[18] Sudrie, L., Franco, M., Prade, B. and Mysyrowicz, A., *Opt. Commun.*, **1999**, *171*, 279.
[19] Yamada, K., Watanabe, W., Nishii, J. and Itoh, K., *J. Appl. Phys.*, **2003**, *93*, 1889.
[20] Yamada, K., Watanabe, W., Kintaka, K., Nishii, J. and Itoh, K., *Jpn. J. Appl. Phys.*, **2003**, *42*, 6916.
[21] Florea, C. and Winick, K.A., *J. Lightwave Technol.*, **2003**, *21*, 246.
[22] Li, Y., Watanabe, W., Yamada, K., Shinagawa, T., Itoh, K., Nishii, J. and Jiang, Y., *Appl. Phys. Lett.*, **2002**, *80*, 1508.
[23] Kawamura, K., Ogawa, T., Sarukura, N., Hirano, M. and Hosono, H., *Appl. Phys. B*, **2000**, *71*, 119.
[24] Kawamura, K., Sarukura, N., Hirano, M. and Hosono, H., *Appl. Phys. Lett.*, **2001**, *78*, 1038.
[25] Kawamura, K., Hirano, M., Kamiya, T. and Hosono, H., *Appl. Phys. Lett.*, **2002**, *81*, 1137.
[26] Li, Y., Watanabe, W., Itoh, K. and Sun, X., *Appl. Phys. Lett.*, **2002**, *81*, 1952.
[27] Li, Y., Yamada, K., Ishizuka, T., Watanabe, W., Itoh, K. and Zhou, Z., *Opt. Express*, **2002**, *10*, 1173.
[28] Glezer, E.N., Milosavljevic, M., Huang, L., Finlay, R.J., Her, T.-H., Callan, J.P. and Mazur, E., *Opt. Lett.*, **1996**, *21*, 2023.
[29] Glezer, E.N. and Mazur, E., *Appl. Phys. Lett.*, **1997**, *71*, 882.
[30] Watanabe, M., Sun, H.B., Juodkazis, S., Takahashi, T., Matsuo, S., Suzuki, Y., Nshii, J. and Misawa, H., *Jpn. J. Appl. Phys.*, **1998**, *37*, L1527.

[31] Watanabe, W., Toma, T., Yamada, K., Nishii, J., Hayashi, K. and Itoh, K., *Opt. Lett.*, **2000**, *25*, 1669.

[32] Watanabe, W. and Itoh, K., *Opt. Express*, **2002**, *10*, 603.

[33] Watanabe, W., Kuroda, D., Itoh, K. and Nishii, J., *Opt. Express*, **2002**, *10*, 978.

[34] Bricchi, E., Mills, J.D., Kazansky, P.G., Klappauf, B.G. and Baumberg, J.J., *Opt. Lett.*, **2002**, *27*, 2200.

[35] Yamada, K., Watanabe, W., Li, Y., Itoh, K. and Nishii, J., *Opt. Lett.*, **2004**, *29*, 1846.

[36] Kondo, Y., Qiu, J., Mitsuyu, T., Hirao, K. and Yoko, T., *Jpn. J. Appl. Phys.*, **1999**, *38*, L1146.

[37] Marcinkevicius, A., Juodkazis, S., Watanabe, M., Miwa, M., Matsuo, S., Misawa, H. and Nishii, J., *Opt. Lett.*, **2001**, *26*, 277.

[38] Masuda, M., Sugioka, K., Cheng, Y., Aoki, N., Kawachi, M., Shihoyama, K., Toyoda, K., Helvajian, H. and Midorikawa, K., *Appl. Phys. A*, **2003**, *76*, 857.

[39] Li, Y., Itoh, K., Watanabe, W., Yamada, K., Kuroda, D., Nishii, J. and Jiang, Y., *Opt. Lett.*, **2001**, *26*, 1912.

[40] Iga, T., Ishizuka, T., Watanabe, W., Li, Y., Nishii, J. and Itoh, K., *Jpn. J. Appl. Phys.*, **2004**, *43*, 4207.

[41] Chan, J.W., Huser, T.R., Risbun, S. and Krol, D.M., *Opt. Lett.*, **2001**, *26*, 1726.

[42] Chan, J.W., Huser, T.R., Risbun, S. and Krol, D.M., *Appl. Phys. A*, **2003**, *76*, 367.

[43] Streltsov, A.M. and Borrelli, N.F., *J. Opt. Soc. Am. B*, **2002**, *19*, 2496.

[44] Efimov, O.M., Gabel, K., Garnov, S.V., Glebov, L.B., Grantham, S., Richardson, M. and Soileau, M., *J. Opt. Soc. Am. B*, **1998**, *15*, 193.

[45] Sudrie, L., Couairon, A., Franco, M., Lamouroux, B., Prade, B., Tzortzakis, S. and Mysyrowicz, A., *Phys. Rev. Lett.*, **2002**, *89*, 186601.

[46] Nguyen, N.T., Saliminia, A., Liu, W., Chin, A.L. and Vallée, R., *Opt. Lett.*, **2003**, *28*, 1591.

[47] Ashcom, J.B., Schaffer, C.B. and Mazur, E., *Proc. of SPIE*, **2002**, *4633*, 107.

Chapter 11

Laser Crystallization of Silicon Thin Films for Flat Panel Display Applications

A.T. Voutsas

Sharp Laboratories of America, 5700 NW Pacific Rim Blvd. Camas, WA 98607, USA

Abstract

This chapter presents the science and technology of Si crystallization by laser annealing. Laser crystallization has been widely accepted to provide the best Si-crystal quality for the fabrication of high performance Thin-Film-Transistors for advanced flat-panel displays. The chapter begins with a historical, brief background in crystallization technology and the fundamentals of laser–matter interaction, as it applies to annealing. An overview of laser equipment technologies is presented. The future outlook on such technologies is discussed, with particular emphasis on new, promising equipment for laser crystallization. The physics of Si transformation by laser annealing is discussed in detail. The evolution of Si microstructure is explained in the context of super-lateral-growth (SLG). Approaches to control SLG are discussed and a model is used to simulate the thermal environment during crystallization and the concomitant mechanisms of nucleation and lateral growth. Based on experimental data and model predictions, a number of practical implementations to achieve lateral growth with high productivity are proposed. An extensive review of characteristics of poly-Si TFTs, fabricated by various laser annealing technologies, is presented and used to draw conclusions between Si material quality and device performance. Such conclusions point to the advantages and disadvantages of current laser annealing schemes and elucidate the remaining challenges in the area of laser crystallization. Such challenges include the development of methods and processes to enable precise control of poly-Si material characteristics (such as defect density and crystallographic orientation) to achieve SOI-like, poly-Si TFT device performance.

Recent Advances in Laser Processing of Materials
J. Perrière, E. Millon and E. Fogarassy (Editors)

Keywords: Classical nucleation theory (CNT); Flat panel displays (FPDs); Gaussian profile; Interface response function; Laser Beam Shaping; Laser Crystallization; Liquid–Solid Interface; Location Control Crystallization; Metal induced crystallization (MIC); MOSFET; Nucleation and Crystal Growth Modeling; Pixel switching; Polycrystalline Silicon Thin-Film-Transistor (poly-Si TFT); Top-hat profile.

11.1. Introduction

Polysilicon thin-film-transistors (TFT) are the key building blocks for active-matrix-driven flat panel displays (FPDs). Many studies have demonstrated the ability of poly-Si based transistors to support a variety of functions beyond pixel switching, which has been the traditional role of TFTs in FPD applications. Poly-Si material enables the design of smaller TFTs that offer higher current and faster switching characteristics. As a result, pixel-driving circuits can be monolithically integrated on the display substrate. Such integration not only reduces the amount of external interconnections to the panel, but also improves the form-factor of the resulting display. The improved performance of poly-Si TFTs is further expected to yield even deeper levels of component integration that will enable the fabrication of unique display systems. In other words adoption of poly-Si material and technology could mark a paradigm shift in the display industry from commodity products to value-added display systems.

The fabrication of poly-Si TFTs bears many similarities to MOSFET process flow. One fundamental difference, however, exists with regards to the maximum processing temperature allowable during fabrication. Whereas the melting point of Si is the approximate upper limit in MOSFETs, the temperature in poly-Si TFT fabrication is typically constrained to less than 600–650°C, due to the heat-sensitive nature of commercially available, display grade, glass substrates. This temperature constraint has significant implications on the quality of the device active layer (poly-Si layer). The quality of a poly-Si film can be generically described as a function of the poly-Si grain size, the defect density (both within grains and at grain boundaries) and the uniformity of the microstructure over a large area. Such attributes are strongly affected by the process by which the poly-Si layer is formed (crystallization process). In addition, the complexity of the crystallization process relates directly to its cost (simpler processes being the most desirable).

High quality poly-Si microstructure is needed for the fabrication of high-quality poly-Si TFTs. The crystallization process is a very critical step of the TFT fabrication process, as it needs to satisfy conflicting requirements on material quality and cost and, at the same time, comply with the thermal-budget constraints imposed by the display substrate. Historically, solid-phase crystallization (SPC) was the first technology to produce poly-Si films for display applications, followed by the development of laser-annealing crystallization (LC) [1]. Both of the technologies evolved significantly over the past 20 years with a variety of spin-offs that aimed at improving different features of the poly-Si crystallization process and/or the poly-Si microstructure.

This chapter discusses primarily Si crystallization by laser irradiation, as this technology has been widely accepted to provide the best Si-crystal quality. Section 11.2 begins with a historical, brief background in SPC and continues with the fundamentals of laser–matter interaction, as it applies to annealing. An overview of laser equipment technologies is further presented. The future outlook on such technologies is discussed, with particular emphasis on new, promising equipment for laser crystallization. Section 11.3 discusses in detail the physics of Si transformation by laser annealing. Both conventional and advanced laser-crystallization processes are discussed in this section. The concept of controlled super-lateral-growth (SLG) is presented and used to explain the details of the laterally grown poly-Si microstructure. A method to model the thermal environment during laser annealing and simulate lateral growth and the resulting microstructure is presented and used to evaluate the impact of various processing options on lateral growth. Section 11.3 closes with the discussion of practical implementation schemes for lateral crystallization. Section 11.4 discusses the characteristics of poly-Si TFTs fabricated with advanced laser-annealing process. A correlation between material quality and device performance is established and used to assess the impact of various material engineering options. Finally, in Section 11.5 remaining challenges in the area of Si laser annealing are presented and discussed. Current challenges include the development of methods and processes to enable precise control of poly-Si material characteristics (such as defect density and crystallographic orientation) to achieve SOI-like, poly-Si TFT characteristics.

11.2. The Evolution in Poly-Si Crystallization Technology

Within the context of poly-Si TFT technology, 'crystallization' refers to the step (or steps, if more than one) that are required to prepare the poly-Si material, which will be used as the TFT active layer in the final device, from the as-deposited, thin Si film. Solid-phase crystallization (SPC) has served that purpose, before laser-based crystallization was recognized as a promising alternative to achieve substantially improved crystal quality. A brief introduction to SPC is presented first to provide the context and the motivation behind the development of laser crystallization technology.

11.2.1. Solid-Phase Crystallization (SPC)

The most direct method of obtaining poly-Si films, from initially amorphous precursor-Si films, is via SPC in a furnace environment. Amorphous silicon is a thermodynamically metastable phase, possessing a driving force for transformation to polycrystalline phase, given sufficient energy to overcome the initial energy barrier. Solid-phase crystallization can be accomplished within a wide annealing temperature range that requires a similarly wide range of annealing times (i.e. time required for complete transformation of the precursor Si film to poly-Si). The relationship between annealing temperature and annealing time, however, is not unique. In other words, depending upon the microstructural details of the precursor-Si film, different annealing times have been observed at the same annealing temperature [2]. A key factor, affecting crystallization, is the nucleation rate in the precursor-Si film. The nucleation rate is strongly influenced by the selected deposition method and conditions, which impact the structural order in the precursor film and its ability to form supercritical nuclei (i.e. stable nuclei), when subjected to thermal annealing [3,4].

Based on the above transformation scenario, larger grain size relates to longer crystallization time and vice-versa. For practical applications, the crystallization time corresponding to average grain size exceeding 0.5–1.0 μm may be prohibitingly long. The typical SPC poly-Si microstructure is characterized by a large density of structural defects (typically twin-boundaries and intra-grain microtwin defects). The result of this high grain-defect density is a saturation in the electrical performance of poly-Si TFTs, fabricated with such poly-Si films, with grain size larger than approximately 0.3–0.5 μm. Therefore, standard SPC technology can only produce poly-Si TFTs of mediocre performance. An improvement in both the SPC poly-Si microstructure, and the temperature–time parameters of the phase-transformation process can be obtained by employing SPC process in the presence of a metal catalyst [5]. In that case, the enhancement in the grain growth is attributed to an interaction of the free electrons of the metal with covalent Si bonds at the growing interface [6]. The growth enhancement has been reported for both elemental metals and metallic silicides. The nickel silicide system has received considerable attention for this application, presumably due to the very close lattice parameter match of the cubic crystal disilicide structure to c-Si (Δ of -0.4%). In addition to Ni, other metals have been investigated as far as their effectiveness in enhancing Si-crystal growth. These include Au, Al, Sb, and In, which form eutectics with Si or Pd and Ti which form silicides with Si [7–9]. However, in all of these cases several issues were reported pertaining to the actual enhancement of the growth rate at sufficiently low-crystallization temperatures (i.e. 500–550°C), and the incorporation of metal impurities in the TFT active layer. As a result, today

Ni remains the undisputed metal of choice for silicide-assisted crystallization. In recent years, another variation to Metal Induced Crystallization (MIC) process has been proposed, to boost the Si-crystal growth rate and allow the additional reduction of the crystallization temperature [10]. In this case an electric field, superimposed on the sample during MIC process, was found to enable reduction of the crystallization temperature to as low as 380°C. Different models have been invoked to explain the increase in the crystal growth rate under the applied electric field [11].

Figure 1 shows a plot of the crystallization time versus crystallization temperature, as a function of various types of SPC technology. The mobility of the resulting poly-Si TFTs are also shown on the plot. These data illustrate that the advances in SPC, made possible by the enhanced crystal growth, due to trace amount metals in silicon, has enabled a tremendous improvement in poly-Si TFT performance, as well as a reduction in the crystallization temperature–time requirements. Mobility values in the order of ~100 cm^2/Vs are sufficient to enable partial integration

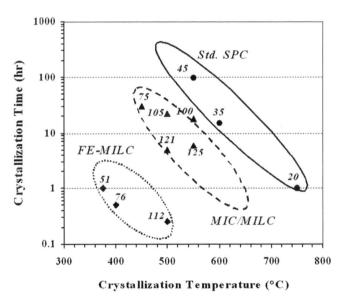

Figure 1: Plot of the crystallization time versus crystallization temperature, as a function of the type of SPC technology. The numbers, within the plot area, denote the mobility of poly-Si TFTs made by the relevant SPC technology. Acronyms are defined as follows: SPC stands for 'Solid-Phase Crystallization', MIC stands for 'Metal Induced Crystallization', MILC stands for 'Metal Induced Lateral Crystallization' and FE-MILC stands for 'Field-Enhanced MILC'.

of driving circuits on the display substrate. However, additional improvements in performance are required to increase the level of on-panel integration and enable the fabrication of advanced display systems. Laser annealing has been investigated as an alternative crystallization technology with the potential of meeting such performance demands.

11.2.2. Laser Crystallization Technology

The coupling of laser radiation to a material is very sensitive to the laser wavelength and the state of the material. The state of the material, in this case, refers to its optical properties, which for energy deposition are more conveniently expressed by the reflectivity (R) and the absorption coefficient (α), or its inverse, the absorption depth (d). These variables relate to the more common optical parameters n and k by well-known relations (see Equations (1) and (2) for normal incidence – λ is the incident light wavelength).

$$R = \left[(n - 1)^2 + k^2 \right] \Big/ \left[(n + 1)^2 + k^2 \right] \tag{1}$$

$$\alpha = 1/d = 4\pi k / \lambda \tag{2}$$

If light of intensity I_0 (in W/cm^2) is normally incident on the surface of a sample, then the power density P (in W/cm^3) absorbed at depth z (from the surface) is given by Equation (3). Laser fluence F (J/cm^2), defined as the time integral of the intensity I_0, is also often used to characterize a laser beam.

$$P(z,t) = I_0(t) \cdot \left[(1 - R) \cdot \alpha \right] \cdot e^{-az} \tag{3}$$

Figure 2 shows the absorption coefficient of Si as a function of wavelength. Silicon absorbs strongly in the UV region, whereas its absorption capacity declines by more than two orders of magnitude in the near-IR. For thin film applications, involving exposing a thin Si film to laser irradiation, high absorption coefficient is preferable to promote effective utilization of the incident energy. This is easily understood by comparing the absorption depth (d^{-1}) to the heat diffusion length $(2D\tau_p)^{0.5}$ ($D = k/\rho c_p$, where k is thermal conductivity, ρ is density, c_p is heat capacity, τ_p is the laser pulse duration).

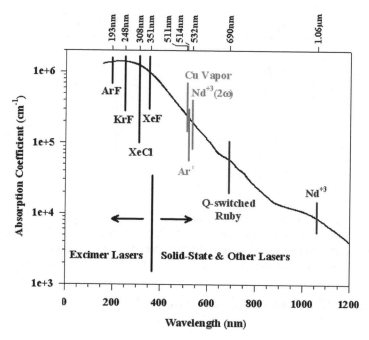

Figure 2: Si absorption coefficient versus wavelength. Light absorption in Si is maximized in the UV region, which contains most of the excimer-laser wavelengths of commercial interest. Solid-state and other types of lasers that have been proposed for Si crystallization (featuring wavelengths in the visible range) are also noted in the figure.

Figure 3 illustrates the temperature and pulse intensity profiles within the laser irradiated volume, corresponding to the two limiting cases (i.e. absorption depth much smaller or much larger than the thermal diffusion length). Small absorption depth implies heavy concentration of the laser energy very close to the irradiated surface. This is desirable for thin film heating, as in the case of poly-Si TFT technology. The thickness (t_f) of poly-Si films is less than 100 nm and typically 30–50 nm. Therefore, the laser energy should be ideally absorbed within this thin slice of Si material for effective utilization of the laser output power.

It is instructive to estimate the approximate surface melt threshold (i.e. laser fluence required to bring the surface temperature of the irradiated sample to the melting point) for the two limiting cases discussed in Fig. 3. The general heat conduction equation (Fick's law), which describes the heating dynamics, can be simplified to express the laser fluence at the surface melt condition as (for more

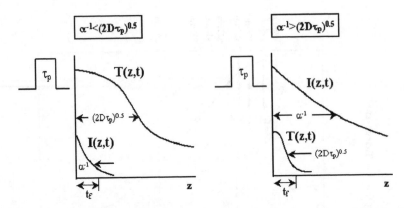

Figure 3: Schematic of the temperature and pulse intensity profiles within the laser irradiated material for the case of (left) absorption depth smaller than the thermal diffusion length and (right) absorption depth larger than the thermal diffusion length. In both cases, (t_f) represents the thickness of the film closest to the surface.

mathematical details, the reader may consult [12]):

$$F_M = \frac{\Delta T \cdot \rho \cdot c_p \left(2D\tau_p\right)^{1/2}}{\left(1 - R\right)} \quad \text{when } \alpha^{-1} < \left(2D\tau_p\right)^{1/2} \tag{4a}$$

$$F_M = \frac{\Delta T \cdot \rho \cdot c_p \cdot \alpha^{-1}}{\left(1 - R\right)} \quad \text{when } \alpha^{-1} > \left(2D\tau_p\right)^{1/2} \tag{4b}$$

where, $\Delta T = T_M - T_o$ (T_M, melting temperature; T_o initial temperature) and R is surface reflectivity (the rest of the symbols as defined before). Table 1 shows typical thermophysical parameters for silicon. Note that for accurate computations, the effect of temperature need be considered, as it will be discussed in later sections.

Table 1: LTPS TFTs whose channels were crystallized by laser processing

Parameter	Units	Fujitsu	Stuttgart	LG Philips	Sony
Laser type	–	DPSS	DPSS	Excimer	Excimer
P-type V_{th}	volts	0.1	−1.8	−3.4	–
P-type mobility	cm²/(Vsec)	200	217	140	–
N-type V_{th}	volts	−0.1	0.0	1.2	~3
N-type mobility, μ_n	cm²/(Vsec)	566	467	290	~250
N-type S-slope	dec/volt	0.14	0.5	0.41	0.16
N-type on-off ratio	–	~10^7	~10^7	>10^7	~10^7

However, for the first order calculations presented here, the temperature effect has been neglected.

Figure 4 shows the relationship between the surface melt threshold fluence and the Si absorption coefficient, as computed from Equations (4a) and (4b). As can be intuitively expected, the fluence requirement decreases significantly with increasing absorption coefficient. This is true in the region of validity of Equation (4b). In contrast, the effect saturates for relatively high absorption coefficient values – i.e. in the region of validity of Equation (4a). In the latter region, the laser pulse duration is the parameter of importance. In this case, the threshold fluence scales proportionally to $(\tau_p)^{1/2}$ – i.e. shorter pulse duration corresponds to lower surface

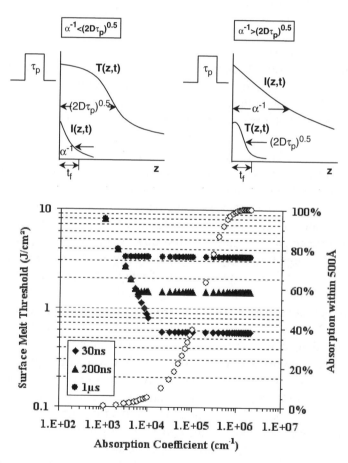

Figure 4: Relationship between Si absorption coefficient and surface melt threshold (left y-axis), as a function of pulse duration. The right y-axis (curve with open symbols) shows the corresponding efficiency in laser absorption, within the typical 500 Å-thick Si film.

melt threshold fluence. Minimizing the critical surface fluence is important from the practical point of view of laser beam size, as will be later discussed. However, given the thickness constrain in thin-film technology, a further consideration is the efficiency of the light absorption process. This point is illustrated on the right y-axis of Fig. 4, which plots the (%) of beam absorption, within a 500 Å-thick Si film (typical thickness in TFT technology), as a function of the Si absorption coefficient. Based solely on arguments of surface melt threshold fluence, it can be argued that absorption coefficients of as low as 10^4 cm^{-1} are usable (i.e. corresponding to near-IR wavelengths). However, when the efficiency factor is added, it becomes apparent that the absorption coefficient needs to exceed 10^5 cm^{-1} to guarantee an absorption efficiency of >50–60%. Ideally, the absorption coefficient should be $>7 \times 10^5$ cm^{-1} to enable efficiency of $>95\%$.

11.2.2.1. Excimer Lasers

Typical excimer lasers operate in pulse mode, at frequencies around 300 Hz, with pulse duration in the range of 10–50 ns. The energy output of TFT production excimer lasers (i.e. Lambda Physiks STEEL series) is in the order of 0.6–2 J [13]. It should be noted that very powerful XeCl excimer lasers, developed by the European maker SOPRA, have also been applied to poly-Si annealing [14]. In that case, the energy of the laser is ~15 J, the pulse duration ~220 ns and the discharge frequency ~1–5 Hz. It is now well understood that the pulse-to-pulse repeatability of the excimer laser is the most important key attribute of the laser equipment, followed by the discharge frequency, output energy and pulse duration. It is convenient to use a simple classification scheme to understand the impact of each laser parameter to (a) productivity and (b) film quality. In that sense, pulse-to-pulse repeatability, discharge frequency and output energy affect primarily productivity, whereas pulse duration affects primarily the film quality. As productivity tends to be the number one priority in manufacturing operations, laser equipment featuring excellent repeatability and high discharge frequency (with reasonable output energy) are preferable. The pulse-to-pulse repeatability of state-of-the-art laser equipment configured for Si annealing application is currently in the 2–3% total range. This number is expected to decline further to ~1–2% in new equipment releases.

A major challenge to crystallization by any laser is extending beam coverage uniformly across large areas. An excimer laser beam comprises of many spatial modes, so it can only be focused to hundreds of microns. Excimer lasers are also spatially non-uniform. Homogenization optics partially helps, but it introduces high transmission losses and interference fringes from overlapping beams. Another limitation is that when projection system is used to image the laser beam, the projection system optics can practically be made only so large (largest projection lens currently made commercially for a laser-annealing application is ⌀30 mm).

The primary wavelength of most excimer lasers resides in the near-ultraviolet (UV) spectrum, where silicon absorption is nearly independent of wavelength, temperature and state of matter (solid or liquid). The main disadvantage is that glass optics cannot clearly transmit near-UV light, so expensive optical elements must be used.

11.2.2.2. Diode-Pumped Solid-State Lasers

The expense of excimer laser processing is a large burden for manufacturers of commodity devices. The large costs are mostly associated with the massive size of excimer systems and their poor stability. Consequently, it is natural to consider if diode-pumped solid-state (DPSS) lasers might cut costs substantially [15].

A distinctive characteristic of excimer lasers is its medium, a mixture of inert and halogen gases. Therefore, a large volume is required to generate ample light flux, and high voltage is needed for photon excitation. Being a disordered material, the gas contributes a partial incoherence to the laser beam. In contrast, the lasing medium of DPSS lasers is typically a polycrystalline ceramic. The solid density means smaller volumes and voltage thresholds, whereas the crystallinity of the medium improves coherence. For quasi-CW (continuous wave) applications, high gain and fleeting excitation lifetimes produce repetition rates up to the order of 100 kHz. The main doubts about DPSS lasers are small power output, which suggests slow throughput, and challenges in stretching and homogenizing the initial beam. Throughput costs can be estimated for a commercial laser system by accounting for the time needed to scan a panel plus time for various motion adjustments. Using typical parametric values, the result indicates that a combination of two, 30 W, DPSS lasers could compete with a typical excimer system. An alternative scheme involves the utilization of multiple DPSS laser units in a configuration that enables crystallization of narrow Si strips on the glass substrate, where high-end circuits are to be subsequently fabricated [16].

Beam shaping in DPSS lasers is particularly problematic. Although its high coherence enables focus to a few microns (using diffractive optics), it impedes beam shaping. The initial output beam of a DPSS laser has a circular cross section and its intensity distribution is Gaussian. Simply stretching it along one axis maintains the Gaussian profile all around, producing in an oval-shaped spot, which is difficult to utilize effectively. To improve efficiency, forming optics are used to make the beam more rectangular. Along the long axis, homogenization is necessary and a top-hat profile is ideal.

The peak-to-peak variation for a typical DPSS laser is roughly 1 to 2%. However, as stated previously, for current excimer lasers, the variation is also typically better than of 3%. Hence, the original advantage of DPSS lasers in pulse-to-pulse stability seems to have been almost neutralized by newer excimer lasers, developed specifically for annealing applications.

For DPSS lasers, its primary wavelength is typically near 1.06 μm, so its frequency needs to be doubled (~532 nm green light) or tripled (~355 nm near-UV light) to affect silicon significantly. Unfortunately, each multiple increment in frequency approximately halves power efficiency. Absorption at visible wavelengths is strongly temperature dependent and changes substantially during melting. Therefore, process variation may be larger for green beams. Conversely, green beams are safer to use, and the output power would be easier to increase (than frequency-tripled beams).

11.2.2.3. Outlook in Laser Technology

Representative TFT results from devices with channels crystallized by either an excimer or DPSS laser are shown in Table 1. The evidence indicates that DPSS lasers can produce TFTs that match or exceed general target specifications for LCD backplanes. Comparison of approximate baseline costs between DPSS and excimer laser systems favors the former, although in these calculations only the laser itself was considered, as opposed to the complete system [17]. Based on such simplified analysis DPSS lasers appear more cost effective than excimer lasers, provided that beam shaping and power output assumptions could be met.

Beyond cost, the most significant advantage of DPSS lasers is the high repetition rate, while their most fateful disadvantages are the high beam coherence and the low output power. On the other hand, excimer lasers favor good beam properties for homogenization and reasonable power output for sufficient beam coverage. Therefore, the opportunity to combine high repetition rate with an incoherent beam and a high lasing-power output will be very advantageous to increase laser productivity for crystallization process. This is the solution that has been proposed by Cymer, Inc. [18]. This company has recently entered the laser-annealing equipment arena by adopting their unique excimer-laser technology, which has been almost exclusively applied in advanced, projection-type, photolithography systems. Table 2 provides a comparison between excimer-laser technology by Cymer and by the current market leader in laser-annealing equipment (JSW). Cymer's main strengths appear to be its remarkable pulse-to-pulse stability, high power output, high repetition rate and long pulse duration (using an integrated pulse-extension system that minimizes optical losses). The last two attributes enable the utilization of this laser tool in a *modus operandi* conducive to high-quality lateral growth, without a concomitant loss in productivity.

11.3. Si Transformation by Laser Crystallization (LC)

Analysis of the transformation scenarios associated with laser annealing shows that one can categorize crystallization of Si films in terms of three transformation regimes (occurring at low, intermediate and high laser fluence, respectively) [19,20].

Table 2: Comparison of excimer-laser annealing equipment technology
by JSW and Cymer

	JSW	Cymer
Wavelength	308 nm (XeCl)	351 nm (XeF)
Power output	500 W	600 W
Repetition rate	500 Hz	6000 Hz
Pulse-to-pulse stability	±2%	±1%
Intrinsic pulse duration	28 ns	200–240 ns
Final pulse duration	200–240 ns (use external pulse extender)	Same
Beam formation	Use mask and 5:1 projection optics	Line beam/No mask
Optical efficiency	30%	40%
Beam size	1.2 mm width × 25 mm length (length limited by diameter of available projection lens)	5–10 μm width × 450–730 mm length (no size limitation on long axis in the absence of projection optics)
Stepping pitch	0.6 mm	2 μm
Throughput (4th-gen sub/h)	18	30
Target poly-Si material quality	Variable, but primarily configured for medium quality poly-Si ('2-shot' process)	Variable, but primarily configured for high quality poly-Si ('directional' process)

The low-laser fluence regime describes a situation where the incident laser fluence is sufficient to induce melting of the Si film, but it is low enough that a continuous layer of Si remains at the maximum extent of melting. For this reason, this regime is also referred to as 'partial-melting' regime. For irradiation of a-Si films, this regime is characterized by a combination of explosive crystallization and vertical solidification. Explosive crystallization can be triggered at the onset, or near the end of melting, respectively, depending upon the presence or absence of microcrystallites in the Si film [21].

The high laser fluence regime corresponds to the situation encountered when the laser fluence is sufficiently high to completely melt the Si film. For this reason, this regime is also referred to as 'complete-melting' regime. The mechanism of transformation in this regime relates to the nucleation of solids within the liquid for the formation of a stable solid–liquid interface that can be used to 'accommodate' the liquid-to-solid conversion. The nucleation in this case takes place as a result of the unusually deep undercooling that occurs in the molten Si film. The term 'undercooling' refers to the degree of deviation of the temperature of the molten Si from its melting point. In that sense, deep undercooling implies a liquid existing at a temperature substantially lower than its melting point. As a result of the copious nucleation that occurs within the undercooled molten Si, the grain sizes obtained in this regime are very small (typically in the order of tens of nanometer in diameter).

In addition to these two regimes, a third regime has been found to exist, within a very narrow experimental window, in-between the two main regimes (super-lateral-growth region, SLG). Despite of the small extent of this third region, it is nonetheless one with great technological significance, as the poly-Si films formed within this region features large-grained polycrystalline microstructures (i.e. grain sizes of several multiples of the film thickness). The transformation scenario associated with this regime has been modeled by Im *et al.*, in terms of near-complete melting of Si films [19,22]. Particularly, it was argued that at the maximum extent of melting, the unmelted portion of the underlying Si film no longer forms a continuous layer, but instead consists of discrete islands of solid material separated by small regions that have undergone complete melting. For this reason, this regime is referred to as the 'near-complete-melting' regime. The practical implication of this model is that the unmelted islands provide solidification 'seeds', from which lateral growth can ensue, thus propagating the solid–liquid interface within the surrounding undercooled molted Si (for this reason this regime has been also coined as 'super-lateral-growth' or SLG regime). In the ideal case, the lateral growth fronts coalesce and form a continuous matrix of similarly sized grains. This, however, is very difficult to control in practice as the laser fluence leading to SLG regime needs to be precisely controlled. Small variations in the laser fluence (i.e. due to pulse-to-pulse energy variations) lead to either partial-melting or complete-melting

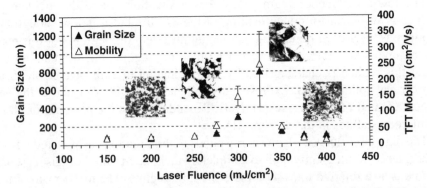

Figure 5: Grain Size and corresponding poly-Si TFT Mobility versus Laser Fluence for conventional ELC process. The inset TEM photographs illustrate the poly-Si microstructure corresponding to (left to right) surface-melting, partial-melting, near-complete-melting and complete-melting regime.

conditions, with grave implications to the poly-Si microstructure as discussed before. Therefore, within the SLG regime, the high quality of the resulting poly-Si material is usually compromised by difficulties in achieving this quality uniformly, across the irradiated film. Figure 5 shows the average grain size and corresponding poly-Si TFT performance, associated with each of these three laser-crystallization regimes.

11.3.1. Conventional LC

In conventional laser crystallization, the substrate is scanned under the laser beam, which is typically homogenized to a 'top-hat' profile. Other beam profiles have also been employed, in efforts to improve the uniformity of the resulting microstructure [23], but the top-hat profile is generally considered as the industry standard. The aim of this process is to induce near-complete-melting in the film and, by irradiating the same region multiple times, improve the size and uniformity of the initially developed grains. This scenario has similarities to conventional normal or secondary grain-growth process and has been previously discussed in detail [24]. In practice, this scenario can be achieved either by static irradiation of a region for multiple times (before moving to the next area), or, on-the-fly, by employing a sufficiently wide overlap between successive shots. The first approach is more suited to a large-area beam, whereas the second approach is commonly used with small-area, high-aspect-ratio beams [25,26]. Naturally, a limitation exists in the practical range of the number of shots (or the overlap distance) dictated in the low end by material quality considerations and in the high end by productivity (i.e. throughput) considerations. Figure 6(a,b) shows the variation of grain size with the number of shots on a given area.

Development of conventional laser crystallization method has been able to provide poly-Si material of higher quality than conventional SPC method. This is primarily attributed to the melt-induced poly-Si growth. As a result of such growth, substantially fewer intra-grain defects form within the grains of laser-annealed poly-Si films. This improvement seems to be more important than the improvement in the grain size itself, as evidenced by the trend in the TFT mobility of poly-Si films as a function of grain size and annealing method (SPC or LC) [27] (see also Fig. 7). However, the stable grain size of LC poly-Si films is typically limited to 0.3–0.6 μm. Larger grain size is possible within the SLG window, but this regime is intrinsically unstable within the context of conventional LC process. For conventional LC, a laser equipment attribute that improves the grain size, within the stable regime, is pulse duration. Longer pulse duration has been found to yield larger grain size (Fig. 6(b)). Theoretical arguments, coupled with simulation results, have attributed this to an increase in the degree of undercooling in the

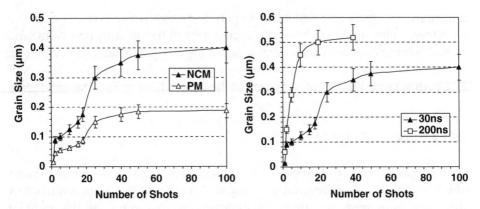

Figure 6: (left) Poly-Si grain size versus number of shots. NCM corresponds to irradiations at the near-complete-melting regime. PM corresponds to irradiations at the partial-melting regime. The number of shots (NS) relates to overlap (OL) by the simple relationship: OL = (NS − 1)/NS. (right) Poly-Si grain size versus number of shots and laser pulse duration (all irradiations at the NCM regime).

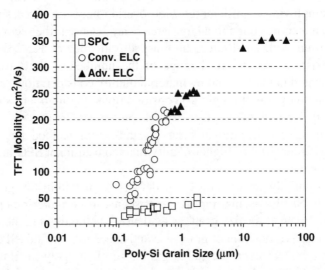

Figure 7: TFT mobility as a function of poly-Si grain size. For a given grain size, ELC (Excimer-LC) process enables superior TFT performance due to the significant reduction of defects within the poly-Si grains. Advanced ELC schemes (discussed in the text) enable customization of the crystal domain size. When the crystal domain size exceeds that of the device channel length very high mobility values can be obtained (independent upon additional increases in the domain size).

molten Si film that enables longer solidification time before the onset of copious nucleation.

One disadvantage of conventional laser crystallization process relates to the difficulty in maintaining a proper balance between performance and process uniformity. This comes about because the pulse-to-pulse repeatability (which is equipment-related parameter) defines the process window that the laser fluence needs to be centered to avoid crossing over to other ELC regimes during annealing. This fluence window, in turn, defines the range of performance that can be expected from TFTs fabricated on such poly-Si microstructure. In manufacturing operations, this kind of compromise, typically, translates to only mediocre TFT performance (i.e. 100 cm^2/Vs) at the benefit of wider process window. A second disadvantage of conventional LC process is the relative disparity between the average grain size and the TFT channel size (length). This means that unless the TFT channel becomes extremely small (i.e. <0.5 μm) it will be impossible to imagine the TFT channel consisting of a single grain. This problem is further exacerbated by the lack of tight grain size distribution in conventional LC and the inability to precisely control the location of grain boundaries with respect to the TFT channel. Therefore, the application of conventional LC process is limited due to these issues and cannot be expected to provide a technology path for future devices. However, this is not to say that laser crystallization is not capable of resolving these issues and improving device performance. By rethinking the physics of laser crystallization and cleverly exploiting previous results, a number of elegant solutions have been indeed identified and developed over the past few years, giving rise to a new generation of laser-annealing crystallization technologies that will be discussed next.

11.3.2. Advanced Lateral Growth (LC) Technology

The key point that drives development of advanced LC concepts is the manipulation of the intrinsically unstable SLG phenomenon in a manner that permits flexible design of the resulting material microstructure, while eliminating all the caveats that are associated with conventional LC processing in the SLG regime [28]. In that sense, all such concepts can be classified under the general term of 'Controlled SLG (C-SLG)' [29].

Super-lateral-growth occurs as a consequence of the lowering in the free energy of the solid-Si/molten-Si system, by the growth of the former (solid-Si) into the undercooled liquid (molten-Si) region. By exploiting this concept, artificial situations can be devised that allow precisely controlled regions of the silicon film to be melted and left in contact with solid regions. These solid regions can then act as seeds for lateral growth of material into the undercooled liquid. The most obvious advantage of C-SLG process is the relative insensitivity of this process to variations in the laser fluence. In the case of C-SLG, the requirement is that the laser fluence is at least sufficient to completely melt the irradiated region, which is in contact with the solid region that will serve as the seed for lateral growth. Note that higher fluence, than this minimum requirement, will work equally well. This implies a wider process window than that of the naturally occurring SLG regime, where the objective is to leave sufficient number of isolated solid islands by precisely controlling the fluence level within the irradiated area (Fig. 8).

By controlling the shape and physical dimensions of the regions that undergo complete melting through various optical, photolithographic and/or other means, the resulting microstructure can be tailored to yield controllable, predictable and uniform grain size and structure. This point is the direct consequence of the relationship between the lateral growth distance (extending in the direction of lateral growth) and the width of the irradiated (completely molten) region. The growth distance of the lateral crystals is limited either by the onset of nucleation in the molten zone, or by collision with the lateral grown crystals from the opposite side of the molten region. Therefore, by tailoring the width of the molten region that is

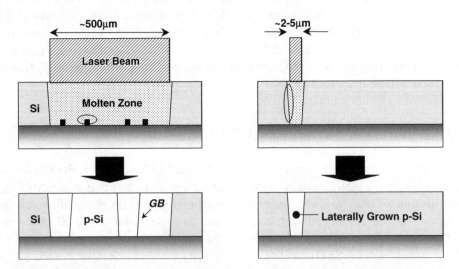

Figure 8: Comparison of SLG and C-SLG modes of growth. In SLG (left panels) growth initiates from seeds, which survived the melting process, at the Si–SiO$_2$ interface. In C-SLG (right panels) the poly-Si microstructure forms by lateral epitaxy from seeds generated at the liquid–solid interface. Notice the difference in the width of the molten zone between SLG and C-SLG.

formed during the laser irradiation to the lateral growth distance, the copious nucleation can be avoided. In this manner, a very uniform and large-grained microstructure can be generated over large areas.

Several techniques can be identified to carry out the task of preselecting the regions to undergo complete melting. These include varying the thickness or contour of the Si film, coating the Si film with a patterned overlayer that can act as an antireflective, reflective, beam-absorbing or heat-sink medium and shaping the intensity profile of the incident laser beam via proximity or projection irradiation involving a shadow mask, by inducing interference on the film surface, or by using diffractive phase elements [30–34]. Figure 9 presents illustrations for a number of published lateral-ELC concepts, including projection irradiation, phase-shift masking and modification of heat diffusion pattern by overlayer/underlayer engineering [29,35,36]. In other related studies, researchers have also demonstrated the formation of high-quality, laterally grown poly-Si films via solid-state laser irradiation in the visible range [37,38].

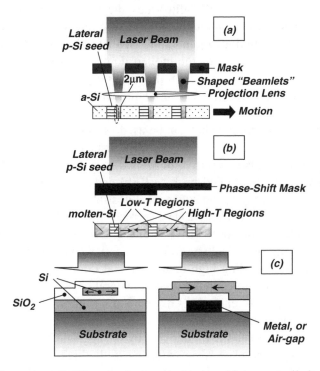

Figure 9: Illustration of different schemes aiming to achieve controlled super-lateral-growth by (a, top) projection irradiation of the laser beam through a mask pattern (SLS), (b, middle) projection through phase-shift mask (PMELA) and (c, bottom) modification of heat flow through overlayer or underlayer engineering (P^2-ELA, Air-gap structure, etc.).

11.3.3. Laterally Grown Poly-Si Microstructure

The concept of lateral growth relies on the generation of Si-crystal seeds that can be then 'dragged' epitaxially in the desired, lateral direction. In the case of laser-based lateral growth a key concept is molten-zone width. The molten-zone width should be precisely controlled to ensure that solidification proceeds via lateral growth and not via copious nucleation, within the undercooled, molten solid. For example, in sequential-lateral-solidification (SLS) method, such control is achieved by precisely defining the extent of the molten-zone, using optical projection technology similar to projection-lithography [29]. It follows that process parameters affecting the heat flow within the laser-irradiated area will also affect the molten-zone width and, possibly, the resulting poly-Si microstructure. Therefore, appraisal of the effects of individual parameters and those of their interactions is critical for understanding the dynamics of the lateral crystallization process and for optimizing the resulting poly-Si microstructure.

Figure 10(a)–(e) schematically illustrates the process of lateral growth via sequential-lateral-solidification method. During the first shot (Fig. 10(a)), the a-Si film melts at the irradiated area generating two lateral growth fronts that recede inward (i.e. toward the center of the molten domain), as the temperature at the solid–liquid interface decreases below the melting point. Ideally, the two melt-fronts will collide at the center of the domain, before the temperature within the melt becomes sufficiently low to trigger copious nucleation (Fig. 10(a)). The crystal seeds generated by the first shot can be used as 'template' for the second shot, thus allowing the grains to grow laterally by a distance equal to the pitch of the advancing substrate (substrate pitch). In this manner the grains can be epitaxially extended indefinitely, as implied by Fig. 10(b) and (c). Such uninhibited lateral growth is achieved as long as the advancing substrate pitch is less than the lateral growth distance achieved in a single shot (termed 'Lateral Growth Length', LGL). If the substrate pitch surpasses the LGL, a different type of poly-Si microstructure is formed, consisting of periodical grain boundaries (with a period equal to the advancing pitch) surrounding high-quality crystal domains (Fig. 10(d) and (e)). Figure 11 shows the relationship between the lateral growth length (LGL) and the laser fluence and film thickness. LGL increases with fluence, as long as the fluence remains within a certain range, defined at the lower end by the full-melt-threshold of the film and, at the higher end, by the agglomeration limit of the film [39]. This process window can be extended with thicker Si films. Therefore, slightly longer LGL can be achieved with thicker films (as Fig. 11 suggests), before hitting the film agglomeration limit.

Although thicker films yield only a small improvement in LGL, the film thickness has been found to affect critically the quality of the poly-Si microstructure.

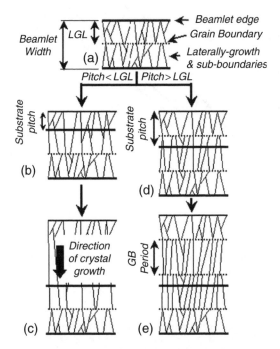

Figure 10: Schematic illustration of lateral growth by sequential-lateral-solidification (SLS) process. (a) Irradiated area after the first shot (LGL denotes the lateral growth length). (b) and (c) Development of continuous, uninterrupted lateral growth, under the condition that the advancing substrate pitch is less than LGL. (d) and (e) Development of periodic-grain and boundary poly-Si microstructure, under the condition that the advancing substrate pitch is greater than LGL (in that case, the grain boundary period is equal to the advancing substrate pitch). The arrow in (c) indicates the direction of lateral growth, which, in this case, coincides with the direction of substrate motion.

Figure 12(a)–(c) shows SEM images of directionally solidified poly-Si films, as a function of the film thickness. The 25 nm-thick poly-Si film (Fig. 12(a)) demonstrates significantly higher density of growth defects (within the crystal domains) than that of either the 50 nm-thick (Fig. 12(b)) or the 100 nm-thick (Fig. 12(c)) film. Such structural defects are primarily identified as microtwins and stacking faults by TEM studies of the poly-Si microstructure [40,41]. In addition to the reduction in defect density, increased film thickness also favors the formation of wider crystalline domains, as is clearly illustrated by the increased sub-boundary width, with film thickness in the SEM images of Fig. 12. Particularly, we estimated the average sub-boundary width as 0.15 ± 0.01, 0.23 ± 0.02 and 0.41 ± 0.06 µm for 25, 50 and 100 nm-thick poly-Si films, respectively. The improvement in crystal quality with film thickness, manifested by the reduction in defect density and increase

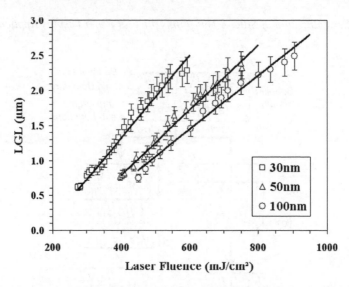

Figure 11: Plot of lateral growth length (LGL) as function of excimer-laser-fluence and Si-film thickness. Single points indicate experimental measurements from poly-Si films irradiated at room temperature. Solid lines indicate model predictions (see later section on numerical simulation of lateral growth).

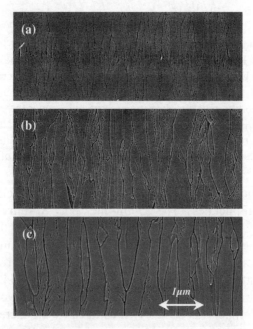

Figure 12: SEM images of directionally solidified poly-Si films, as a function of the film thickness. (a) 25 nm-thick Si film, (b) 50 nm-thick Si film and (c) 100 nm-thick Si film. All films were irradiated in air and at RT. The laser fluence was approximately: (a) 350 mJ/cm², (b) 500 mJ/cm² and (c) 650 mJ/cm².

in the sub-boundary spacing has important implications on the electrical characteristics of the directional poly-Si material, as will be discussed in a later section.

Texture analysis of the SLS-processed material has been carried out as a function of film thickness [41]. Grain orientations were measured at 0.07 μm intervals over a large (e.g. ~7 × 20 μm²) area to obtain orientation maps of the microstructure. Inverse pole figures (IPF) showing the grain orientations in the normal (001) and in-plane directions (lateral growth direction (100) and transverse direction (010)) for crystallized films are presented, as a function of the film thickness, in Fig. 13(a)–(c). In all cases, a random orientation was observed in both the normal (001) direction and transverse (010) in-plane directions. In the lateral growth direction, occlusion of slower growing grains led to a moderate texture development, which for thicker films (i.e. ≥50 nm) was manifested by the predominance of the <100> growth direction. In contrast, for thinner films (i.e. ≤30 nm) a strong (110) orientation was observed in the lateral growth direction. The change in the predominant film texture can be assigned to interfacial breakdown, which commonly

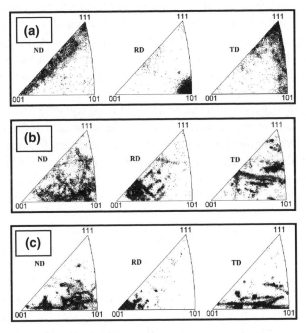

Figure 13: Inverse pole figures (IPF) showing the grain orientations in the normal (001) direction – ND – lateral growth (100) direction – RD – and transverse (010) direction – TD – for: (a) 30 nm-thick Si film, (b) 50 nm-thick Si film and (c) 100 nm-thick Si film. In all cases the advancing substrate pitch was 0.5 μm. Each film was irradiated in air and at RT, at the optimum laser fluence range for the specific thickness.

occurs in ultra-rapid epitaxial growth of very thin Si films [42]. This phenomenon occurs when the solid–liquid interface velocity is greater than a threshold value (that depends upon the orientation in the growth direction) and leads to the formation of defects that assist in the growth of the crystalline phase. At even higher interfacial velocity, the defect-mediated growth breaks down and interfacial amorphization of the film occurs [43]. The window for defect-mediated growth, before interfacial amorphization sets in, has been reported to be very small in the <100> direction by comparison to <110> direction [20]. Therefore, development of (110) orientation in thin, laterally grown films is consistent with this model. The film texture is known to affect the electron and hole mobility and, in that sense, development of (100) texture is preferable for optimum electron mobility [44].

11.3.3.1. Numerical Simulation of Lateral Growth

The simulation of lateral growth in a laser-irradiated thin Si film requires some basic pieces of information. These include: (a) the temporal and spatial temperature distribution within the irradiated volume, (b) a method to track the location of the solid–liquid interface and its evolution over time and (c) a method to enable the triggering of random nucleation, within the irradiated Si volume. Simulating and monitoring random nucleation in the film is an important and technically challenging feature. However, such nucleation represents an alternative and strongly undesirable mechanism for transforming molten Si to solid Si, which significantly impacts the final crystal quality.

Temperature distribution. The basic heat diffusion equation (Fick's law) needs to be modified to capture the distinct features of the laser-based film heating. In this case, there are two distinct heat sources that drive the development of a temperature gradient in the film: (a) the absorption of laser energy and (b) the absorption or release of latent heat as a result of phase change (solid to liquid and liquid to solid, respectively). Accounting for these processes, Fick's law may be written as:

$$\rho c_p \frac{\partial T}{\partial t} = \frac{\partial}{\partial x}\left(k \frac{\partial T}{\partial x} \right) + \frac{\partial}{\partial y}\left(k \frac{\partial T}{\partial y} \right) + \frac{\partial}{\partial z}\left(k \frac{\partial T}{\partial z} \right) + S_{\text{Laser}} + S_{\Delta H_m} \tag{5}$$

where ρ is the density, c_p is the specific heat and k is the thermal conductivity of Si, T is the absolute temperature, S_{Laser} is the source term for the absorbed laser energy and $S_{\Delta H_m}$ is the source term for the latent heat released/absorbed during phase changes. If the source terms in the above equation were known, we could solve for T by using a finite-element method. It is the main task of the model to compute the two source terms in the above equation, for each time step Δt, in the simulated time period. At each time step, the model is called to compute the laser energy absorbed (S_{Laser}) to determine which cells undergo phase change and to compute the latent heat that is released or absorbed ($S_{\Delta H_m}$).

Using values for the absorption coefficient and reflectivity of silicon at $\lambda = $ 308 nm, the laser energy term can be readily computed by applying Lambert's law for the energy distribution in an absorbing medium [45]. The evaluation of the latent heat term, on the other hand, is quite challenging. Phase change only occurs either at the advancing/retreating solid–liquid interface or wherever a solid phase nucleus is created. This means that away from the interface, the liquid phase persists at temperatures well below the equilibrium melting point (undercooling) until a solid-phase critical nucleus appears. This is unlike the thermodynamic formulation where phase change occurs whenever the equilibrium melting temperature is reached.

Simulation of interface motion. To simulate the motion of the solid–liquid interface, a form of an 'interface response function' (IRF) is used. Such function relates the normal velocity of the interface, v_{int}, to the deviation of the interface temperature, T_{int}, from the equilibrium melting point, T_{melt}. A linear IPF form is typically used.

$$v_{\text{int}} = K \left| T_{\text{int}} - T_{\text{melt}} \right|, \tag{6}$$

where K is a proportionality constant. The value of this proportionality constant can be determined by comparing simulated and experimental LGL values under various irradiation conditions. Using this approach the value of K is determined in the range of 6.7–7.2 cm/s·K [46]. The linear form is valid for the degree of under-cooling seen in liquid metal systems, such as silicon [47]. Although the kinetics of solidification at the solid–liquid interface are simple, the non-isothermal environment and the merging of interfaces, which occur as solid-phase regions grow and coalesce which mean that interface position, as a function of time, will not have an analytical form. Instead of having a simple geometric shape, in general the interface will be an irregularly shaped surface that moves with time. This difficulty can be partly addressed by a cellular automata (CA) algorithm that determines phase change without *a priori* knowledge of the shape of the interface. The algorithm works on a cell per cell basis to decide whether phase change will occur, based solely on the state of the cell and of its immediate neighbors. It also determines, from various possibilities, the way by which the interface migrates from a given cell to its neighbors. Another part of the difficulty is addressed by a set of functions that compute the displacement of the interface in the cell and the latent heat that is released or absorbed. To this end, a published model defines six parameters that are needed to completely describe the state of a cell [48].

A set of rules is applied to update these parameters as the simulation progresses from one time step to the next. The rules allow the cell a certain level of 'intelligence' to determine its future state. An example would be for a liquid cell with temperature above the melting point, in which case phase change does not occur. Another would be a cell that contains an interface within it, or in other words, is partly solid and liquid. The interface will then move according to the interface response function [Equation (6)] and, depending on the original position and the direction of the interface, the interface may or may not migrate to adjacent cells. A key result is that, although the evolution in each cell is determined by local parameters, over time, the configurations of the entire system resembles the propagation of contiguous surfaces seen in crystal growth.

Simulation of random nucleation. In accordance with classical nucleation theory (CNT), the crystallization model should account for nucleation that occurs in the bulk liquid (homogeneous), as well as, on catalytic surfaces such as the Si–glass interface (heterogeneous). In the general model, homogeneous nucleation may occur in any fully liquid cell, resulting in the creation of a solid nucleus (hence a new solid–liquid interface) at the center of the cell. Heterogeneous nucleation, on the other hand, is restricted to the liquid cells at the Si–glass interface to which a new nucleus would be attached. Nucleation is then followed by the isotropic growth of the solid nucleus in which the spherically shaped solid–liquid interface expands in all direction. Vertical growth, however, ends after the interface reaches the top and bottom surfaces of the silicon and, henceforth, growth proceeds laterally.

To simulate the random nature of nucleation, a Monte-Carlo-based algorithm can be used, adapted from Leonard and Im [49]. For each cell and each time step the algorithm calculates the probability of the absence of nucleation by using homogeneous Poisson statistics. This formulation treats a sequence of time steps as a series of homogeneous Poisson trials through which a critical solid nucleus may be formed. The duration of a time step, Δt, is typically 0.5 ns. In the interval between t and $t + \Delta t$, the probability of the i-th cell remaining in liquid phase, $P_{i,t}^{LIQUID}$, is given by Equation (7).

$$P_{i,t}^{LIQUID} \cong \exp\left[- \int_{t}^{t+\Delta t} \Gamma_i(\tau)\cdot d\tau \right] \cong \exp\left[-\Gamma_{i,t}\cdot\Delta t \right], \qquad (7)$$

where $\Gamma_i(\tau)$ is the instantaneous nucleation frequency of the cell which, at time t, is approximated by $\Gamma_{i,t}$, defined by the cell temperature which is assumed to be constant within the small time step Δt. Simultaneously, the algorithm generates and assigns a random fraction to each cell at each time interval Δt for comparison with $P_{i,t}^{LIQUID}$. When the assigned number is larger than $P_{i,t}^{LIQUID}$, nucleation is said to occur within the cell. Since the probability (for a cell to remain liquid) decays as the temperature decreases, it will become more likely for nucleation to occur as quenching proceeds.

For homogeneous nucleation, $\Gamma_{i,t}$ is given by the product $[\Delta V \cdot I_v(T_{i,t})]$, where ΔV is the cell's volume and $I_v(T_{i,t})$ is the homogeneous nucleation rate (events/m^3·s) corresponding to the cell's temperature $T_{i,t}$. For heterogeneous nucleation, its frequency will be given by the product $[\Delta A \cdot I_a(T_{i,t})]$, where ΔA is area of the catalytic surface bounding the cell and $I_a(T_{i,t})$ is the heterogeneous nucleation rate (events/m^2·s). From CNT, the homogeneous and heterogeneous rates of nucleation are given by Equations (8a) and (8b), respectively.

$$I_v\left(T_{i,t}\right) = I_{ov}\left(T_{i,t}\right)\exp\left[\frac{16\pi\sigma^3}{3k_B T_{i,t}\Delta G_{ls}\left(T_{i,t}\right)^2}\right] \tag{8a}$$

$$I_a\left(T_{i,t}\right) = I_{oa}\left(T_{i,t}\right)\exp\left[\frac{16\pi\sigma^3}{3k_B T_{i,t}\Delta G_{ls}\left(T_{i,t}\right)^2} \cdot f(\theta)\right] \tag{8b}$$

where k_B is Boltzmann's constant, I_{ov} and I_{oa} are kinetic prefactors, σ is surface energy and $\Delta G_{ls}(T_{i,t})$ is the difference in Gibbs free energy between liquid and solid phases. The heterogeneous rate includes the correction factor $f(\theta) = [2 - 3\cos(\theta) + \cos^3(\theta)]/4$ which accounts for the catalytic surface's role in lowering the energy barrier to nucleation as characterized by the contact angle θ. Defined for θ between 0 to 180°, $f(\theta)$ has minimum value of 0 at $\theta = 0°$ (activation energy is eliminated) and maximum value of 1 at $\theta = 180°$ (no effect) [47]. Typically, heterogeneous nucleation is the predominant mechanism because the catalytic surface reduces the energy barrier to nucleation, allowing it to occur at higher temperatures. In simulations using parameters in Table 3, heterogeneous nucleation accounts for most (if not all) of the nucleation events, since heat from the earliest heterogeneous events would quickly prevent deeper undercooling required for homogeneous nucleation to occur.

The simulation provides detailed information about the solidification process that occurs under the non-equilibrium conditions of laser annealing. As an example, Fig. 14(a) shows the temperature history of a cell within a simulated volume that undergoes 'crystallization' [48]. The temperature of the cell is seen to rise to about 1500 K as the film melts. It then rises even further for the remainder of the pulse duration, after which the film quenches. At about 50 ns, the undercooling initiates random nucleation in the film with subsequent rise in temperature due to the released latent heat. This is in contrast to the equilibrium case (shown by

Table 3: Thermophysical parameters used in simulations of lateral growth
in poly-Si films

Density, ρ	c-Si, a-Si	2.33 g·cm^{-3}
	l-Si	2.53 g·cm^{-3}
	Glass	2.54 g·cm^{-3}
Enthalpy, ΔH_m	c-Si	1799 J·g^{-1}
	a-Si	1148 J·g^{-1}
Equilibrium melting point, T_m	c-Si	1683 K
	a-Si	1420 K
Thermal conductivity, k	c-Si, a-Si, l-Si	$k(T)$
	Glass	1.2 W·m^{-1}·K^{-1}
Specific heat, c_p	c-Si, a-Si, l-Si	$c_\mathrm{p}(T)$
	Glass	1.0 J·g^{-1}·K^{-1}
Reflectivity, R (at 308 nm)	a-Si	0.58
	l-Si	0.72
Absorption coefficient, α (at 308 nm)	a-Si	2×10^6 cm^{-1}
	l-Si	2×10^6 cm^{-1}
Kinetic prefactor (homogeneous nucleation), I_ov	l-Si	10^{39} m^{-3}·s^{-1}
Kinetic prefactor (heterogeneous nucleation), I_oa	l-Si	10^{27} m^{-2}·s^{-1}
Contact angle (heterogeneous nucleation), θ	l-Si	70°
Surface energy, σ	l-Si	0.40 J·m^{-2}

the dashed line), which does not take into account undercooling and nucleation. Figure 14(b) shows the distribution of temperatures at which nucleation events occur, demonstrating the random nature of nucleation. Figure 15(a) is an example of the simulated, evolving microstructure: a snapshot of a re-solidification process showing the expanding lateral growth areas and nucleated grains. From a similar image of a completely re-solidified domain, the LGL can be measured as the distance from the edge of the laterally crystallized region to the beginning of the nucleated region. Figure 15(b) is a plot of the corresponding temperature field vividly showing the exact locations where phase change is occurring, i.e. at the moving solid–liquid interfaces at the edge of the laterally crystallized region and at the perimeter of the growing nuclei.

With the aid of realistic numerical simulation of the crystallization process, screening of process conditions can be readily conducted, in terms of their effect to the computed solidification rate, melt time, maximum film temperature and ultimately, the material microstructure. Table 3 lists the thermophysical parameters

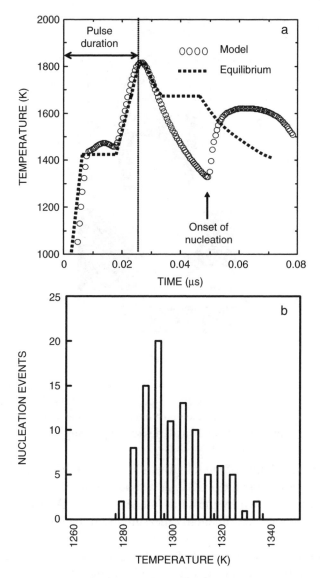

Figure 14: (a) Temperature history on the Si–glass interface at the beam center (50 nm-thick film irradiated at RT and with 400 mJ/cm²). (b) Distribution of temperatures at which nucleation occurs.

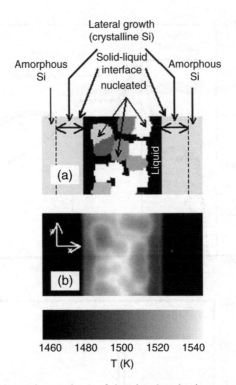

Figure 15: (a) Snapshot (planar view) of the simulated microstructure showing lateral growth and distinct grains arising from stochastic nucleation. (b) Bright areas in temperature field indicate localized release of latent heat at the solid–liquid interface and correspond to the boundaries of the growing grains.

required for the simulation of lateral growth in a Si film that is typically deposited on a SiO_2 layer [48]. Simulated LGL, as a function of excimer-laser-fluence and Si-film thickness is shown by the solid lines in Fig. 11. An excellent agreement between predicted and experimentally measured LGL can be readily observed.

11.3.3.2. Improving Lateral Growth Length and Poly-Si Microstructure
Figure 16(a) and (b) shows the effect of the advancing substrate pitch on the resulting film texture. Comparing the orientation maps for the samples shown in Fig. 16(a) and (b) and Fig. 13(b) we first note that the substrate pitch has no effect on the poly-Si orientation in the normal and transverse directions, which remain random for all three pitch settings. On the other hand, the orientation in the lateral growth direction is observed to remain predominantly (100) for a pitch of 0.25–0.5 μm, but become bimodal (i.e. two peaks at (100) and (110)) as the pitch further increases to 0.75 μm. Increasing pitch (for a given beam width) means that

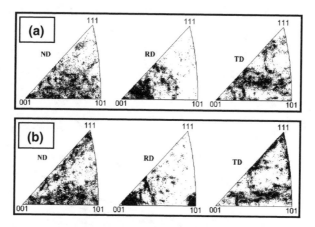

Figure 16: Inverse pole figures (IPF) showing the grain orientations in the normal (001) direction – ND – lateral growth (100) direction – RD – and transverse (010) direction – TD – for 50 nm-thick films annealed with an advancing substrate pitch of: (a) 0.25 μm and (b) 0.75 μm. Samples crystallized in air and at RT.

the lateral growth is 'seeded' from crystal material that was formed under a deeper interfacial undercooling. Such deeper undercooling increases the material propensity for defect generation. Under such conditions, it is plausible that defect-mediated crystallization also contributes to lateral growth, thus promoting growth in the <110> direction. As a result of the particular setup, in experiments performed with 0.75 μm pitch, each shot sampling crystal was very close to the center of the irradiated domain; i.e. crystal that was formed under relatively deep undercooling. Therefore, to increase the substrate pitch (which may be beneficial for process throughput), the degree of interfacial undercooling has to be controlled. This kind of consideration will be necessary to define the optimum beam-shaping aperture-width for a given set of irradiation process parameters (laser fluence, film thickness, etc.). Additional control on the degree of undercooling is expected through the use of pulse extension and/or substrate heating [50].

Figure 17 illustrates the effect of substrate temperature on LGL. The lateral growth length, obtained from 100 nm-thick Si films, has been plotted as a function of excimer-laser-fluence and substrate temperature. For a given laser fluence, a significant increase in LGL can be observed by increasing the substrate temperature. The solid lines represent a fit to numerical simulations of LGL at the respective conditions. Based on the excellent agreement between experimental and simulated LGL, the developed model was used to extract LGL at temperatures approaching the melting point of a-Si material. The computed trend is shown in Fig. 18. LGL is shown to initially increase linearly with substrate temperature in

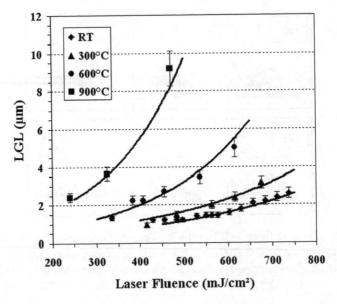

Figure 17: Trend in LGL as a function of excimer-laser-fluence and substrate temperature. Single points indicate experimental measurements. Solid lines indicate model predictions.

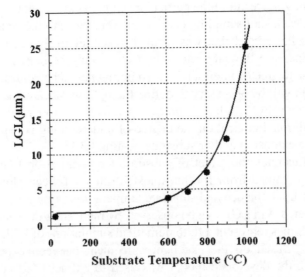

Figure 18: Simulated LGL as a function of substrate temperature for a 50 nm-thick Si film. Simulations were run at the estimated optimum fluence for each temperature setting.

the range of RT–700°C and then exponentially when the substrate temperature becomes greater than ~800°C and up to the melting point. This second region is of significant interest, as it enables a wide range of processing capabilities. Unfortunately, this region is not practically accessible by conventional means due to the temperature sensitivity of the underlying glass substrate. One possibility to enable operation in this regime appears to be the use of temporal pulse extension for the laser beam. Figure 19 shows the trend in LGL as a function of laser fluence and laser pulse duration at two substrate temperatures (RT and 600°C). Based on the simulated LGL at 600°C, with 210 ns pulse duration, LGL of >4 μm can be readily obtained. However, there is a caveat in the use of pulse extension in the sense that it typically restricts the useful process window, in terms of laser fluence (see Fig. 20). This imposes some restrictions in the number of shots that a given volume of the film can experience before reaching the limit of agglomeration. Therefore, processing with significantly 'extended' pulses seems more appropriate for low-shot-type schemes (i.e. <4 shots per area). Nonetheless, the combination of moderate substrate heating with moderate pulse extension appears attractive as a method to increase the lateral growth length and improve the quality of the generated crystal. For example, as Figure 21 demonstrates, the crystal domain width (within the laterally grown crystal) increases both by employing higher substrate temperature and/or longer pulse extension. The so-called

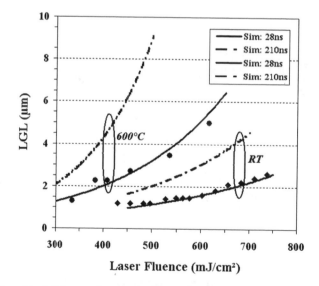

Figure 19: Trend in LGL as a function of laser fluence and laser pulse duration at two substrate temperatures (RT and 600°C). Simulation was used to generate the long pulse duration (210 ns) curves.

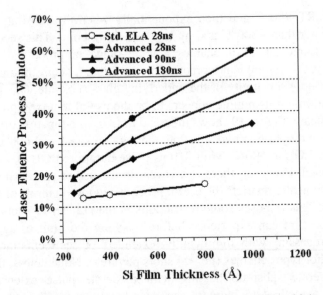

Figure 20: Laser-fluence process window as a function of Si film thickness and laser
pulse duration for advanced irradiation process (lateral growth). The process window is
determined by the minimum and maximum fluence that the Si film can be processed at
without agglomeration.

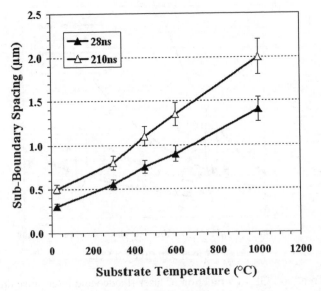

Figure 21: Sub-boundary spacing in laterally grown poly-Si film as a function of
substrate temperature and laser pulse duration.

'sub-boundary spacing' has important implications in the quality of the poly-Si film and its electrical characteristics, as will be discussed in a later section.

11.3.3.3. Effect of Laser Beam-Profile on Lateral Growth Characteristics

An important – and independently controlled parameter – for lateral growth is the beam profile. The actual beam profile on the sample surface is primarily determined by the relationship between the critical size of the beam-shaping pattern (i.e. the width in the case of a large aspect ratio rectangle) and the resolving power of the projection optics (i.e. the numerical aperture of the projection lens). This relationship determines the edge 'sharpness' of the beam-profile, i.e. the spatial requirement for the intensity to rise from 10 to 90% of full intensity. Efficient SLS process requires an abrupt beam-edge profile to minimize the inefficient utilization of laser energy. In practice, this profile is dictated by the diffraction limit of the projection optics due to the compromise between numerical aperture and depth-of-focus requirements. Figure 22 shows the spatial intensity distribution (on the substrate) for a 9 μm-wide beamlet generated from a 5:1 projection lens with NA = 0.05. The intensity range corresponding to the SLS process window is also shown in this figure. For intensity values outside this range, partial melting of the film occurs with the subsequent formation of fine grain poly-Si material. The sharpness of the beam profile will define the extent of this regime at each edge of the beam (d_e). Depending upon the width of the beamlet, random nucleation may occur at the center of the beam, if the LGL under the irradiation conditions is smaller than half of the beam width. We define as d_L the lateral growth length and as d_c the width of the central, nucleated region. If follows that: $w = 2d_e + 2d_L + d_c$, where w is the beam width. Therefore, for optimum utilization of the beamlet, it is required that $d_e \rightarrow 0$ and $d_c \rightarrow 0$. Although the nucleated, 'center' region can be effectively eliminated (i.e. by decreasing the beamlet width), the beam edge region can be restricted but never, practically, eliminated.

Beyond the limitations of the projection optics, another (equivalent) source of beam-profile distortion is focusing. For a given numerical aperture for the projection optics, the depth-of-focus can be determined, defining the degree of variation in the distance between the projection lens and the surface of the irradiated sample that results in negligible changes in the imaging capability of the lens (i.e. its resolving power). If, for a fixed projection lens, the location of the sample plane exceeds this limit (depth-of-focus), distortion in the imaged beam profile will occur, manifesting itself as 'blurring'; in other words, the imaged beam cannot be fully resolved. Both underfocus and overfocus conditions will produce this result. Under these conditions, the maximum beam intensity decreases and the edge becomes more diffuse. In other words, d_e increases and the probability of nucleation at the center of the beamlet also increases. Figure 23(a) shows the variation

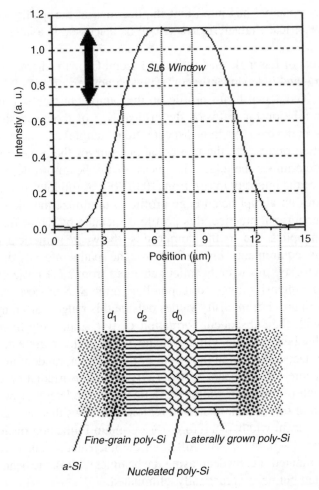

Figure 22: Simulated spatial intensity distribution for a 9 μm-wide beamlet projected with a 5:1 projection lens having NA = 0.05. The intensity range corresponding to the SLS process window is shown in the figure. The schematic at the bottom, matching the intensity distribution, illustrates the regions that may form upon irradiation of a silicon film with this beamlet. The region with width d_e corresponds to the transition between amorphous silicon and fine-grain poly-Si, as a result of the increasing beam intensity (but still below the SLS process window). The region with width d_L corresponds to laterally grown poly-Si material. Finally, the region with width d_c corresponds to poly-Si material crystallized at the center of the irradiated domain by copious nucleation and growth.

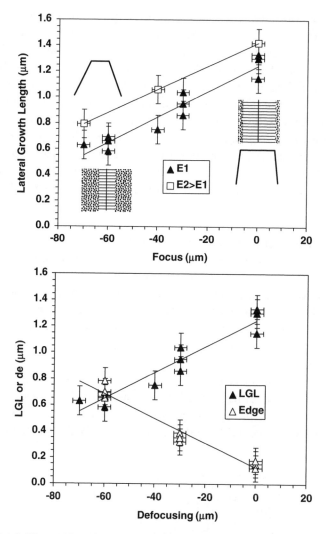

Figure 23: (a) LGL as a function of the degree of defocusing (in micrometer) and the level of laser fluence (for 50 nm-thick films). The inset sketches illustrate the distortion in the beam profile, with increasing defocusing, and its qualitative effect on the poly-Si microstructure. (b) Edge length (d_e) and LGL variation as a function of defocusing (for irradiation at fluence level E_1).

of LGL as a function of the degree of defocusing (in micrometer) from the best focal plane (denoted as 0 μm). The lateral growth length is reduced with defocusing, due to the increased edge diffusion in the beam profile, which renders an increasing part of the beam ineffective for lateral growth. This is better shown in Fig. 23(b), which shows the edge length variation as a function of defocusing. As shown, the decrease in LGL is accompanied by a concomitant increase in the edge length, at increased defocusing. Increasing the laser fluence can somewhat compensate for the defocusing losses on LGL. However, agglomeration poses a limitation as to the extent of such compensation schemes. These results imply that distortions in the beam profile (i.e. due to defocusing) will have to be accounted for the determination of the optimal substrate pitch to maintain lateral growth continuity over large areas.

11.3.3.4. Practical Implementation Schemes

Beam shaping by masking and projection. As mentioned earlier, the concept of lateral growth relies on the precise control of the width of the molten zone, formed during laser irradiation. A generally accepted method for this purpose shapes the intensity profile of the incident laser beam via projection irradiation through a shadow mask, in a manner equivalent to projection lithography. Depending on the design of the crystallization mask pattern, a number of different process schemes have been proposed [51,52].

One of the simplest mask patterns involves two rows of slits, offset to each other, with each row covering approximately half of the beam width. A schematic of this pattern is shown in Fig. 24. Such pattern is used on a very rapid, lateral growth crystallization scheme, dubbed '2-shot' process. With this scheme, Si material

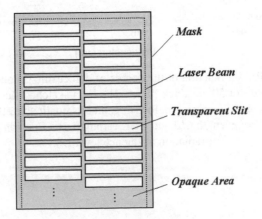

Figure 24: Schematic illustration of the '2-shot' mask design.

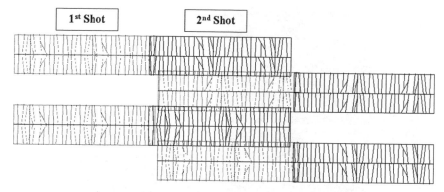

Figure 25: Illustration of '2-shot' process. Within two shots, the area within one column of slits is completely crystallized. The resulting periodical microstructure consists of laterally grown poly-Si 'strips' separated by grain boundaries. The period of the grain boundaries is precisely determined by the geometry of the mask (i.e. slit width and overlap).

within the area of any of the two columns is completely annealed within two shots (see illustration in Fig. 25). The substrate continuously scans under the masked beam with a velocity that allows its exact translation by a distance equal to the width of one row of slits, within one discharge period. Typical throughput of this scheme is in the range of 12–18 panels/h for fourth generation mother glass (i.e. 730×920 mm). The resulting poly-Si microstructure consists of a periodical pattern of laterally grown Si 'strips' separated by grain boundaries. The strength of this scheme is its rapid processing time. On the other hand, its main weaknesses are the modest crystal quality and the stochastic nature of grain boundary inclusion within the active layer of devices fabricated on such periodic microstructure. At grain boundaries, substantial surface roughness is typically observed, which is undesirable in terms of device performance and yield control. A final point on 2-shot process is that it also suffers from the same issue that plagues all lateral growth schemes, namely the intrinsic anisotropy in crystal quality. This means that TFT device performance will be quite different between devices oriented parallel and devices oriented perpendicular to the direction of lateral growth.

Improvements on the basic 2-shot mask-pattern and process have been proposed to alleviate some of its disadvantages, while preserving its fast throughput. Such improvements include the bi-directional mask design [53], the so-called '2 + 2-shot' process and its derivative '2^N-shot' process [54]. The bi-directional process capitalizes on an optimized geometrical arrangement of slits to achieve

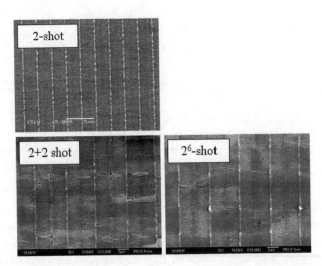

Figure 26: Microstructure detail (by SEM) in poly-Si films crystallized by 2-shot,
2 + 2 shot and 2^6-shot laser annealing process.

bi-directional lateral growth and increase the width of the laterally grown
poly-Si strips tacked in-between grain boundaries. 2^N-shot process uses a combination of horizontal and vertical slits to refine the crystal, formed by the basic
2-shot process, and decrease the anisotropy in crystal quality [55]. Figure 26
shows examples of poly-Si material formed by 2-shot, 2 + 2-shot and 2^6-shot
process.

Line-beam type. Borrowing from conventional laser-crystallization, line-beam
scheme aims at a mask-less, projection-less lateral growth process. In this case, the
laser beam is optically shaped into a high-aspect-ratio, approximately rectangular
shape. The width of this beam should be typically in the range of the LGL – for
example, < 10 μm. In contrast, the length of the beam, being free of limitations
imposed by projection optics, should be made as long as technologically possible.
Given that the moving pitch for a line-beam process is very small – i.e. < 5 μm – it
follows that the throughput of this process will be practical only when the
discharge frequency of the laser becomes sufficiently large. This is possible with
'green' DPSS lasers. In addition, a recent introduction of a high repetition-rate
laser has opened this avenue in the excimer laser category as well [18].

Beyond process throughput, the weakness of the line-beam scheme is the inherent anisotropy of the 'directionally' formed Si crystals. However, the superb crystal quality, along the lateral growth direction, and the lack of grain boundary
formation perpendicular to the lateral growth front are significant advantages
from the point of view of device performance. The issue of anisotropy can be

addressed partially by process improvements (as discussed in previous sections) and partially by adjusting the circuit design to match the specific characteristics of this material.

11.4. Characteristics of Advanced Poly-Si TFTs

The performance of various n-channel TFTs fabricated with laterally crystallized, poly-Si active layer is summarized in Fig. 27. TFT mobility is affected by the presence of grain boundaries (perpendicular to the direction of carrier transport) in the device channel. When no such boundaries are present, the mobility values are seen to approach that of SOI material.

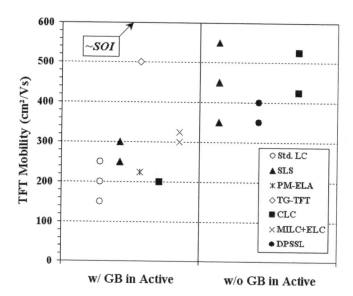

Figure 27: Comparison of the mobility of n-channel poly-Si TFT prepared by lateral growth, laser-crystallization techniques (std. LC performance is provided as a reference). 'w/GB in Active' means active layer including grain boundaries perpendicular to the direction of conduction. Conversely, 'w/o GB in Active' means active layer without such grain boundaries. The acronyms listed in the legend are defined as follows: SLS = 'sequential-lateral solidification', PM-ELA = 'phase-mask excimer-laser-annealing', TG-TFT = 'twin-grain TFT', CLC = 'CW-laser lateral crystallization', MILC = 'metal induced lateral crystallization', DPSSL = 'Diode-pumped solid-state-laser crystallization'.

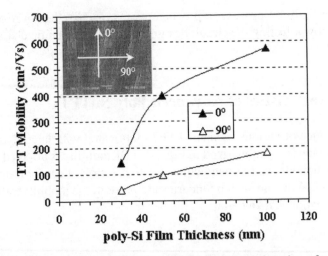

Figure 28: Room-temperature mobility of LCP-TFTs as a function of active layer thickness and device channel orientation. The channel orientation is defined with respect to the lateral growth direction. We denote as '0°' LCP-TFTs with channels parallel to the lateral growth direction and as '90°' LCP-TFTs with channels perpendicular to the lateral growth direction (see inset SEM photograph).

Figure 28 shows the relationship between the mobility of laterally crystallized, poly-Si TFTs (LCP-TFTs) and the thickness of the LC poly-Si active layer [56]. The trend is shown for two different channel orientations, with respect to the lateral growth direction. We denote as '0°' LCP-TFTs with channels parallel to the lateral growth direction and as '90°' LCP-TFTs with channels perpendicular to the lateral growth direction. As shown in Fig. 28 both the film-thickness and the device orientation have significant effects on the device mobility. The root of both effects reduces to the details of the poly-Si microstructure that develops after lateral crystallization. This microstructure is typically composed of long crystal domains that run parallel to the direction of lateral growth and are separated by grain boundaries termed 'sub-boundaries'. For '0°' TFTs sub-boundaries are oriented parallel to the direction of current conduction, while for '90°' such sub-boundaries are oriented perpendicular to the direction of current conduction. The existence of sub-boundaries can explain the severe anisotropy in mobility demonstrated by this type of LC poly-Si material [57,58]. It is also known that the sub-boundary spacing increases with the film thickness [59]. The increase in sub-boundary spacing is one reason for the improvement in the mobility of 90°

LCP-TFTs with increasing poly-Si thickness. At the same time, the improvement in the mobility of 0° LCP-TFTs implies that the mobility enhancement cannot be solely attributed to the widening of the crystal domains, but it is rather a manifestation of an improvement in the quality of the Si crystal itself with increasing thickness. However, even for the thickest poly-Si films utilized in previous studies, the resulting mobility is still far from that of SIMOX material of the same thickness (i.e. ~770 ± 40 cm²/Vs). This implies that some form of correlation exists between the sub-boundary spacing and the crystal quality itself. In that sense, the crystal quality will tend to approach that of SIMOX at the extreme limit of infinite sub-boundary spacing.

It has been suggested that the temperature–mobility relationship can be used to infer information on the crystal quality of the active layer [60]. This method has been also used in a recent study to compare the trend in LPC-TFT mobility as a function of the measurement temperature [56]. Figure 29 shows a series of normalized mobility–temperature curves as a function of the LC poly-Si thickness and device channel orientation. Data for SIMOX-TFT are also shown as a reference. Although in all cases the normalized mobility is eventually seen to decline with temperature, it is clear that the rate of mobility decrease in LPC-TFTs is not the same as in SIMOX-TFTs. Furthermore, it is also clear that different types of lateral crystallized poly-Si 'material' demonstrate different responses.

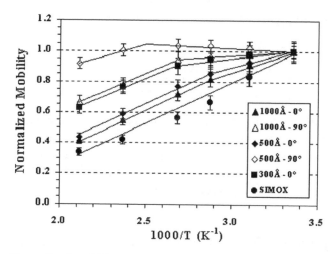

Figure 29: Normalized mobility versus temperature as a function of the LC poly-Si thickness and device channel orientation. Data for SIMOX-TFTs are also shown as a reference. The mobility is normalized with respect to the room-temperature (RT) value.

11.4.1. Material-Quality to Device-Performance Correlation

The variation of the crystal quality with film thickness in laterally crystallized poly-Si films gives credence to the idea that increasing film thickness must amount to a reduction in the crystal defect concentration in the Si layer. Typically, it has been observed that thin laser-crystallized poly-Si films are very rich in various types of defects (i.e. micro-twins and stacking faults [20]). Such defect concentration decreases with the film thickness, but even for 100 nm-thick poly-Si films defects are still readily observable (most notably stacking faults). The apparent mobility of LCP-TFTs is observed to increase with film thickness, reaching a maximum level in the case of practically defect-free crystal (i.e. SIMOX). This trend suggests that, beyond lattice scattering (observed in c-Si and SIMOX MOSFETs), an additional scattering mechanism exists, lumping, in the simplest case, the effects of various crystal defects within the active layer. This is the so-called 'crystal-defect scattering' mechanism [56].

Crystal-defect scattering seems to be primarily responsible for the reduction in mobility (from the ideal case of SIMOX) observed in devices with channels oriented parallel to lateral growth. For such channel orientation, carriers do not cross grain boundaries in their path from source to drain. In contrast, 90° devices encounter an additional scattering mechanism due to the existence of sub-boundaries that oppose the free flow of carriers. This third mechanism ('grain-boundary scattering') has, therefore, to be added to explain the mobility trend of 90° LPC-TFTs. Applying Mathiessen's rule, the complete mobility model can be written as:

$$\frac{1}{\mu_{\text{app}}} = \frac{1}{\mu_{\text{SIMOX}}} + \frac{1}{\mu_{\text{Defect}}} + \frac{1}{\mu_{\text{GB}}} \tag{9}$$

where: μ_{app} is the apparent (measured) mobility, μ_{SIMOX} is the equivalent SIMOX mobility accounting for lattice scattering, μ_{Defect} is a mobility term attributed to defect-scattering and μ_{GB} is a mobility term attributed to grain-boundary scattering. Expressions for the three contributions to the mobility model of Equation (9) have been developed on a semi-empirical basis and are summarized in Table 4 [56].

11.4.2. Performance Predictions for Advanced Poly-Si TFTs

It is instructive to investigate the apparent TFT mobility as a function of the angle between the device channel-length direction and the lateral growth direction in the film. In terms of this angle, two limiting cases correspond to the 'parallel' and

Table 4: Formulations for the three scattering mechanisms included in the mobility model of Equation (9) and best-fit parameters in each case (L expresses sub-boundary spacing in micrometer, T is the absolute temperature in deg. Kelvin and k is Boltzmann's constant)

Scattering mechanism	Formulation	Best-fit parameters
Lattice scattering	$\mu_{SIMOX} = \mu^* \cdot T^a$	$\mu^* = 5.83 \times 10^8$; $a = -2.38$
Crystal-defect scattering	$\mu_{Defect} = \mu_o \cdot L^b$	$\mu_o = 31761.2$; $b = 2.71$
Grain boundary scattering	$\mu_{GB} = \mu^{**} \cdot \left(L - L_{crit}\right) \cdot \exp\left(-\dfrac{E_B}{kT}\right)$	$\mu^{**} = 2384.7$; $L_{crit} = 0.1009$
		$\dfrac{E_B}{kT} = 7.662 \cdot 10^{-25} \cdot N_t^2$
		$N_t \sim 1.16 \times 10^{12}$ cm^{-2}

'perpendicular' devices discussed previously. To remedy the intrinsic material anisotropy in directionally crystallized poly-Si films, previous works have suggested the intentional modification of the channel angle to intermediate values (i.e. 30–60°) [58]. To model the behavior of 'tilted' TFTs, one needs to recognize that crystal-quality (expressed by the magnitude of the crystal-defect scattering term) is constant. On the other hand, the 'apparent' sub-boundary spacing varies as the device orientation, with respect to the lateral growth direction, changes from 0 to 90°. Using a simple geometrical approach, one can model the 'apparent' sub-boundary spacing as $L_{app} = L/\sin\theta$, (where θ is the angle between the device channel and the lateral growth direction) and replace 'L' with 'L_{app}' in the expressions of Table 4. Figure 30 shows the computed mobility as a function of the angle θ and compares it to experimental measurements. A good agreement is observed between the model predictions and the actual data.

11.4.2.1. Sub-Boundary Spacing and Uniformity

Using the mobility model, embodied by Equation (9), it is possible to predict the effect of different material engineering options on the resulting TFT performance. The simplicity of the model is one key advantage, but, at the same time, the semi-empirical nature of the model is one of its main drawbacks in correctly assigning material-related causes to device-observed effects.

One key area of interest is the impact of material uniformity (i.e. uniformity in sub-boundary spacing) to device mobility uniformity. The relationship between

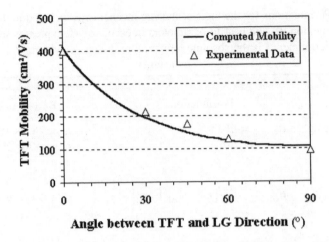

Figure 30: Calculated versus measured TFT mobility as a function of the angle (θ) between the TFT channel length direction and the lateral growth direction in the poly-Si film. For this calculation it was assumed that the crystal-defect scattering contribution is constant as a function of the angle, whereas the grain-boundary scattering contribution becomes increasingly smaller as the angle becomes increasingly closer to 0°.

the two can be readily assessed by the mobility model and compared to experimental data. The method to make this assessment has been explained in detail before [56]. Figure 31(a,b) shows this comparison for parallel and perpendicular TFTs. The computed μ_{app}-range was calculated based on the experimentally measured sub-boundary spacing range (also plotted on Fig. 31). A very good agreement is observed between the computed and the actual mobility distributions. The observed trend demonstrates a strong correlation between the uniformity in sub-boundary spacing and the concomitant mobility distribution.

Figure 32 shows the calculated LCP-TFT mobility (for 0 and 90° devices) as a function of the sub-boundary spacing. Based on this calculation, we recognize an interesting trend for 0° devices. As shown, the device mobility is initially quite sensitive to the sub-boundary spacing, which implies significant variation in device performance for rather small variation in sub-boundary spacing. This situation, however, is considerably improved, once the sub-boundary spacing becomes sufficiently large (i.e. >0.5 μm). Under these conditions, significant variations in sub-boundary spacing are predicted to cause only small changes in device mobility. Therefore, operation in this regime is desirable to optimize uniformity. Note that a similar trend is not observed for 90° devices as, in that case, μ_{GB} mobility, which is directly proportional to sub-boundary spacing, is the

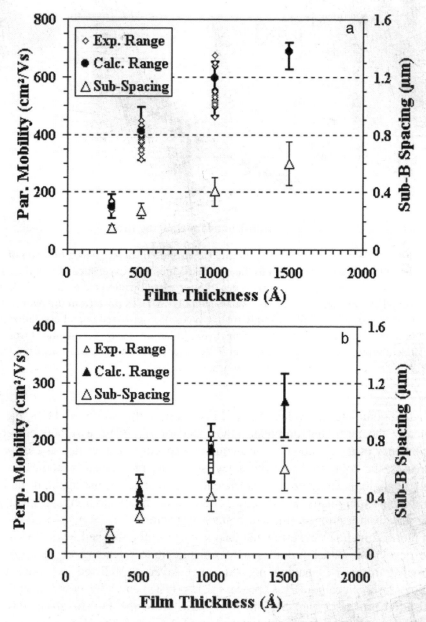

Figure 31: Comparison of computed and actual TFT mobility distribution as a function of LC poly-Si film thickness. The computed mobility range was calculated based on the experimentally measured sub-boundary spacing range, corresponding to the particular film thickness (also plotted in the figure). Panel (a) shows data for parallel TFTs and panel (b) shows data for perpendicular TFTs.

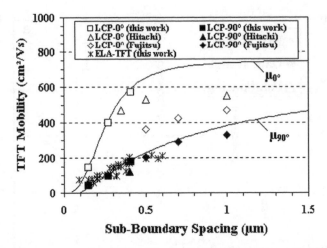

Figure 32: Calculated LCP-TFT mobility (for 0 and 90° devices) as a function of the sub-boundary spacing. Overlaid to the calculated mobility, experimental mobility data are shown corresponding to several types of laser-crystallization technology. ELA data correspond to standard excimer-laser-annealed poly-Si TFTs (also from this work). Data from other groups, working with alternative types of laser-based lateral crystallization technologies, include SELAX data (Selectively Enlarging Laser X-tallization) from the Hitachi group (ref. 24) and CLC data (CW-Laser Lateral Crystallization) from the Fujitsu group (ref. 18).

term that dominates the value of μ_{app}. In contrast, in the case of 0° TFTs, μ_{Defect} is typically the term that dominates the value of μ_{app}. The assumed power-law dependency of μ_{Defect}, upon the sub-boundary spacing, ensures that as the spacing increases, the contribution of the defect-scattering mechanism becomes increasingly negligible. Therefore, for sufficiently large spacing, the apparent mobility will be only marginally sensitive to variations of the material crystal quality (as expressed by variations to the sub-boundary spacing).

Figure 32 includes experimental data from several published works [56,60,61]. The developed model provides a reasonable fit to a variety of mobility data extracted from TFTs having poly-Si active layer crystallized by conventional excimer laser annealing (ELA) process (ELA-TFTs) and laser-based, lateral crystallization processes (i.e. SLS in [56], CLC from the Fujitsu group [60] and selectively enlarging laser X-tallization (SELAX) from the Hitachi group [61]). It is noted that the mobility model embodying grain-boundary scattering provides a very good fit to the experimental data, over a wide spectrum of crystallization processes. We remind that in the case of lateral growth processes, the model applies to devices that are oriented perpendicular to the direction of lateral growth.

The mobility predictions, for devices oriented parallel to the direction of lateral growth, are found to exceed the experimental measurements for some of the lateral growth technologies included in Fig. 32. This could be interpreted as the manifestation of the existence of additional crystal defects in the active layer, beyond the concentration level allowed by the current model. In that sense, some detail of the lateral crystallization process may have a bearing on the level of defects formed in the crystallized layer. It is further noted that this simple model does not account for texture effects. As the sub-boundary spacing increases (i.e. crystal domains become wider), texture effects will start to modulate mobility [44]. Therefore, without a means to control crystallographic orientation in the film, a substantial increase in the sub-boundary spacing may be, in fact, counterproductive in terms of uniformity of TFT characteristics. This point will be discussed in more detail next.

11.4.2.2. Texture Issues in Wide Sub-Boundary Poly-Si Films
Some methods exist to increase the sub-boundary spacing in laterally grown poly-Si material – for example, film thickness and substrate temperature among others. Although such methods may not always be practical in terms of manufacturing, increasing the sub-boundary spacing in such manner is, nonetheless, instructive. Figure 33 shows an example of the microstructure of a laterally grown poly-Si film exhibiting very wide sub-boundary spacing. Using such material as the active layer, poly-Si TFTs have been fabricated and characterized [62]. Figure 34 shows the statistical distribution of mobility and threshold voltage based on the measurement of 400 devices having $W \times L = 8 \times 1.3$ µm. Variation in both mobility and

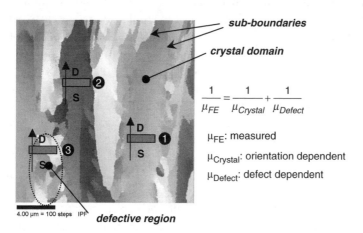

Figure 33: Electron-Backscattering Diffraction Pattern from a 50 nm-thick poly-Si film exhibiting very wide sub-boundary spacing (>4 µm).

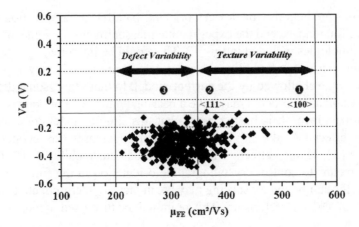

Figure 34: Mobility and threshold voltage distribution from (W × L) 8 × 1.3 μm TFT fabricated with poly-Si active layer having similar microstructure to that shown in Fig. 32. See text for the explanation of the regimes shown on the plot.

threshold voltage is clearly seen, although the variation in mobility seems to be particularly wide. The wide range in mobility reflects the combination of two effects: (1) the natural variation in sub-boundary spacing, which implies that, occasionally, defective regions are included in the device channel region and (2) the variation in the 'effective' channel texture, resulting when TFTs end up positioned on a single-crystal domain. These two effects can be summed-up using Mathiesen's rule as shown in Fig. 33. For example, 'lucky' devices, which do not contain defective regions, have high mobility (i.e. ❶ and ❷), but only devices oriented closely to <100> (i.e. ❶) demonstrate mobility in excess of 450 cm²/Vs. On the other hand, devices that include defective regions (crossing the conduction path – i.e. ❸) demonstrate lower mobility, in the range of 250–300 cm²/Vs. We note that the concomitant variation in the threshold voltage is consistent with the aforementioned model of material defects and crystallographic texture. This model has been confirmed by the detailed study of the channel-region microstructure of TFT devices demonstrating various levels of mobility. The result of this study is summarized in Fig. 35, which shows the cumulative probability of the mobility of TFTs fabricated with poly-Si active layer similar to that shown in Fig. 33. It was concluded that low-mobility devices correlate with heavily defective channel region (i.e. including multiple crystal domains and/or other defects), whereas high mobility devices correlate with typically mono-domain channel microstructure (see insets in Fig. 35).

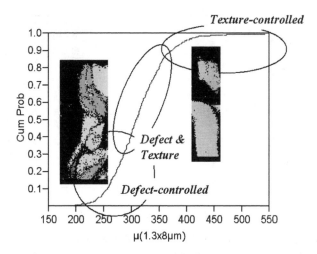

Figure 35: Cumulative probability of TFT mobility based on wide sub-boundary poly-Si active layer. Examples of device channel microstructure corresponding to the two main influence regimes (defect controlled and texture controlled performance) are shown in the inset.

11.4.2.3. Impact of Defects to Performance

Figure 36 shows the impact to the mobility and threshold voltage of poly-Si TFTs by the inclusion of a single grain boundary to the device channel. The average mobility decreases by ~120 cm^2/Vs and the average threshold voltage increases by ~0.5 V, when a grain boundary is included within the active layer and oriented perpendicular to the direction of carrier transport. This result agrees with the overall view of the literature in this subject [63,64] and points out a serious issue in controlling the uniformity in TFT characteristics. Crystallization schemes, which result in microstructures with periodic grain boundaries, are very vulnerable to this limitation, especially as the device channel scales down in accordance to increasingly demanding design rules for poly-Si TFTs. To avoid the undesirable inclusion of a grain boundary in the active layer, special and costly measures have to be taken, including the intentional degradation of overall performance (i.e. all TFTs will include a grain boundary) or the necessity of precise alignment to ensure that all TFTs fall within laterally grown 'stripes'.

Figure 37 compares the density-of-states function for various poly-Si microstructures, formed by laser crystallization [64,65]. Compared to conventionally crystallized material (std. LC), all laterally grown poly-Si materials show reduced density of states, both close to the conduction bands ('tail' region) and close to the mid-gap ('deep' region). Within the lateral growth family of microstructures, the benefit of thicker active layer is confirmed, in terms of the observed

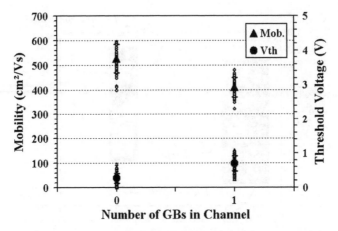

Figure 36: Impact to TFT mobility and threshold voltage by the inclusion of a single grain boundary (GB) in the device channel. The GB is perpendicular to the direction of carrier transport.

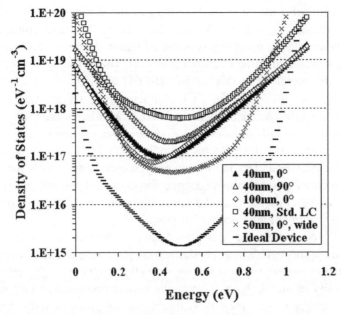

Figure 37: Density-of-State (DOS) function for various poly-Si microstructures.

decrease in the density of deep-states. The inclusion of grain boundaries, perpendicular to the direction of carrier transport, is shown to be quite detrimental to the material quality, increasing the levels of defects, particularly around mid-gap and also tail-state on the donor side. This implies that the mobility of p-channel devices is more strongly affected by the inclusion of grain boundaries, a fact that has been also experimentally observed. The DOS for a wide sub-boundary spacing poly-Si material is also shown in Fig. 37. A striking feature of this material is its low-deep-state density. The tail-state density is equivalent, or slightly higher, to other poly-Si materials, although the rise in DOS appears to be much sharper for this type of microstructure.

Tail states are typically connected to on-current (transconductance and mobility), whereas deep states to threshold voltage and subthreshold slope. The density of states function is described by Equation (10), written for donors and acceptors.

$$N_{acc}(E) = N_{deep} \exp\left(-\frac{E_c - E}{w_{deep}}\right) + N_{tail}\left(-\frac{E_c - E}{w_{tail}}\right) \quad (10)$$

TFT simulation has been used to assess the impact of the four parameters of interest in Equation (10) to device characteristics (μ, *Vth* and *S*) [65]. The relevant simulation results are summarized in Fig. 38 (simulations for $W \times L = 8 \times 1.3$ μm, with a 50 nm-thick active layer and 50 nm-thick SiO_2 GI layer). Based on the simulated performance, one can select values for the parameters of interest in Equation (10) describing an 'ideal' device and compute the corresponding DOS function. The values selected were: N_{tail} ~2 × 10^{18} cm^{-3}, N_{deep} ~1 × 10^{17} cm^{-3}, w_{tail} ~0.01 eV and w_{deep} ~0.10 eV. The corresponding, computed DOS trace is shown in Fig. 37. This result clearly indicates the necessity for further improvements on the material crystal quality to enable consistent, SOI-like TFT performance for devices with near-micrometer and sub-micrometer channels.

11.5. Remaining Challenges in Si Crystallization

Laser-based crystallization for TFT has progressed significantly, since its introduction in the late 1980s. Both in terms of process uniformity and crystal quality, significant strides have been made by the progress in laser source technology and the introduction of lateral growth process technology. As a result of the improvements in poly-Si microstructure, TFT performance rivaling that of SOI

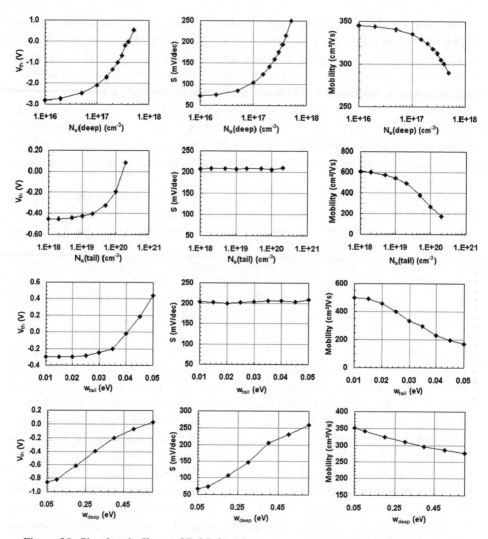

Figure 38: Simulated effects of DOS function parameters on TFT mobility, threshold voltage and subthreshold slope.

has been reported by some research studies. However, upon careful evaluation of the statistical performance of TFT characteristics it has become clear that the variation in TFT data is still present, albeit clustered around better average numbers. The main reasons behind the variation in TFT performance are the occasional inclusion of defective regions within the active layer and the lack of precise control of the crystallographic orientation of Si material within the device channel.

Defect formation in the poly-Si film is intimately related to the stress build up in the film due to the extreme thermal gradients experienced during the course of lateral growth. The onset of defect formation is difficult to predict, as it is ultimately dependent on the local thermal environment and microstructure. Even for a perfectly seeded material (i.e. SOI) exposed to laser irradiation, defect formation has been observed to occur within a few microns from the origin point. Therefore, any crystallization scheme that enables continuous lateral growth over long distance is bound to suffer from stochastic defect incorporation, within the laterally grown material. When such defective regions are incorporated into a device, significant deviation in performance will be experienced. This problem is further exacerbated by the decreasing trend in channel dimensions for next generation TFTs. Under these conditions, no sufficient 'averaging' can occur to moderate the device-to-device variation.

From this point of view, both control of the extent of lateral growth and the crystallographic orientation within the laterally grown region appear necessities to improve TFT performance and, most importantly, curtail variation. These are not trivial pursuits and present substantial challenges for the next step in improving silicon crystallization technology.

References

[1] Mimura, A., Konishi, N., Ono, K., Ohwada, J.I., Hosokawa, Y., Ono, Y.A., Suzuki, T., Miyata, K. and Kawakami, H., *IEEE Trans. Electron Dev.*, **1989**, *36*, 351.

[2] Voutsas, A.T. and Hatalis, M.K., *J. Electrochem. Soc.*, **1992**, *139*, 2659.

[3] Voutsas, A.T. and Hatalis, M.K., *J. Electrochem. Soc.*, **1993**, *140*, 871.

[4] Voutsas, A.T. and Hatalis, M.K., *J. Electron. Mat.*, **1994**, *23*, 319.

[5] Wagner, R.S. and Ellis, W.C., *Appl. Phys. Lett.*, **1964**, *4*, 89.

[6] Spaepen, F., Nygren, E. and Wagner, A.V., *NATO ASI Series E: Applied Sciences*, **1992**, *222*, 483.

[7] Hultman, L., Robertsson, A., Hentzell, H.T.G. and Engstrom, I., *J. Appl. Phys.*, **1987**, *62*, 3647.

[8] Radnoczi, G., Robertsson, A., Hentzell, H.T., Gong, S.F. and Hasan, M.A., *J. Appl. Phys.*, **1991**, *69*, 6394.

[9] Nemanichi, R.J., Tsai, C.C., Thompson, M.J. and Sigmon, T.W., *J. Vac. Sci. Technol.*, **1981**, *19*, 685.

[10] Park, S.J., Cho, B.R., Kim, K.H., Cho, K.S., Yoo, S.Y., Kim, A.Y., Jang, J. and Shin, D.H., *SID Symposium Digest*, **2001**, *XXXII*, 562.

[11] Yoon, S.Y., Oh, J.Y., Kim, C.O. and Jang, J., *J. Appl. Phys.*, **1998**, *84*, 6463.

[12] Poate, J.M. and Mayer, J.W., *Laser Annealing of Semiconductors*, Academic Press, New York, **1982**.

[13] Fiebig, M., Stamm, U., Oesterlin, P., Kobayashi, N. and Fechner, B., *SPIE Proceedings*, **2001**, *4295*, 38.

[14] Fogarassy, E., Prevot, B., de Unamuno, S., Prat, C., Zahorski, D., Helen, Y. and Mohammed-Brahim, T., *Mat. Sci. Res. Symp. Proc.*, **2001**, *685E*, D7.1.1.

[15] Ready, J.F., *Industrial Applications of Lasers*, Academic Press, New York, 2nd edition, **1997**.

[16] Hara, A., *IEDM Tech. Digest*, **2001**, 34.2.1.

[17] Flores, J., Droes, S., Joshi, P., Crowder, M.A. and Voutsas, A.T., *Laser Focus World*, July, **2002**.

[18] *Private communication with Cymer, Inc.*

[19] Im J.S., Kim H.J. and Thompson, M.O., *Appl. Phys. Lett.*, **1993**, *63*, 2969.

[20] Crowder, M.A., PhD Dissertation, Columbia University, **2001**.

[21] Sinke, W. and Saris, F.W., *Phys. Rev. Lett.*, **1984**, *53*, 2121.

[22] Im, J.S. and Kim, H.J. *Appl. Phys. Lett.*, **1994**, *64*, 2303.

[23] Brotherton, S.D., McCoulloch, D.J. and Edwards, M.J., *Solid State Phenomena*, **1994**, *37–38*, 299.

[24] Kuriyama, H., Sano, K., Ichida, S., Nohda, T., Aya, Y., Kuwahara, T., Noguchi, S., Kiyama, S., Tsuda, S. and Nakano, S., *Mat. Res. Soc. Symp. Proc.*, **1993**, *297*, 657.

[25] Voutsas, A.T., Prat, C. and Zahorski, D., *IDRC '00 Conference Record*, **2000**, 451.

[26] Brotherton, S.D., McCulloch, D.J., Clegg, J.B. and Gowers, J.P., *IEEE Trans. Electron Dev.*, **1993**, *40*, 407.

[27] Hatalis, M.K. and Voutsas, A.T., *Proceedings of RTP 2001 Conference*, **2001**.

[28] Im, J.S. and Sposili, R.S., *MRS Bulletin*, **1996**, *21*(3), 39.

[29] Im, J.S., Crowder, M.A., Sposili, R.S., Leonard, J.P., Kim, H.J., Yoon, J.H., Gupta, V.V., Song, H.J. and Cho, H.S., *Phys. Stat. Sol. A*, **1998**, *166*, 603.

[30] Makihira, K. and Asano, T., *Proceedings of AMLCD '00*, **2000**, 33.

[31] Kim, H.J. and Im, J.S., *Appl. Phys. Lett.*, **1996**, *68*, 1513.

[32] van der Wilt, P., Ishihara, R. and Bertens, J., *Mat. Res. Soc. Symp. Proc.*, **2000**, *621*, Q7.4.1.

[33] Sposili, R.S. and Im, J.S., *Appl. Phys. Lett.*, **1996**, *69*, 2864.

[34] Matsumura, M., *SPIE Proceedings*, **2001**, *4295*, 1.

[35] Kim, C.-H., Song, I.-H., Nam, W.-J. and Han, M.-K., *IEEE Electron Dev. Let.*, **2002**, *23*, 315.

[36] Matsumura, M., *Proceedings of IDRC '99*, **1999**, 351.

[37] Helen, Y., Dassow, R., Nerding, M., Mourgues, K., Raoult, F., Köhler, J.R., Mohammed-Brahim, T., Rogel, R., Bonnaud, O., Werner, J.H. and Strunk, H.P., *Thin Solid Films*, **2001**, *383*, 143.

[38] Sasaki, N., Hara, A., Takeuchi, F., Mishima, Y., Kakehi, T., Yoshino, K. and Takei, M., *SID Symposium Digest*, **2002**, *XXXIII*, 154.

[39] Sposili, R.S., PhD Dissertation, Columbia University, **2001**.

[40] Crowder, M.A., Limanov, A.B. and Im, J.S., *Mat. Res. Soc. Symp. Proc.*, **2000**, *621E*, Q9.6.

[41] Voutsas, A.T., Limanov, A. and Im, J.S., *J. Appl. Phys.*, **2003**, *94*, 7445.

[42] Cho, H.S., Kim, D.-B., Limanov, A.B., Crowder, M.A. and Im, J.S., *Mat. Res. Soc. Symp. Proc.*, **2000**, *621*.

[43] Cullis, A.G., Webber, H.C., Chew, N.G., Proate, J.M. and Baeri, P., *Phys. Rev. Lett. B*, **1982**, *49*, 219.

[44] Sato, T., Takeishi, Y., Hara, H. and Okamoto, Y., *Phys. Rev. B*, **1971**, *4*, 1950.

[45] Ditchburn, R.W., *Light*, Chapter 15, Dover, **1991**.

[46] Kisdarjono, H., PhD Dissertation, Oregon Graduate Institute, **2004**.

[47] Jena, A. K. and Chaturvedi, M.C., *Phase Transformation in Materials*, Prentice Hall, **1992**.

[48] Kisdarjono, H., Voutsas, A.T. and Solanki, R., *J. Appl. Phys.*, **2003**, *94*, 4374.

[49] Leonard, J. and Im, J., *Appl. Phys. Lett.*, **2001**, *78*, 3454.

[50] Voutsas, A.T., Kisdarjono, H., Solanki, R. and Kumar, A., *Proceedings of SISPAD 2001,* Springer-Verlag Wien, **2001**, 132.

[51] US Patent 6,368,945.

[52] US Patent 6,563,077.

[53] Personal communication with Prof. James Im of Columbia University.

[54] Crowder, M.A., Voutsas, A.T., Droes, S., Moriguchi, M. and Mitani, Y., *IEEE Trans. Electron Dev.*, **2004**, *51*, 560.

[55] Crowder, M.A., Moriguchi, M., Mitani, Y. and Voutsas, A.T., *Thin Solid Films*, **2003**, *427*, 101.

[56] Voutsas, A.T., *IEEE Trans. Electron Dev.*, **2003**, *50*, 1494.

[57] Brotherton, S.D., Crowder, M.A., Limanov, A.B., Turk, B. and Im, J.S., *Proceedings of Asia Display/IDW '01*, **2001**, 387–390.

[58] Jung, Y.H., Yoon, J.M., Yang, M.S., Park, W.K., Soh, H.S., Cho, H.S.,Limanov, A.B. and Im, J.S., *Mat. Res. Soc. Symp. Proc.*, **2000**, *621E*, Q8.3.1.

[59] Crowder, M.A., Limanov, A.B. and Im, J.S., *Mat. Res. Soc. Symp. Proc.*, **2000**, *621E*, Q9.6.1.

[60] Hara, A., Takeuchi, F., Takei, M., Suga, K., Yoshimo, K., Chida, M., Sano, Y. and Sasaki, N., *Proceedings of AM-LCD'02 Conference*, **2002**, 227.

[61] Hatano, M., Shiba, T. and Ohkura, M., *Proceedings of SID 02 Digest*, **2002**, *XXXIII*, 158.

[62] Voutsas, A.T., *to be published in IEEE Electron Dev. Lett.*

[63] Mizuki, T., Matsuda, J., Nakamura, Y., Takagi, J. and Yoshida, T., *IEEE Trans. Electron Dev.*, **2004**, *51*, 204.

[64] Bonfiglietti, A., Valletta, A., Gaucci, P., Mariucci, L., Fortunato, G. and Brotherton, S.D., *to be published in IEEE Trans. Electron Dev.*

[65] Afentakis, T. and Voutsas, A.T., *to be published in IEEE Trans. on Electron Dev.*

Chapter 12

Long Pulse Excimer Laser Doping of Silicon and Silicon Carbide for High Precision Junction Fabrication

E. Fogarassy[a] *and J. Venturini*[b]

[a]*CNRS/ULP – InESS, BP 20, 67037 Strasbourg Cedex 2, France;* [b]*SOPRA Laser, 13-21 Quai des Grésillons, 92230 Gennevilliers, France*

Abstract

As an alternative to classical thermal heating, laser annealing (LA) appears very suitable for junction formation both in silicon (Si) and silicon carbide (SiC). In this work, we demonstrate the possibility to anneal ion-implanted damage and activate the dopants into Si and SiC by laser processing using an XeCl excimer source of 200 ns-pulse duration. Simulation calculations highlight the added value brought by the relatively long thermal cycle induced in Si and SiC from the long duration laser pulse. Physical characterizations of the junctions show a very efficient dopant activation and defects annealing while the electrical properties measured on real device structures confirm the great potential of the 200 ns-pulse duration excimer laser annealing for dopant activation both in Si and SiC.

Keywords: Annealing; Dopant activation; Excimer laser; Ion implantation; Junctions; Silicon; Silicon carbide; Solid phase; Thermal simulation.

12.1. Introduction

In the last 20 years, laser processing of semiconductors has been regarded as one of the most promising technique for manufacturing microelectronic devices both in silicon (Si) and silicon carbide (SiC) according to the International Technology Roadmap of Semiconductors (ITRS) requirements [1].

In silicon, the Ultra-Shallow Junction (USJ) formation for the sub-65 nm CMOS (Complementary Metal-Oxide Semiconductor) node is a major challenge.

Recent Advances in Laser Processing of Materials
J. Perrière, E. Millon and E. Fogarassy (Editors)

As an alternative to Rapid Thermal Processing (RTP), various nanosecond-pulsed laser doping techniques, including laser annealing of ion-implanted dopants [2–4] and laser induced diffusion of dopants from gas (GILD) [5–7] and solid sources (Spin-On-Glass) [8–10] were investigated recently in order to achieve such requirements. These approaches offer the possibility to suppress Transient Enhanced Diffusion (TED) phenomena observed in RTP [11,12] by applying ramp up and down times in the order of hundred of nanoseconds, which enables USJ (less than 20 nm in depth) formation, very abrupt profiles and a solubility limit higher than that usually encountered with any conventional rapid thermal process.

Silicon carbide is a wide-gap semiconductor which presents unique material properties especially suitable for high temperature, high power and high frequency applications. However, SiC device fabrication has to face various technological difficulties. Among them, one of the more crucial appears to be the doping step. The high melting point and limited diffusion of impurities into SiC have greatly restricted the use of ion implantation and furnace annealing commonly employed in the silicon microelectronics industry to incorporate and activate the dopants. As for Si, excimer laser processing was demonstrated to be suitable for the doping of SiC in various experimental conditions.

In Fig. 1, we show the time evolution over years of the annealing tools used in semiconductors industry as followed by ITRS. We draw attention here on the fact that several generations of technologies are co-existing in the same manufacturing site. Pulsed laser tools will begin to be used in production from the year 2006 and are inducing thermodynamics 10^6 times faster than present annealing tools.

Both short (~20 ns) and long pulse (~200 ns) duration excimer lasers were used to deposit a large amount of energy in short time into the near-surface region of the semiconductor, while maintaining the substrate essentially at room temperature. Under suitable conditions, the irradiation could lead to surface melting to a depth ranging from a few tens to a few hundred nanometers and its rapid solidification from the bulk, allowing the dopant to be incorporated by liquid phase diffusion. In addition, by taking advantages of the highly non-equilibrium nature of the melt/regrowth process, high level electrical activation of the dopant can be achieved both in Si and SiC.

In addition, in order to overcome certain limitations related to the liquid phase, such as uncontrolled diffusion phenomena, surface degradation and stoechiometry change (in the case of SiC), recent studies suggested working in the sub-melt regime.

Here we review the most recent works performed in the long pulse duration regime for laser doping of silicon and silicon carbide by comparison to the more classical short pulse duration excimer laser processing. We focus here on the specificity of the thermal regime induced by the long laser pulse. We show here

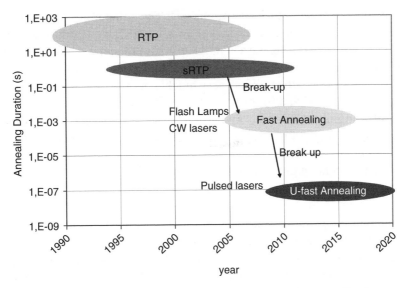

Figure 1: Roadmap of hot process tools for semiconductors dopant activation according to ITRS.

that there is a trade-off between the high speed and the high temperature of heating required to activate the dopants over the solid solubility limit and the relative longer time necessary to anneal properly the defects from the implant.

We show that in a melt regime the long pulse is well adapted to the formation of ultra-shallow junction in amorphized silicon where the laser process window is enlarged noticeably thanks to the explosive crystallization process.

We also demonstrate the possibility to anneal ion-implanted damage into Si and SiC by solid phase laser processing using an XeCl excimer source of 200 ns-pulse duration.

Finally, the electrical properties measured on real device structures confirm the great potential of this novel source (model VEL from SOPRA) for high precision junction fabrication both in Si and SiC.

12.2. Laser Doping of Silicon

12.2.1. Introduction: CMOS Technology

As the International Technology Roadmap of Semiconductors (ITRS) plans to introduce in production increasingly small CMOS devices, features size of the electrical junctions of these devices reach depths in the 10 nm range with dopant activation concentration upto 1.10^{21} at/cm^3 [1]. The structure of a present CMOS device at the step of dopant activation is shown in Fig. 2. Non-controlled diffusion of the dopant in silicon during the activation process is a major problem with p-type CMOS where small boron (B+) ions are commonly implanted before annealing. Boron atoms in silicon lattice diffuse very quickly at the solid phase when temperature overpasses 800°C and for duration bigger than millisecond (ms) range. One of these diffusion phenomenon is known as Transient Enhanced Diffusion (TED) and is a complex chain of processes involving several types of defects induced by the boron implant [11,12].

None of the technological solution used today in CMOS industry is able to achieve ultra-shallow extensions with a proper device operation and a good control of the manufacturability.

By applying temperature ramp up and ramp down in the order of hundreds of nanoseconds, pulsed laser can heat up a very shallow silicon surface layer while reaching higher solid state dopant solubility limit than those featured by other standard thermal process technologies. As pictured in Fig. 1, ITRS predict that ultra-fast annealing featured by pulsed laser tools will begin to be used in semiconductor fabs from today until at least two decades later. These predictions are

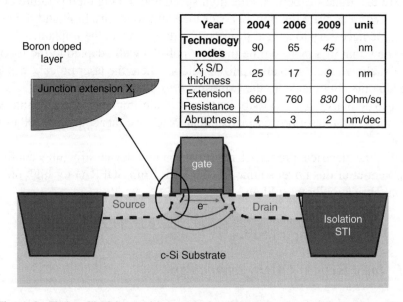

Year	2004	2006	2009	unit
Technology nodes	90	65	45	nm
X_j S/D thickness	25	17	9	nm
Extension Resistance	660	760	830	Ohm/sq
Abruptness	4	3	2	nm/dec

Figure 2: Today CMOS transistor structure. The table shows the corresponding ITRS specifications of junction depth and resistance for present and future CMOS technology nodes.

justified by the very broad capabilities of processes allowed by pulsed UV lasers. Indeed, the strong absorption of Si, SiC or even Ge, which is today integrated as a new high-performance material, induce a very well controlled process where the depth of the annealed layer ranges around the absorption depth of the material (7 nm for Si).

As we will see in the next paragraph, there is a trade-off in the annealing duration. The temperature ramp up shall be fast enough to reach a high activation rate of the dopant and avoiding dopant diffusion, while lasting enough in time to anneal the defects properly.

12.2.2. Thermal Simulations

The precise control of the laser doping process requires a good knowledge of the basic mechanisms governing the interaction between the incident laser beam and the irradiated material. This mechanism involves heat generation at the surface of the irradiated area by light absorption, followed by cooling through heat conductivity into the substrate [8].

Heating and cooling stages were determined by solving numerically the heat flow equation and using the appropriate optical and thermal parameters for silicon [13]. Simulations have been calculated with FIDAP® software from FLUENT Inc. Details on the model used for calculations are provided in ref [14–15].

We remind here the equation used:

$$\frac{\partial T(x, y, z, t)}{\partial t} = \nabla(D\nabla T(x, y, z, t)) + \frac{S_{\text{laser}}(x, y, z, t)}{C_{\text{p}}(T)}$$

where T is the temperature, $D = k/C_{\text{p}}$ is the thermal diffusivity of Si or SiC, C_{p} is the heat capacity, k the heat conductivity and $S_{\text{laser}}(x, y, z, t)$ is the laser source term induced by the laser-matter interaction defined by:

$$S_{\text{laser}}(x, y, z, t) = I_0(x, y, t)(1 - R)f(z)$$

where $I_0(x, y)$ is the laser intensity at the material surface ($z = 0$), R the reflection coefficient at 308 nm, α the absorption coefficient and where $f(z) = \alpha e^{-\alpha z}$ describe the variation of the laser intensity in the material taking into account the absorption versus z [16].

As shown in numerical data of Fig. 3 corresponding to silicon, the surface melting threshold (E_{m}, in J/cm²)) strongly depends on both the XeCl pulse duration (ranging from 20 to 200 ns) and properties of the irradiated material. It has to be

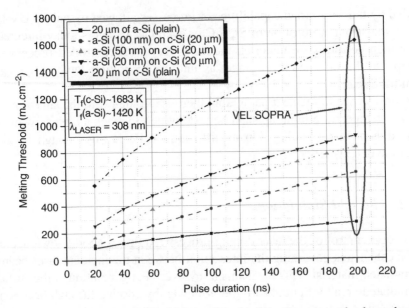

Figure 3: Simulated melting threshold for c-Si and a-Si/c-Si versus pulse laser duration at 308 nm.

noticed that, whatever the pulse duration, the heating process is more efficient in silicon than in silicon carbide. In addition, a diminution of E_m is measured when irradiated surface is non-crystalline (amorphous). For a pulse of 200 ns duration (VEL laser) the melting threshold in c-Si (Fig. 3) is close to 1.7 J/cm², while in a-Si E_m is ranging between 1.3 and 0.7 J/cm², depending on the thickness (20 to 100 nm) of the amorphized surface layer.

Temperature dynamics are fundamental in defects annealing, dopants diffusion [17] and in dopants activation rates. This parameter has a major effect on the rate of activation of the dopants and the diffusion of small dopant atoms (like boron in Si).

Time duration and also ramp-up and -down rates of the thermal process in the solid regime (just below the melting threshold) can also be extracted from the simulations.

Figure 4 shows the evolution of c-Si surface temperature versus time for different laser pulse duration (from 20 to 200 ns). In Fig. 5 we compare the simulated ramp-up rate (dT/dt) at the silicon surface under heating from different annealing tool available today. The temperature ramp-up rates of the pulsed laser processes are much larger than for the RTP tool ranging only around 10^3 °C/s in the best case. One can notice the break-up in the temperature dynamics between the classical annealing technology tools used in the present CMOS silicon technology, like

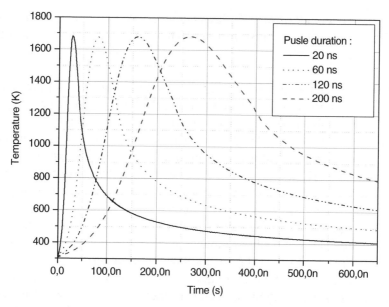

Figure 4: Evolution of the surface temperature of c-Si (at the melting threshold) versus laser pulse duration (from 20 to 200 ns).

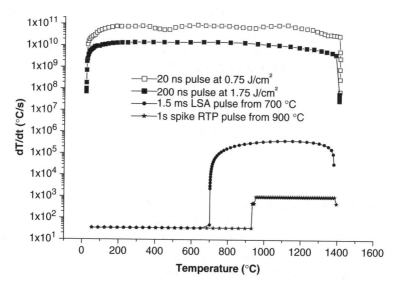

Figure 5: Simulated temperature ramp-up rates in silicon as a function of the temperature for two different laser pulses (25 ns and 200 ns) and for two other classical annealing tools (LSA: Laser Spike Annealing and Spike RTP: Spike Rapid Thermal Processing).

Laser Spike Annealing (using CW lasers comparable to the temperature dynamic of the Flash Lamps tools) or Spike Rapid Thermal Processing tools, and the two pulsed laser technologies. The rates of temperature dynamics are jumping from 10^3 and 10^6 up to 10^{10}.

Nevertheless, although very high ramp-up and ramp-down rates are required to avoid Transient Enhanced Diffusion (TED) of the boron in Si, their values should be kept under a value ensuring a good rate of defects annealing.

Indeed, contrary to the 200 ns laser annealing process, the 25 ns laser annealing process is moving further from an equilibrium thermodynamic process in a regime called adiabatic [18]. In this regime, the system made by the upper atomic layers of the silicon surface does not have enough time to exchange efficiently the heat with the underlying Si substrate during the time of the irradiation process. Particularly, in the case where melting occurs, lower solidification velocities are likely to improve crystal quality following recrystallization in the full melt and in the explosive regime for PAI (Pre-Amorphization Implant) samples [19].

We summarize in Fig. 6 the thermal processes currently used in semiconductor industry for dopant activation and compare them, on a thermodynamic scale, to the ultra-fast processes induced by laser thermal processing (LTP). In this diagram, we call adiabatic regime, the regime where the time constant of the heat diffusion in the material (Si or SiC) begin to be longer than the duration of heating of the surface of the sample from the annealing tool. This regime induce the risk to reach the melting (or even the damage through ablation or superheating of the melting phase) of the material before any efficient annealing of the materials occurs through reduction of the rate of defects or proper activation of the dopants [18].

12.2.3. Laser Annealing of Implanted Silicon (Liquid and Solid Phase)

12.2.3.1. Introduction
Both solid and liquid phase laser induced activation in silicon has been studied. When one deal with such fast silicon thermodynamics like those induced by pulsed excimer lasers, the original crystalline state of the implanted silicon before laser annealing plays an important role on the activation process and efficiency.

When the silicon sample is directly implanted with boron atoms, the rate of defects from implant in the silicon lattice is not inducing amorphization of Si. On the other hand, when the sample is pre-implanted by germanium or silicon atoms to avoid short channel effect during the post-boron implantation, an upper layer of the silicon sample is amorphized.

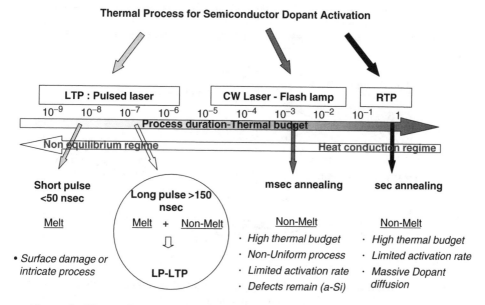

Figure 6: Thermodynamic diagram of the physical processes and thermal regime involved with different activation annealing tool.

The work presented here summarizes two types of laser processes which depend on the original crystalline state of the sample. The first deals with direct boron implantation where both solid and melting phase processes occur in a crystalline structure close to the c-Si one. The second concerns pre-amorphized samples. In this case, the melting phase is required in order to anneal properly the amorphized Si layer and the end of range defects lying beyond the a-Si/c-Si interface.

In the first case, when the sample is not amorphized and the laser energy density is higher than the melting threshold, the melting front propagates through the depth of the sample in a single liquid layer until, following a fast cooling phase, it solidificates back to the surface. In the second case, when an amorphized Si layer is present, a more complex process occurs where several liquid layers co-exist and propagate in the a-Si layer. The different mechanisms involved versus the implantation condition are summarized in Fig. 7.

We present here the experimental results and modelisation of the long-pulse laser activation process of Si wafers implanted under several different conditions and at different laser energy density including solid phase annealing. In terms of process integration, solid phase appears well adapted to the present process flow requirements for CMOS manufacturing while melting phase offers the possibility

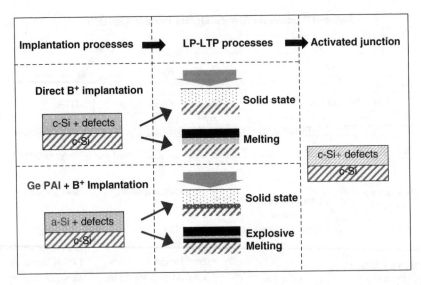

Figure 7: Physical processes induced by Long Pulse Laser Thermal Processing of a silicon implanted layer.

to raise the best activation rate and appears more adapted in terms of laser process window to pre-amorphized implanted wafers.

The experimental results are also discussed and compared to the results we obtained by simulation calculations.

12.2.3.2. Laser Experimental Setup

Laser irradiations were carried out at ambient temperature and in Ar atmosphere using the XeCl (308 nm) excimer source from SOPRA (VEL Model). Pulses of 200 ns in duration can be delivered at maximum output energy of 15 J. The maximum energy density (fluence) is 3.5 J/cm^2 over an area of 4 cm^2. X-ray pre-ionization enables this laser to reach a pulse-to-pulse stability better than ±1%.

The experiments done with the long pulse laser were compared to those performed with a 25 ns-duration excimer source (COMPEX model) from LAMBDA PHYSIKS.

The homogenizing optics employed with the VEL SOPRA are designed to cover a whole chip area in a single pulse. It is composed of a double fly eyes matrix and a multiple lens objective ensuring uniformity over the whole area of ±1.5%. This value is calculated as $(E_{max} - E_{min})/(E_{max} + E_{min})$ where energies E_{max} and E_{min} are defined on the top hat part of the beam. The design and coatings of the optics are providing, at the same time, high uniformity and high power delivery. The experimental setup used in SOPRA is shown in Fig. 8. An *in situ* reflectivity CW He-Ne laser probe has been setup to evaluate melting of the sample surface and to follow the surface state of the sample during the excimer laser irradiation [20–21].

The dynamic reflectivity signal not only gives the onset of the melting of the surface samples but is also modelised and fitted to retrieve the dynamic of complex melting processes occurring when a thin upper layer (10–50 nm) is pre-amorphized (in the case of Si).

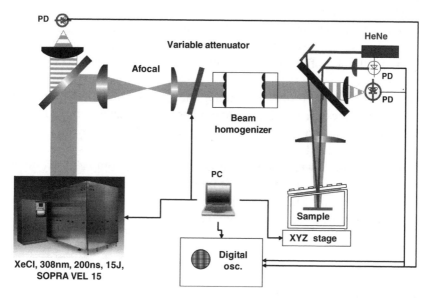

Figure 8: Long Pulse Laser Thermal Processing experimental setup with the *in situ* reflectivity setup. PD: photodiodes.

12.2.3.3. Wafers Implantation, Experimental Conditions and Reflectivity Modelisation

Boron was implanted in (100) Si wafers at 0.5 and 3 keV at a dose of $2.10^{15}/cm^2$ and $1.10^{15}/cm^2$ with or without germanium PAI. Ge PAI were done at 10 keV at a dose of 1.10^{15} which gives a a-Si layer of ~20 nm over the c-Si substrate. BF2 implant were also performed at 3 and 5 keV at a dose of 3.10^{15} /cm².

Secondary Ion Mass Spectroscopy (SIMS) characterization has been done with a CAMECA 5F tool using an O_2^+ primary beam. Spread sheet resistances (R_s) have been measured with a CDE ResMap 4-point probe R_s mapping tool. The Spreading Resistance Profile (SRP) was done by Cascade Europe. Scanning (SEM) and Transmission (TEM) Electron Microscopy pictures were also performed.

Samples have been laser annealed with the two excimer lasers previously described. Single and triple-shot irradiations were conducted at different laser energy densities in N_2 environment at room temperature. The reflectivity curves versus time have been fitted through a MATLAB® code to retrieve the different melting dynamics when an upper silicon layer has been amorphized.

12.2.3.4. Results

a) SIMS profile. Figure 9 shows a typical SIMS profile obtained on a Ge pre-amorphized sample and irradiated with the 200 ns laser pulse over the melting threshold of the amorphous layer. One can appreciate a 2.4 nm/decade steep abruptness of the resulting profile which has been obtained with a low SIMS primary ion energy (500 eV). This ion beam energy of the SIMS tool is crucial to get the real abruptness of the dopant profile, as probing the sample with 1 keV primary beam will lead to abruptness profile 2 or 3 time higher for the same sample [22].

This SIMS profile shows also that the dopants are homogeneously distributed over the a-Si pre-amorphized layer (~20 nm) without overpassing the a-Si/c-Si interface. Indeed, the lower melting threshold of the a-Si compared to c-Si allows the melting front to stop at the a-Si/c-Si interface, annealing and melting the End Of Range (EOR) region where lies a high rate of defects and bouncing back to the surface. Through this, a full re-epitaxy takes place from the interface back to the interface. The details of this mechanism are made explicit in 12.2.3.4d.

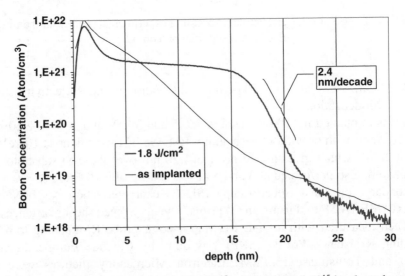

Figure 9: SIMS profile of a Ge 10 keV 1.10^{15} + B 0.5 keV 5.10^{15} implanted sample annealed at 1.8 J/cm² with SOPRA VEL laser compared to the as-implanted profile.

SIMS profile gives the distribution of both the electrically activated and non-activated dopant. In the case where the melt front overpasses the as-implanted profile, the depth of the activated profile follows the SIMS profile. The maximum rate of activated dopants can reach 60% for doses up to 5.10^{15}. In the contrary, when the melt front does not overpass the implantation profile or when the laser energy density lie under the melting threshold, the SIMS profile does not reflect anymore the carrier profile accurately. For this, Spreading Resistance Profiling (SRP) is necessary to follow the real profile of the dopants that have been activated by the laser annealing process. Figure 10 shows the SRP profile of a BF_2 implanted wafer after laser annealing at 1.6 J/cm^2, a value lying just under the melting threshold of the sample [23].

We superimposed over the SRP profile, the SIMS profiles of the as-implanted and laser annealed samples to show that no melting and neither dopant diffusion occurred. A 30% activation rate in a *non-melt* regime has been obtained through a single laser annealing process. Others studies on non-melt annealing processes

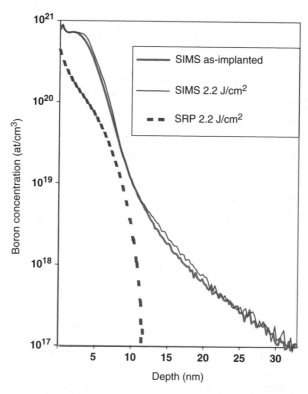

Figure 10: SIMS and SRP (Spreading Resistance Profiling) results showing the diffusionless (SIMS) and abrupt activation profile (SRP) process from the 200 ns irradiation LTP. Implantation condition is BF_2 5 keV 1.10^{15}.

involving shorter laser pulses brought the necessity to use a complementary 'slow' annealing tool (RTP) before the laser process [24]. This non-melt regime obtained here is extremely important in terms of integration of the pulsed laser process in the CMOS fabrication process flow.

b) Sheet resistance results. To compare the effect of the different laser pulse duration on the activation rate of the dopants, we have performed melt and non-melt irradiation on different implanted wafers on the two laser setups with 25 ns and 200 ns laser pulse durations, respectively. We summarize on Fig. 11 the sheet resistances obtained at different laser energy densities for both lasers and for a direct boron implantation without amorphizing the silicon wafers.

To assess when the R_s data measured correspond to a solid or liquid silicon phase process, we have plotted the surface melting durations obtained from the *in situ* He-Ne reflectivity setup on the same graph.

It is again here interesting to compare the results in terms of electrical activation for the melting regime and the non-melting regime. The breaks in the melting

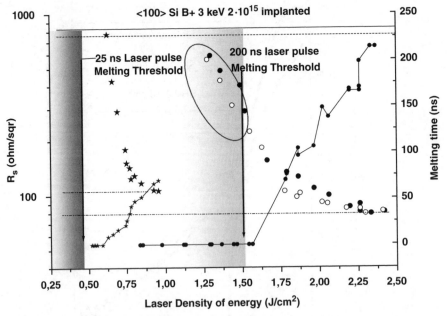

Figure 11: R_s and melting duration versus energy density for B^+ 3 keV/2.10^{15} implanted samples, laser annealed at the two pulse durations of (★) 25 ns and (●) 200 ns.
(o) triple-shot processing. (★-★ and ●-●) corresponding melting duration; Horizontal: maximum R_s-values for the 65 nm (----) and 45 nm (——) ITRS nodes;
(·· — ··) saturation of activation for both laser anneals.

duration curves occur at the melting thresholds. We can first observe that, as expected, the energy necessary to melt c-Si with a 25 ns is smaller than with a 200 ns pulse process. This difference confirms the non-equilibrium thermodynamics induced in silicon by shorter laser pulses. A large amount of R_s data points are gathered under the maximum R_s required for the 65 nm and 45 nm ITRS nodes. But what is particularly interesting to stress here is that, in contrast to the 25 ns laser process, non-melt relevant R_s were obtained with the 200 ns laser pulse irradiations. We also observe that subsequent shots (up to three in the present experiment) on the same location induce improvement in R_s for low laser energy densities, particularly in the non-melt regime, where up to 20% improvement in R_s is achieved. The extension junction depth (X_j) values in this non-melt regime are identical to or smaller than the as-implanted ones since no diffusion of the dopants occurs at these temperature ramp rates [14].

As for the short 20 ns pulses, previous work reported that more than 100 pulses in a non-melt regime were required to reduce the R_s down to 800 Ohm/sq and to anneal properly the implant damage [25].

We see that the maximum activation rates in the melting regime (saturation of the lowest R_s-value while increasing energy density) are higher in the case of the 200 ns laser. This behavior is not expected since the activation should be better with higher temperature ramp-up and -down rates. On the other hand, if we look at the diffusion of boron in the liquid phase (10^{-4} cm^2/s) and the extremely fast solidification front induced by the 25 ns laser pulse suggested by the simulations (5 to 10 m/s), fall in the boron concentration at the extend of the melting front could be expected. This result shall be studied further since the segregation coefficient k is supposed to be close to one at these melting front speeds [26].

Finally, Fig. 12 shows a comparison of the 200 ns laser activation for pre-amorphized wafers. Figure 12 also shows the relevant X_j values, when available from SIMS profiles in the melting regime.

c) Comparison with the simulations. In Fig. 13 we show the computer simulations of the melt depth versus the incident laser energy density and compare them to experimental results. The melt depths are extrapolated from the SIMS profiles.

The correlation between simulations and experimental results is good for the 200 ns laser pulse for all considered implantation and PAI conditions while for the 25 ns, the fit is clearly shifted for the bulk sample. At this step of the simulation work, the 25 ns pulse is not reproduced by our model based on equilibrium heat diffusion equation, while the 200 ns pulse experimental results almost fit an equilibrium thermodynamic model. These results show how, contrary to the 200 ns laser annealing process, the 25 ns laser annealing process is moving further from an equilibrium thermodynamic process.

Table 1 shows the simulation of the deviation of the melt depth due to pulse to pulse laser energy instabilities for both laser melting processes. For a process energy

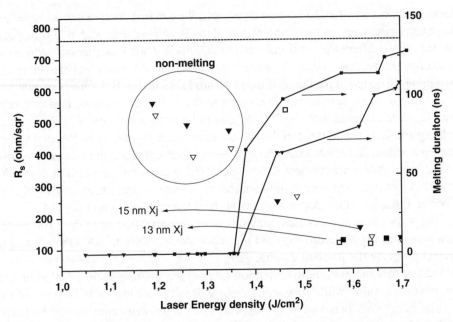

Figure 12: R_s value from SOPRA 200 ns annealed samples showing the sub-melt R_s lying under the maximum R_s-values for the 65 nm and 45 nm ITRS nodes. (■) Ge PAI + B+ 0.5 keV/5.10^15. (▼) Ge PAI + B+ 3keV/ 2.10^15. Relevant X_j are indicated.

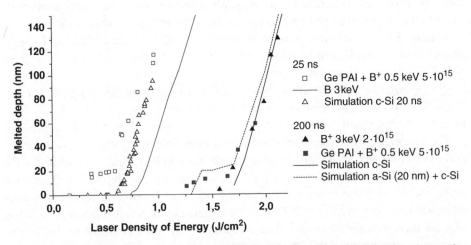

Figure 13: Simulated (dashed: 25 ns laser pulse, continuous: 200 ns laser pulse) and experimental results of melt depth as a function of the melting time. (□) 25 ns irradiation of Ge PAI + B+ 3 keV. (o) 25 ns irradiation of B+ 3 keV. (■) 200 ns irradiation of Ge PAI + B+ 3 keV. (●) 200 ns irradiation of B+ 3 keV.

Table 1: Simulation of ΔX_j during the process of a 5 and 13 nm junction depth by the different laser pulses applying the $3\sigma = 1.5\%$ energy stability of SOPRA VEL15 Laser. The values reported for the 25 ns laser pulse are theoretical minimum since 3σ stability is larger for this laser in the present study

Pulse duration	Energy density of the process during melting	ΔX_j for 5 nm X_j and $\sigma_{Elaser} = 0.5\%$	ΔX_j for 13 nm X_j and $\sigma_{Elaser} = 0.5\%$
25 ns	600 mJ/cm^2	0.6 nm (theor. min.)	1.5 nm (theor. min.)
200 ns	1500 mJ/cm^2	0.3 nm	1.2 nm

density next to the melting threshold, the deviation for the 200 ns laser process is around two times better than the 25 ns laser process for a 5 nm X_j and similar while better for a 13 nm X_j. Moreover, these deviations are optimistic for the 25 ns pulse process since they have been calculated with laser energy instabilities of 1.5% at 3σ, which is the value provided by the SOPRA industrial tool.

These last observations confirm the peculiar thermodynamic regime induced by the 200 ns pulse which is fast enough to activate efficiently the dopant without diffusion and slow enough to enable efficient dopant activation in a solid phase regime.

d) Explosive crystallization
(i) Introduction
The case of PAI wafer exhibits a peculiar laser activation regime and is treated independently. Particularly, the long laser pulse induces here a process able to enlarge the laser activation process window.

The amorphous phase of the silicon of the upper layer of a PAI wafer is energetically metastable relative to the crystalline phase. The free energy difference that exist between this two phases can be released through different heating mechanisms. The classical method consists in heating up the a-Si layer in a furnace (~450–700°C) for several minutes or hours. This technique, called Solid Phase Epitaxy Regrowth (SPER), leads to a re-epitaxy of the amorphous layer. The rate of activation of the dopants of this process is high thanks to the metastable mechanism involved in the crystallization, but the rate of defect remaining at the end of range from the PAI implant is not satisfactory.

As heating rates go higher (like those induced by a few tenth of ns laser pulse duration), the enthalpy release from crystallization is able to self-propagate over the a-Si layer through an intermediary liquid-mediated mechanism [27]. Details of this mechanism called 'explosive crystallization' have been widely studied [28–30].

For laser pulse durations longer than 20 ns, a buried liquid layer starts to propagate through the a-Si layer from the primary molten layer existing at the surface.

Although several experiments and models have been proposed in these studies to understand the nucleation and propagation of this explosive buried liquid layer, few studies have addressed the case where a long laser pulse (>40 ns) is irradiating the top a-Si layer [30–32].

(ii) Modelling and *in situ* reflectivity simulation

The nature of the buried liquid layer (nucleation, thickness, propagation velocities) and the way it propagates when generated by a long laser pulse irradiation, depend on several parameters, including the a-Si thickness and the incident laser energy density [33].

The dynamic of the propagation of the co-existing surface molten layer and the buried liquid layer can be pictured with a Depth-Phase-Time (DPT) diagram. This diagram is represented in Fig. 14 for the case of a thin a-Si layer and a laser energy density lower than the melting threshold of c-Si. An important consequence of the latent heat difference between a-Si and c-Si is that, in addition to generating the explosive melting front, this front reverses its propagation when reaching the a-S/c-Si interface. This bounce induces the explosive front to solidify epitaxially from the underlying c-Si substrate towards the surface while re-melting the overlying p-Si remaining from the primary melting front [30].

To follow the dynamic of the crystallization of the a-Si layer in function of the laser energy density, we have developed a calculation code simulating the reflectivity from the He-Ne laser probing the molten silicon surface during the UV laser

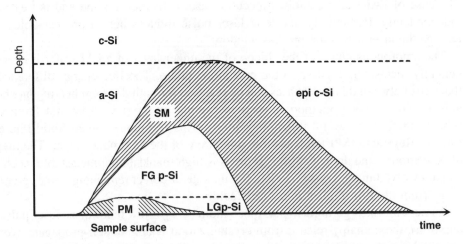

Figure 14: Depth-Phase-Time diagram picturing the a-Si layer melt and crystallization dynamics. Case of a thin a-Si layer (<40 nm) and a long laser pulser (>40 ns).
PM: Primary melt. SM: Secondary melt. FG p-Si: Fine grain poly-silicon.
LG p-Si: Large grain poly-silicon.

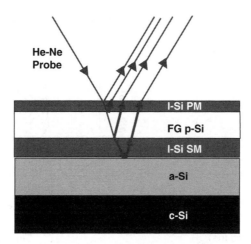

Figure 15: Output interferences from the reflection of the He-Ne beam at the different interfaces.

irradiation (Fig. 15). The code which determines the dynamic reflectance of the surface sample at 630 nm (He-Ne Probe), allows retrieving, at every moment, the structures of the liquid Si layers and finally the corresponding DPT diagram which describes completely the crystallization dynamic.

The variable parameters adjusted by the code to fit the experimental reflectivity are: the melting and solidification velocities, the maximal thickness and times of the onsets of the two melting fronts.

We bounded the variation of the adjusted parameters of the simulation with the possible solutions provided by the DPT model of the Fig. 14. This model is adapted to the case of a very thin a-Si layer irradiated by a long laser pulse [30]. A more complete study of the explosive crystallization under long pulse laser irradiation is published elsewhere [33].

(iii) Results

In Fig. 16a we report the adjustment of the simulation on the experimental reflectivity performed under a UV laser irradiation at 1.8 J/cm². We limited the fit (bold line on the figure) to the part of the curve where a liquid layer still exists. The fit takes place adjusting the velocities and onset of the different melting front. Those adjusted parameters are reported on the DPT diagram of the Fig. 16b. We see here that the primary melting front is very small (few nm) and that the main crystallization process is driven by the secondary explosive melting front. The explosive front is solidifying back to the surface from the a-Si/c-Si interface, re-epitaxying the whole underlying a-Si layer. The energy for melting the Large Grain p-Si zone

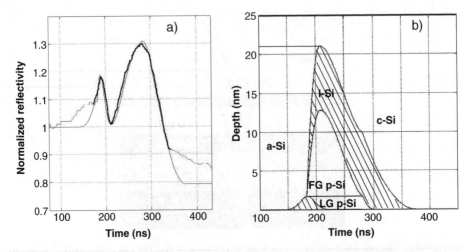

Figure 16: (a) Fit of the experimental surface dynamic reflectivity (Bold line) and (b) the corresponding Depth-Phase-Time diagram calculated. Laser energy density is 1.8 J/cm².

remaining from the primary melt is obtained from both the enthalpy of crystallization and the continuous deposition of energy by the laser pulse [29,33].

As shown in Fig. 17, the TEM picture confirms the DPT diagram of Fig. 16. The HR-TEM on Fig. 17b shows that the a-Si zone has been totally re-crystallized in c-Si and that few defects are remaining.

In Fig. 18 is reported the sheet resistance of a wafer pre-amorphized by Ge 10 keV 1.10^{15} at/cm² and implanted with BF_2 2.2 keV 2.10^{15} at/cm². The thickness

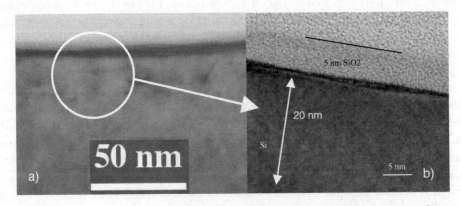

Figure 17: (a) TEM of a 20 nm pre-amorphized layer irradiated at 1.8 J/cm². (b) HR-TEM of the same sample.

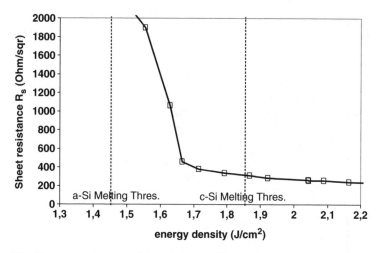

Figure 18: Sheet resistance of a Ge 10 keV 10^{15} at/cm^2 PAI sample implanted with BF2+ 2.2 keV 2.10^{15} at/cm^2 for various laser energy densities.

of the amorphous layer under such PAI conditions is roughly 20 nm. We see that above 1.6 J/cm^2 the sheet resistance dramatically decreases, correlating the sudden full activation of the implanted layer, although the laser energy density is still under the melting threshold of c-Si.

In terms of integration of the process, the explosive crystallization of the a-Si layer leads to an efficient activation of the dopants, while maintaining the laser energy density under the damage threshold of the p-Si gate standing in between the junctions.

12.2.3.5. Conclusion

We demonstrate here that the pulsed laser activation process is able to build, through a diffusionless process, a shallow activated junction satisfying the next technology nodes requirements of the ITRS for CMOS device fabrication. We also showed that the process is substantially improved by using a long 200 ns excimer laser pulse.

Two main LP-LTP activation regimes were studied through thermal simulation and experimental characterization. The first one concerns crystalline implanted silicon where a non-melt activation regime has been obtained. The second one involves pre-amorphized wafers where an explosive crystallization occurs inducing a full activation of the amorphous layer through its complete c-Si re-epitaxy.

The results are summarized in the Fig. 19 where we report the sheet resistance and junction thickness for different processes. The non-melt LP-LTP X_j values are only depending on the implantation process.

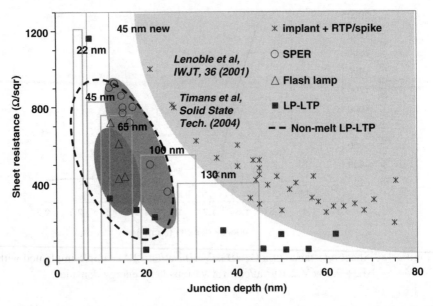

Figure 19: Long Pulse-LTP junctions characteristics compared to those obtained with other classical processes (RTP, Flash-Lamp, SPER). Boxes indicate the ITRS technology nodes requirements.

The two regimes studied here show that we improved the integration of the long laser pulse process in the industrial fabrication of a CMOS device by tuning both implantation process and laser parameters.

12.2.4. Laser Induced Diffusion from Spin-On Glass

12.2.4.1. Introduction

One of the main drawback of the ion implantation technique for dopant incorporation into silicon is the possible generation of implantation tail-related defects which are difficult to anneal and could be detrimental for device performances. This point is particularly critical when using the low energy (<1 keV) ion implantation technique for ultra-shallow junction formation. One possible way to overcome these difficulties is to use as a dopant source a thin film of doped oxide deposited on top of the Si wafer by the spin coating technique [8–10]. Under suitable conditions, the excimer laser irradiation leads to surface melting and dopant incorporation by liquid phase diffusion from the surface. By this way, junction formation of less than 20 nm in depth was successfully achieved.

12.2.4.2. Experimental Conditions

The first stage consisted in depositing doped oxide film on samples which were initially submitted to chemical etching in a hydrofluoric acid (HF) solution diluted to 10%, in order to remove the native oxide layer. BoroSilicate Glass (BSG) thin films are deposited uniformly by the spin-coating method, controlling the spinning speed (rpm) and the volume of solution deposited onto the sample. BSG layers of approximately 200 nm in thicknesses contain 1.10^{20} and 1.10^{21} B at/cm^2, respectively. An annealing step is carried out at a temperature of 100°C for 30 min in order to evaporate the solvents and densify the films. The second stage consisted in building the junctions. The surface of the sample is first irradiated using a laser source with an energy density above the melting threshold of c-Si (as deduced from simulations in Fig. 3). The incorporation of the dopant atoms located near the surface of Si takes place by fast diffusion during the molten phase of the process. Two excimer laser sources were used: A classical 20 ns-duration KrF laser from LAMBDA PHYSIK (EMG 201 MSC) delivering up to 300 mJ per pulses and the 200 ns-duration XeCl laser from SOPRA (VEL 15).

12.2.4.3. 20 ns-KrF Laser Processing

SIMS results reported in Fig. 20a,b show the strong influence of both the laser energy density (a) and number of shots (b) on the final boron distribution profiles into Si. The junction depth increases from 25 to 70 nm for energy densities varying from 500 to 800 mJ.cm^{-2}. A maximum boron concentration of about 1.10^{21} at/cm^3 is recorded at the highest fluence (800 mJ/cm^2). It is also interesting to notice that the junction is deeper and deeper when working under multishot (1 to 100)

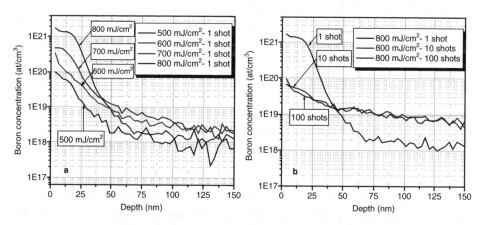

Figure 20: SIMS profiles of boron-doped c-Si, as a function of the 20 ns-laser energy density (a) and the number of laser shots (b).

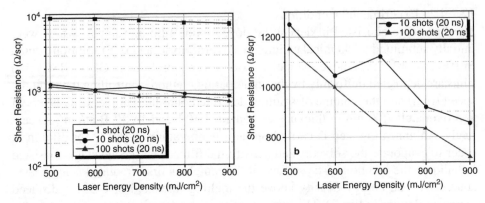

Figure 21: The sheet resistance of boron-doped c-Si samples versus the laser energy density and shot number (1, 10, 100) (a). Enlarged scale for 10 and 100 shots (b).

laser irradiation conditions (Fig. 20b). This behavior is assumed to be related to a modification of the optical and thermal properties of the surface after the first shot irradiation.

As shown in Fig. 21a,b the sheet resistances (R_{sqr}) of the boron-doped layer (after 20 ns-KrF laser irradiation) decreases when increasing the laser energy density mainly because silicon is molten deeper and deeper. When performing multishot laser treatment, the increasing amount of dopant atom incorporated into the irradiated zone explains the observed extra diminution of sheet resistance. Finally, a maximum electrically active boron concentration of 3.10^{19} at/cm^3 can be deduced.

12.2.4.4. 200 ns-XeCl Laser Processing

As shown in Fig. 22, the boron distribution profiles obtain following long pulse (200 ns) irradiation processing present a similar behavior to those corresponding to short pulse (20 ns) treatment.

When working close to the melting threshold of c-Si ($E_{MT} = 1400$ mJ.cm^{-2}, for 200 ns-XeCl) ultra-shallow distributions (ranging from 10 to 20 nm, at the level of 1.10^{18} cm^{-3}) can be achieved with a profile abruptness of about 8 nm/decade. In addition, from sheet resistance measurements reported in Fig. 23 a maximum electrically active boron concentration of $1.2.10^{20}$ at/cm^3 can be deduced.

12.2.4.5. Conclusion

The possibility to fabricate by excimer laser induced diffusion from spin-on-glass source, ultra-shallow junctions ranging between 10 and 20 nm in depth was demonstrated. Using the 200 ns-duration XeCl source, the process can be achieved under control way on very large area (several cm^2) due to the excellent quality of the

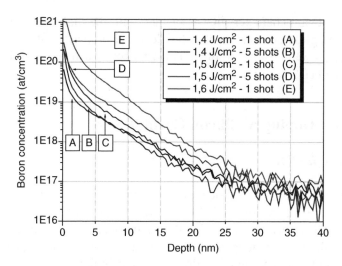

Figure 22: SIMS profiles of boron-doped c-Si samples as a function of the 200 ns-laser energy density and shot number (1 and 5).

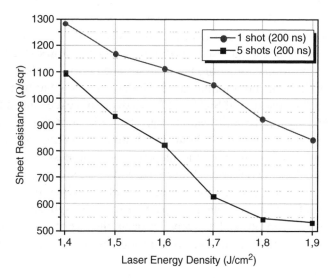

Figure 23: Sheet resistance of boron-doped c-Si samples as a function of the 200 ns-laser energy density and shot number (1 and 5).

SOPRA laser beam characteristics associated with the high thickness uniformity of the spin-on-glass deposit. Finally the electrical properties of the boron-doped layers ($R_{sqr \cdot min} \approx 530$ Ω/sqr) and diode performances carried out by this technique are fitting the ITRS requirements.

12.3. Laser Doping of Silicon Carbide

12.3.1. Introduction

Silicon carbide (SiC) is a wide-gap semiconductor which presents unique material properties especially suitable for high temperature, high power and high frequency applications. However, SiC device fabrication has to face various technological difficulties. Among them, one of the more crucial appears to be the doping step [34,35]. The high melting point and limited diffusion of impurities into SiC have greatly restricted the use of ion implantation and furnace annealing commonly employed in the silicon microelectronics industry to incorporate and activate the dopants. As an alternative to classical thermal heating, excimer laser processing was demonstrated to be suitable for the doping of SiC in various experimental conditions, including laser annealing of ion-implanted SiC [36–41], laser induced diffusion of dopant from gas [42,43] or solid sources [44,45] and laser crystallization of doped-SiC films deposited by CVD [46]. The use of high powerful pulsed laser beams in the nanosecond duration regime allows to deposit a large amount of energy in short time into the near-surface region, while maintaining the substrate essentially at room temperature. Under suitable conditions, the irradiation leads to surface melting of SiC [37,47] to a depth not exceeding a few hundred nanometers and its rapid solidification from the bulk, allowing the dopant to be incorporated by liquid phase diffusion. In addition, by taking advantages of the highly non-equilibrium nature of the melt/regrowth process complete electrical activation of the dopant could be achieved, as shown for silicon [48]. It has also to be mentioned that recent works [49,50] suggested the possibility to suppress the ion-implanted defects and activate the dopant into SiC by high temperature excimer laser annealing in solid phase. For such approach, the use of a long-pulse excimer laser beam such as the VEL appears very promising for fabricating good quality SiC devices [51].

12.3.2. Simulations

As for silicon, heating and cooling stages in silicon carbide were determined by solving numerically the heat flow equation and using the appropriate optical and

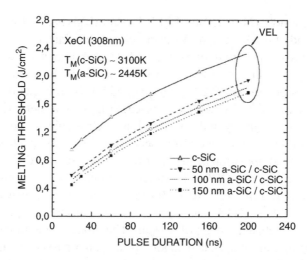

Figure 24: Simulated melting threshold of c-SiC and a-SiC/c-SiC for three amorphous
thicknesses (50, 100, 150 nm) as a function of the XeCl pulse duration.

thermal parameters for crystalline (c) and amorphized (a) SiC [51]. As shown on numerical data of Fig. 24, the surface melting threshold (E_m in J/cm^2) strongly depends on both the excimer pulse duration (ranging from 20 to 200 ns) and structural properties of the near surface layer (c- or a-SiC). It has to be noticed that, whatever the pulse duration, the heating process is more efficient in the amorphous phase due to its stronger absorption coefficient (at 308 nm) and lower thermal conductivity and melting temperature, as compared to the crystalline one [51]. For a pulse duration of 200 ns, E_m is close to 2.3 J/cm^2 in c-SiC, while for a-SiC E_m is ranging between 1.8 and 2 J/cm^2, depending on the thickness of the amorphized surface layer. From these simulations, we can conclude that laser treatments have to be carried out in the 1.8 to 2.3 J/cm^2 fluence range, according to the implantations conditions, in order to work essentially in the solid phase regime.

12.3.3. Experimental Conditions

The 4H-SiC material purchased from Cree Research, consisted of 6-μm-thick n-type epitaxial layers grown on off axis (0001) oriented with a net doping concentration of $\sim 5.10^{15}$ at/cm^3. In order to obtain a square-like aluminum (Al) atom distribution over approximately 0.15 μm, multiple-implantation technique carried out at room temperature was used up to a maximum energy of 120 keV. Samples were implanted at three different doses: 1.5, 15 and 150.10^{13} Al/cm^2, giving uniform Al concentration of 1, 10 and 100.10^{18} Al/cm^3, respectively, at the plateau.

As deduced from the simulations, the VEL laser treatments were carried below 2.3 J/cm^2 energy density, in order to work essentially in the solid phase regime.

The structural properties of Al$^+$ ion-implanted SiC before and after laser annealing were characterized by Rutherford backscattering spectrometry (RBS) of 2 MeV He$^+$ in channeling geometry and transmission electron microscopy (TEM) and Al distribution profiles were deduced from secondary ion mass spectrometry (SIMS). Surface chemical composition of the laser irradiated SiC was deduced from X-ray photoelectron spectroscopy (XPS) and surface morphology was controlled by atomic force microscopy (AFM) from 10×10 μm squares. Finally, forward and reverse current density-voltage (I–V) characteristics were measured on pn diodes at room temperature (RT) with a diameter of 630 μm. The diodes were processed into mesa structure, using optimized annealed conditions and after passivating the surface with deposited oxide. To minimize the contact resistance on implanted layers, Al$^+$ ions were multi-implanted (30 and 60 keV) at 650°C with a total 8×10^{14} at/cm^2 dose through a 50 nm-thick oxide mask in the diode fabrication. Sputtered Al/Ti and TiSi$_2$ annealed at 950°C were employed for ohmic contacts on implanted layers and n$^+$ substrates, respectively. The junction depth was estimated to be approximately 0.4 μm.

12.3.4. Results

Figure 25a shows the RBS spectra of as-implanted and laser-annealed (one shot, 2.2 J/cm^2) SiC samples. The channeling spectra indicate that the efficiency of the laser treatment strongly depends on the structural characteristics of the as-implanted region. When the top surface layer is amorphized (at high implantation dose of 1.5×10^{15} Al/cm^2), we do not observe any structural improvement of the laser irradiated region (top spectra in Fig. 25a). However, TEM observations of the amorphized sample (Fig. 25b) show the transition from amorphous to polycrystalline phase following laser irradiation. By contrast, when working below the amorphization threshold (at 1.5×10^{14} Al/cm^2 mid-implantation dose), laser processing is demonstrated to be efficient to suppress the induced-implanted structural defects (middle spectra in Fig. 25a). Indeed, the normalized backscattering yield χ was estimated as the ratio of the aligned yield in the damaged region and that of the random yield, for as-implanted and annealed samples. The χ value is strongly reduced from 31 to 7% following laser treatment (as compared to $\chi \sim 3\%$ in virgin 4H-SiC). This indicates that a large amount of defects can be removed after a single-shot laser treatment. This result was also confirmed by TEM (Fig. 25c), which shows good crystalline quality of the irradiated layer. It is interesting to notice that

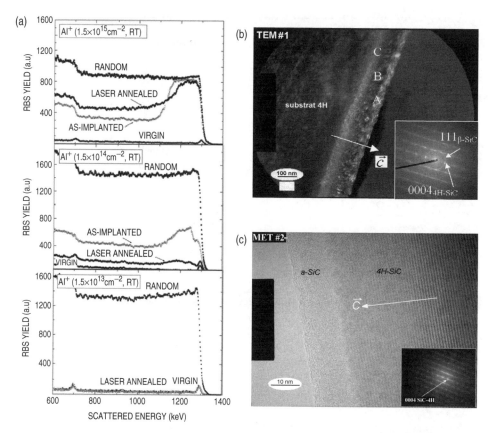

Figure 25: RBS spectra of Al-implanted (dose of : 1.5, 15 and 150 × 10¹³ Al/cm²) bulk 4H-SiC before and after laser annealing (a). TEM picture of the laser annealed high implantation dose (150 × 10¹³ Al/cm²) sample (b). TEM picture of the laser annealed middle implantation dose (15 × 10¹³ Al/cm²) sample (c).

similar results were previously reported by Japanese groups [49,50,52] when working with a classical excimer source (~20–30 ns) in a multishot (>1000) procedure. Finally, at the lowest implantation dose (1.5×10^{13} Al/cm²), for which no related-implantation damage is observed, the high crystalline quality of the SiC layer (bottom spectra in Fig. 25a) is maintained after laser treatment. As deduced from SIMS measurements, the Al distribution profiles are not modified following the laser treatment (Fig. 26 corresponding the mid-implantation dose). This strongly suggests that the process occurs in the solid phase regime. XPS measurements do not show (Fig. 27a) any significant change in the C/Si ratio (measured either in surface or at 20 nm in depth) after VEL laser processing at 2.2 J/cm², excepted

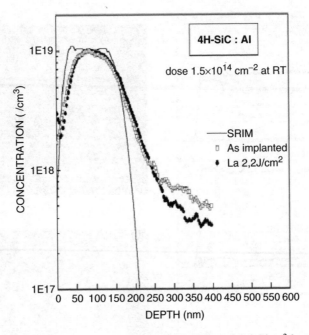

Figure 26: SIMS profiles of Al-implanted 4H-SiC, and 2.2 J/cm^2 laser annealed samples.

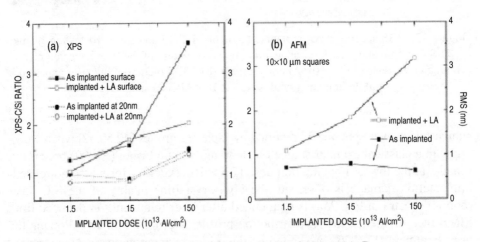

Figure 27: C/Si ratio at the surface and at a depth of 20 nm (a). Root mean square (RMS) value extracted from AFM morphology studies as a function of the implanted doses before and after 2.2 J/cm^2 laser annealing (b).

when the as-implanted sample was amorphized (1.5×10^{15} Al/cm^2 dose). In this last case, the graphitized surface (C/Si = 3.2) of SiC is chemically modified (C/Si = 2) by the laser treatment. AFM measurements (Fig. 27b) reveal that the surface roughness of the as-implanted samples is comprised typically between 0.5 and 0.8 nm, as depending on the implantation dose. After laser annealing a slight increase of the RMS values is observed. At the lowest implantation dose, the surface roughness is close to 1.1 nm, which satisfies surface device quality requirement, whereas a value of 3.2 nm is recorded for the pre-amorphized laser annealed sample. This surface roughness degradation could be related to the different thermal behavior of a-SiC, as compared to c-SiC, leading to a reduction of its surface melting threshold (Fig. 24) and a concomitant enhancement of its surface sublimation.

In Fig. 28 are reported the forward I–V characteristics (measured at 300 K) for a 630 μm mesa pn vertical junction prepared in optimized conditions of implantation and annealing (2.4×10^{14} Al/cm^2 dose, 2.2 J/cm^2 VEL single-shot). The reverse bias I–V curve obtained at 400 K is also shown in the inset. At RT, similar

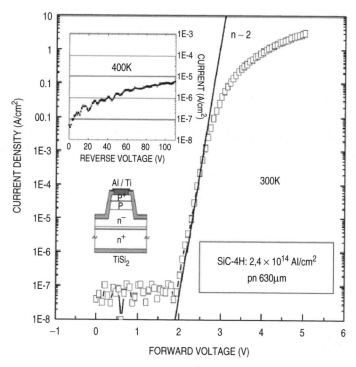

Figure 28: Forward and reverse current density-voltage characteristic for a 630 μm mesa pn vertical junction obtained by RT Al ion implantation and 2.2 J/cm^2 laser annealing.

to the works reported in Ref. [53,54], the forward current essentially relies on the generation-recombination component ($n = 2$) and the saturation current density was extrapolated to be 2×10^{-23} A/cm^2. As also deduced from the I–V data, a high voltage forward drop of 5 V is required to obtain a 3 A/cm^2 current density. The high on state resistance of 0.6 Ω.cm^2 coupled to this poor forward conduction is supposed to be a consequence of the room temperature Al implantation procedure which generates a high concentration of compensated defects and limits the dopant activation efficiency, as previously reported for SiC furnace annealing [55]. At 110 V reverse voltage, a fairly satisfactory leakage current density of 5 μA/cm^2 (@ 400 K) is measured.

12.3.5. Conclusion

Despite the simple technology used, namely the implantation step carried out at RT and without any device periphery protection, these results are quite encouraging for long-pulse duration excimer laser processing of SiC in solid phase. Finally, it has to be noticed that from our most recent device characterizations up to 80% of as-implanted Al$^+$ ($C_m \sim 10^{20}$ Al/cm^3 at the plateau) were electrically active after laser processing.

12.4. General Conclusion

In this chapter, it was investigated in detail the technological potential of excimer laser processing for high precision junction fabrication both in silicon and silicon carbide.

Simulation calculations highlight the added value brought by the relatively long thermal cycle induced in Si and SiC from the long duration (200 ns) pulse provided by the VEL excimer laser from the SOPRA company.

Physical characterizations of the junctions show a very efficient dopant activation and defects annealing while the electrical properties measured on real device structures confirm the great potential of the long pulse (200 ns) excimer laser annealing for high level dopant activation both in Si and SiC.

Acknowledgment

This work has been supported by the French ministry of Research and Technology in the framework of the RMNT DOLAMI Project and ST Microelectronics.

References

[1] The International Technology Roadmap for Semiconductors, **2001**.

[2] Chong, Y.F., Pey, K.L., Wee, A.T.S., See, A., Chan, L., Lu, Y.F., Song, W.D. and Chua, L.H., *Appl. Phys. Lett.*, **2000**, *76*, 3197.

[3] Fortunato, G., Mariucci, L., Stanizzi, M., Privitera, V., Whelan, S., Spinella, C., Mannino, G., Italia, M., Bongiorno, C. and Mittiga, A., *Nucl. Instrum. Methods Phys. Res. B*, **2002**, *186*, 401.

[4] Laviron, C., Séméria, M.N., Zahorski, D., Stehlé, M., Hernandez, M., Boulmer, J., Debarre, D. and Kerrien, G., *International Workshop on Junction Technologies, Tokyo, Japan*, **2001**, 6.1.1.

[5] Sigmon, T.W., *MRS Proceedings*, **1987**, *75*, 619.

[6] Slaoui, A., Foulon, F., Bianconi, M., Correra, L., Nipoti, R., Stuck, R., de Unamuno, S., Fogarassy, E. and Nicoletti, S., *MRS Proceedings*, **1989**, *129*, 591.

[7] Kramer, K.-J., Talwar, S., McCarthy, A.M. and Weiner, K.H., *IEEE Electron. Device Lett.*, **1996**, *17*, 461.

[8] Fogarassy, E., Pattyn, H., Elliq, M., Slaoui, A., Prevot, B., Stuck, R., de Unamuno S. and Mathé, E.L., *Applied Surface Science*, **1993**, *69*, 231.

[9] Kim, C., Jung, S.-H., Jeon, J.-H. and Han, M.-K., *Thin Solid Films*, **2001**, *397*, 4–7.

[10] Coutanson, S., Fogarassy, E. and Venturini, J., *MRS Proceedings*, **2004**, *810*, C4.13.

[11] Jones, K.S., Banisaukas, H., Glassberg, J., Andideh, E., Jasper, C., Hoover, A., Agarwal, A. and Rendon, M., *Appl. Phys. Lett.*, **1999**, *75*, 3659.

[12] Lindsay, R., Pawlak, B., Kittl, J., Henson, K., Torregiani, C., Giangrandi, S., Surdeanu, R., Vandervorst, W., Mayur, A., Ross, J., McCoy, S., Gelpey, J., Elliott, K., Pages, X., Satta, A., Lauwers, A., Stolk, P. and Maex, K., *MRS Proceedings*, **2003**, *765*, D7.4.

[13] de Unamuno, S. and Fogarassy, E., *Applied Surface Science*, **1989**, *86*, 1.

[14] Hernandez, M., Venturini, J., Zahorski, D., Boulmer, J., Kerrien, G., Sarnet, T., Laviron, C., Semeria, M.N. and Santailler, J.L., *Applied Surface Science*, **2003**, *208–209*, 345.

[15] Hernandez, M., PhD Thesis, "Procédé laser de réalisation de jonctions ultra-minces pour la microélectronique silicium: étude expérimentale, modélisation et tests de faisabilité", Paris XI Orsay University, **2005**.

[16] Von Allmen, M., *Laser-Beam Interactions with Materials: Physical Principles and Applications (Updated), Springer-Verlag*, **1995**.

[17] Jain, S.C., Schoenmaker, W., Lindsay, R., Stolk, P.A., Decoutere, S., Willander, M. and Maes, H.E., *J. Appl. Phys.*, **2002**, *91*, 8919.

[18] Lojek, B., *Advanced Thermal Processing of Semiconductors*, RTP, **2004**, 53.

[19] Yang, S. and Thompson, M.O., *MRS Proceedings*, **2001**, *669*, J7.4.1.

[20] Débarre, D., Kerrien, G., Nogushi, T. and Boulmer, J., *IEICE Trans. Electron*, **2002**, *E85-C*, 1098.

[21] Kerrien, G., Hernandez, M., Laviron, C., Sarnet, T., Debarre, D., Noguchi, T., Zahorski, D., Venturini, J., Semeria, M.N. and Boulmer, J., *Applied Surface Science*, **2003**, *208–209*, 277.

[22] Baboux, N., Dupuy, J.C., Prudon, G., Holliger, P., Laugier, F., Papon, A.M. and Hartmann, J.M., Ultra-low energy sims analysis of boron deltas in silicon, *J. Crys. Growth*, **2002**, *245*, 1.

[23] International Workshop on Junction Technology, Nov. 29, **2001**.

[24] Earles, S., Law, M., Jones, K., Talwar, S. and Corcoran, S., *MRS Proceedings*, **2001**, *669*, J4.1.1.

[25] Earles, S., Law, M.E., Jones, K.S., Frazer, J., Talwar, S., Downey, D. and Arevalo, E., *IEEE. RTP*, **2004**, 143.

[26] Wood, R.F. and Giles, G.E., *Phys. Rev. B*, **1982**, *25*, 2786.

[27] Thompson, M.O., Galvin, G. J., Mayer, J. W., Peercy, P. S., Poate, J. M., Jacobson, D. C., Cullis, A. G. and Chew, N. G., *Phys. Rev. Lett.*, **1984**, *52*, 2360.

[28] Bruines, J.J., van Hal, R.P.M. and Boots, H.M., *Appl. Phys. Lett.*, **1986**, *49*, 1160.

[29] Tsao, J.Y. and Percy, P.S., *Phys. Rev. Lett.*, **1987**, *58*, 2782.

[30] Peercy, P.S., Tsao, J.Y., Stiffler, S.R. and Thompson, M.O., *Appl. Phys. Lett.*, **1988**, *52*, 203.

[31] Murakami, K., Eryu, O., Takita, K. and Masuda, K., *Phys. Rev. Lett*, **1987**, *59*, 2203.

[32] Fogarassy, E., de Unamuno, S., Prévot, B., Harrer, T. and Maresch, S., *Thin Solid Films*, **2001**, *38*, 48.

[33] Venturini, J., Huet, K. and Hernandez, M., Proceeding of the 12th IEEE International Conference on Advanced Thermal Processing of Semiconductors, RTP 2004, p. 73.

[34] Harris, G.L., (Ed.), in *Properties of Silicon Carbide*, Emis Datareviews Series No. 13, INSPEC, **1995**, 157.

[35] Dmitriev, V.A. and Spencer, M.G., in *SiC Materials and Devices, Semiconductors and Semimetals*, Vol. 52, Park, Y.S. (Ed.), Academic Press, **1998**.

[36] Pehrsson, P.E. and Kaplan, R., *J. Mater. Res.*, **1989**, *4*(6), 1480.

[37] Chou, S.Y., Chang, Y., Weiner, K.H., Sigmon, T.W. and Parsons, J.D., *Appl. Phys. Lett.*, **1990**, *56*, 530.

[38] Bourdelle, K.K., Chechenin, N.G., Akhmanov, A.S., Poroikov, A.Yu. and Suvorov, A.V., *Phys. Stat. Sol. (a)*, **1990**, *121*, 399.

[39] Ahmed, S., Barbero, C. and Sigmon, T.W., *Appl. Phys. Lett.*, **1995**, *66*(6), 712.

[40] Key, P.H., Sands, D., Schlaf, M., Walton, C.D., Anthony, C.J., Brunson, K.M. and Uren, M.J., *Thin Solid Films*, **2000**, *364*, 200.

[41] Hishida, Y., Watanabe, M., Sekine, K., Sugino, K. and Kudo, J., *Appl. Phys. Lett.*, **2000**, *76*, 3867.

[42] Eryu, O., Nakata, T., Watanabe, M., Okuyama, Y. and Nakashima, K., *Appl. Phys. Lett.*, **1995**, *67*, 2052.

[43] Russell, S.D. and Ramirez, A.D., *Appl. Phys. Lett.*, **1999**, *74*, 3368.

[44] Eryu, O., Kume, T., Nakashima, K., Nakata, T. and Inoue, M., *NIMB*, **1997**, *121*, 419.

[45] Krishnan, S., D'Couto, G.C., Chaudhry, M.I. and Babu, S.V., *Mater. J., Res.*, **1995**, *10*(11), 2723.

[46] Mizunami, T. and Toyama, N., *Jpn. J. Appl. Phys.*, **1998**, *37*, 94.

[47] Baeri, P., Spinella, C. and Reitano, R., *Int. J. Thermophys.*, **1999**, *20*.

[48] Fogarassy, E., Lowndes, D.H., Narayan, J. and White, C.W., *J. Appl. Phys.*, **1985**, *58*, 2167.

[49] Hishida, Y., Watanabe, M., Nakashima, K. and Eryu, O., *Mater. Sci. Forum*, **2000**, *338–342*, 873.

[50] Eryu, O., Aoyama, K., Abe, K. and Nakashima, K., *Mater. Res. Soc. Symp.*, **2001**, *640*.

[51] Dutto, C., Fogarassy, E., Mathiot, D., Muller, D., Kern, P. and Ballutaud, D., *Applied Surface Science,* **2003**, *208–209*, 292.

[52] Tanaka, Y., Tanoue, H. and Arai, K., *Mater. Sci. Forum*, **2002**, *389–393*, 799.

[53] Kimoto, T., Takemura, O., Matsunami, H., Nakata, T., Inoue, M., *J. Electron. Mater.*, **1998**, *27*, 358.

[54] Bluet, J.M., Pernot, J., Camassel, S., Contreras, S., Robert, J.L., Michaud, J.F. and Billon, T., *J. Appl. Phys.*, **1998**, *88*, 1971.

[55] Inoue, N., Itoh, A., Matsunami, H., Nataka, T. and Watanabe, M., in *Proceedings of the silicon carbide and related materials, Conf. Kyoto (Japan, 1995) Inst. Phys. Conf. Ser. 142,* Chap. 3, **1996**, 525.

Chapter 13

Laser Cleaning: State of the Art

Ph. Delaporte[a] and R. Oltra[b]

[a]*Lasers, Plasmas and Photonic Processes Laboratory (LP3), CNRS – Mediterranean university, 163, avenue de Luminy, C. 917, 13288 Marseille Cedex 9, France;* [b]*Laboratoire de Recherches sur la Réactivité des Solides, Université de Bourgogne, B.P. 47 870, 21078 DIJON Cedex, France*

Abstract

Laser cleaning became a real alternative to mechanical and chemical techniques to remove pollution from a large set of materials. The motivations for using such new process are both for environmental considerations, as waste reduction or replacement of organic solvents, and to address cleaning problems which cannot be solved by other techniques. Many studies allowed to identify the interaction mechanisms leading to the laser removal and to understand, and then predict, the surface modifications induced by this process. Laser cleaning is certainly not a universal process, but it has a great potential in many fields of applications. Some of them are already operational, and the development of new ones require a good knowledge of the laser cleaning physic and of the advantages and the limitations of this technique. After a brief description of the laser matter interaction processes, the three main mechanisms of laser ablation, thermal, mechanical and photochemical are discussed through two characteristic cases: particle removal and superficial layer removal. At last, a typical laser cleaning experimental setup is described and some applications will be presented in different fields such as microelectronics, nuclear, art or surface preparation.

Keywords: Laser ablation; Laser cleaning mechanisms; Particle removal; Process efficiency; Pulsed laser matter interaction; Surface damage; UV laser.

13.1. Introduction to Laser Cleaning

13.1.1. Cleaning Processes

The implementation of many industrial processes requires a precise control of the different surfaces involved in the processes. Cleaning is one of the major steps of

Recent Advances in Laser Processing of Materials
J. Perrière, E. Millon and E. Fogarassy (Editors)

preparation of these surfaces. Today, most of the cleaning techniques are based on mechanical or chemical techniques. However, both the new environmental requirements and the necessity to work with pollution-free surfaces in some high technological fields (microelectronics, optics, medical, etc.), imposed the development of new processes like cryogenic (supercritical CO_2, dry ice, etc.) and photonic ones (plasma, laser).

The ideal cleaning process must never generate wastes or modify the surface properties, and must improve them. It must have a higher efficiency for lower cost than the current processes. The aim of this chapter is not to present the laser cleaning as the ideal process, but to give some data on the physics of the process and its performances as a function of the different kinds of pollution and of the surface properties. Some typical applications of laser cleaning will be detailed.

13.1.2. The Basic Laser Material Interaction of the Laser Cleaning Process

Studies of physical processes involved in laser matter interaction have been extensively investigated on both theoretical and experimental points of view [1–3]. The aim of this part is to present the fundamental mechanisms occurring in laser cleaning. It is necessary to specify that the lasers used for the development of cleaning processes are pulsed lasers with pulse durations ranging from few nanoseconds to tens of nanoseconds. Longer pulse durations induce deeper thermal effects and then higher risks of modifications of material properties. Picosecond and femtosecond lasers could give relevant results with some kind of pollution, but their low average powers do not allow their use for industrial cleaning applications. The lasers currently used for cleaning are therefore the UV excimer lasers, the YAG lasers which are powerful in the visible and the near infrared wavelength range, and for some applications the CO_2 TEA lasers emitting at 10.6 μm.

When a laser beam irradiated a surface, its energy is shared in three components: reflection, transmission and absorption. The latter will be responsible of the laser ablation process. For cleaning applications, the materials are not perfect and their surface state (roughness, oxidation, etc.) strongly modify their reflectivity. The absorption coefficients of both pollution and bulk are wavelength dependant. The absorption of the laser beam energy with pulse duration larger than nanosecond by conducting material is done by the free electrons of the material. Free electrons transfer this energy to the lattice in few tens (or hundred) of picoseconds. These thermal relaxation times being short in regard to the pulse duration, we can consider the laser effect as a heating source at the material surface. The resulting phenomena (heating, thermal diffusion, vaporization, etc.) are considered to be purely thermal.

Time and space evolution of the temperature in the absorbing volume and the surrounding space can be simply described by the heat equation:

$$\phi(t,z) + \frac{\partial}{\partial z}\left(K_{th}\frac{\partial T}{\partial z}\right) = \rho C_p \frac{dT}{dt}$$

where ϕ is the absorbed laser flux, K_{th} the thermal conductivity of the material ($J \cdot m^{-1} \cdot s^{-1} \cdot K^{-1}$), C_p the specific heat ($J \cdot kg^{-1} \cdot K^{-1}$), ρ the material density ($kg \cdot m^{-3}$). The flux is calculated from the laser intensity in the volume defined by the Beer–Lambert law:

$$I(z) = I(0)e^{-\alpha z}$$

where α is the absorption coefficient of the material.

For semiconductor or dielectric, processes occurring during laser beam absorption are more complex and their relative importance is strongly wavelength dependant. Laser energy absorption in semiconductor is related to free carrier density which can increase due to thermal or photonic processes. When this density is high enough, some phenomena, like inverse Bremssthralung or ionization avalanche, can produce a thermal interaction. For dielectric materials, which have an optical gap higher than the photon energy, the laser absorption can be done by multi-photonic processes, by structural defects of the material or in two steps, creation of absorbing centers in the bulk followed by photon absorption, as it happens when UV laser irradiates silica.

The main processes induced by laser irradiation of a surface are strongly dependant of the optical properties of the material in regard to the laser wavelength (λ) and pulse duration (τ). Two physical parameters are used to classify the interaction processes:

The optical absorption length: $\delta_a = \alpha^{-1}$

The thermal diffusion length: $\delta_{th} = \sqrt{\dfrac{K_{th}\tau}{C_p\rho}}$

Purely non-thermal ablation processes may exist, as it is encountered, for example UV laser irradiation of polymers. In this case, the photon energy is higher than the binding energy of the monomer chains, and some links can be broken, inducing a volume increase and the material ablation.

Three kinds of phenomena may induce the ablation process: thermal, mechanical and photochemical. They are tightly correlated, but as a function of the material and the laser parameters, one of them prevails. As discussed previously, the laser wavelength influences the ablation process (efficiency, surface modification, etc.) and its choice is an important factor to define a laser cleaning setup.

Whatever the ablation mechanism, the key parameter to characterize the processes is the fluence (J/cm^2). We have also to define the cleaning threshold and the damage threshold. The first one corresponds to the lowest fluence necessary to obtain, after one or more laser shots, a clean surface (for the specific application). The damage threshold cannot be defined in a such universal way than the cleaning one, because it depends both on the material to clean and on the application. Indeed, a surface coloration can be considered as damage for some applications and not for others. Moreover, it has been shown that for some materials, like glass, this threshold could be dependant of the number of laser shots and of the pulse repetition frequency [4]. The fluence definition is easy for a multi-mode laser beam which allows to get a uniform energy repartition in the laser spot (top hat). The monomode beams have a low divergence, but also a gaussian energy profile, and then the fluence in the laser spot is not uniform. If the cleaning and damage thresholds are close, the process requires a very uniform energy profile and optical devices allowing to shape the beam from gaussian to top hat must be used.

13.2. Mechanisms and Performances

There are many laser cleaning applications in a wide range of fields such as cleaning of artworks and antiquities, nuclear and biological decontamination, mold cleaning in aeronautic industry and particle removal in microelectronics and optronics. They can be divided in two domains: particle removal and layer removal. The physics and the performances of these two processes are described hereafter.

13.2.1. Laser Removal of Particles

The miniaturization of devices and structures for application in microelectronics, optics and optoelectronics requires to reduce both the density and the size of defects on the material surfaces. The chemical techniques used at the present time induce an erosion of the substrate which is composed of very thin layers. Soon, these methods will reach their limits and will not allow to fill the requirements of ultra high cleanness of these industries. Some tests, realized with cryogenic techniques did not solve the problem of submicronic particle removal. In this context, the removal of possibly high adhering nanoparticles from surfaces of silicon wafers or other materials remain a challenge. Three laser-based processes are currently investigated:

- Dry Laser Cleaning (DLC): the particles are ejected due to the surface motion induced by the direct laser irradiation;
- Steam Laser Cleaning (SLC): a thin liquid layer, deposited on the surface, is removed by the laser beam and the particles are carried away with the liquid film;
- Laser induced Shock Wave Cleaning (LSC): the laser beam is focused just above the surface and a gas breakdown occurs at the focal point, generating strong shock waves which push the particles outside the surface.

13.2.1.1. Dry Laser Cleaning

The main adhesion forces binding submicronic particles and the surface are the Van der Walls forces [5].

$$F = \frac{Aa}{6h^2}\left(1 + \frac{r_c^2}{ah}\right)$$

where a is the radius of the particle, h the equilibrium distance between the particle and the surface, r_c the radius of contact correlated to the particle deformation and A the Hamaker constant which depends on the properties of the particle, surface and surrounding medium.

For small particles, the second term is predominant ($r_c^2 \gg ah$) and the Van der Walls forces become inversely proportional to the cube of the separation distance. So, the challenge is to be able to initiate the detachment of the particle to increase h and then strongly reduce the attraction forces.

As discussed previously, the laser beam irradiation of the surface induces a fast heating of a thin layer of the substrate leading to a fast expansion of the surface followed by the return to its initial position. The time scale of this motion is in order of pulse duration. During the expansion phase, the particle follows the surface motion with the same acceleration (if we neglect the particle deformation), and when the surface cools down the particle can be ejected far enough to be out of the Van der Walls attraction. So, two criteria must be filled to allow the particle removal:

The ejection force must be higher than the Van der Walls force: $m\gamma > F_{vdw}$ where γ is the acceleration of a particle with a mass m. In some cases of particle–substrate couple, like inorganic particle on hydrophilic surface, one must also take into account the hydrogen bounding [6] and add the corresponding adhesion force.

The second condition is the energetic one. To overcome the attraction potential of the surface, the particle must have energy higher than the adhesion energy.

That gives a condition on the minimum velocity which must be reached by the particle:

$$v > \sqrt{2E_{adh}/m}$$

These two conditions are simple enough to give a clear understanding of the basics mechanisms of the dry laser cleaning. However, they do not allow to take into account all the phenomena occurring during this process. Some very good theoretical descriptions can be found in literature [7–9]. To develop a realistic model, thermal expansion and deformation of both substrate and particle must be considered simultaneously and elastic forces must be taken into account, as well as optical resonance effects [10,11]. The process efficiency is not only related to the fluence, the temporal shape and the duration of the laser pulse have also a significant influence [12]. Indeed, faster is the trailing edge of pulse and higher is the deceleration of the surface. Then, for the same particle acceleration, due to the front edge of the pulse, the distance between the particle and the substrate grows faster for a sharp end of the laser pulse. Figure 1 exhibits a significant reduction of cleaning threshold when shorter and sharper pulses are used [13]. The field enhancement effects must also be considered. The particles with low absorption coefficient at the irradiation wavelength, can act 'like small lenses' (Mie theory) and focus the laser beam on the substrate, leading to a local increase of the fluence. In a first step, these effects are favorable because the surface expansion is enhanced under the particles [14–16]. However, they can also induce surface damages if the fluence becomes locally higher than the damage threshold [17].

The laser beam incidence angle plays also a role on cleaning efficiency, but the published experimental results are apparently not consistent together. On one hand, a drop of cleaning efficiency from 100% to zero has been observed for 2.5 μm SiO$_2$ particles on silicon when the incident angle moves from zero (normal to the surface) to 20° [18]. On the other hand, the cleaning efficiency has been improved by one order of magnitude when the angle moves from zero to 80° (almost grazing) for 10 μm copper particles deposited on copper substrate [19]. These phenomena can be explained by the influence of the absorption coefficient of the particles. For non-absorbing particles (i.e. SiO$_2$), the field enhancement effect discussed previously plays a major role and the irradiation of the substrate with a significant angle induces a shift of the surintensity spot which is no more located under the particle and does not amplify anymore the substrate expansion. For absorbing particles (i.e. metals), there is never field enhancement effect and the particle expansion is important in the removal process. When the incidence angle is increased, the fluence delivered on the substrate is reduced, because the irradiated surface is larger, but the fluence on spherical particles is unchanged, as well as the particle expansion. Moreover, one can expect a focalization of the laser beam

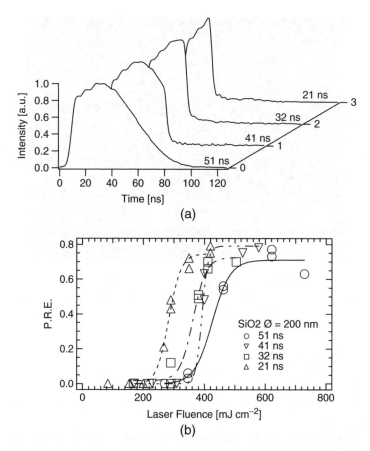

Figure 1: Pulse shapes of an XeCl laser (308 nm) used for laser cleaning of 190 nm SiO₂ particles on silicon (a) efficiencies of particles removal according to the pulse shapes (b) the laser pulse is shortened by KDP crystal-based electro-optic gating system.

under the particle by reflection between substrate and particle for the grazing angle. This example illustrates the complexity of physical mechanisms involved in dry laser cleaning and their relative importance in regard to parameters such as optical properties of material and wavelength. That also makes this technique powerful for many kind of applications.

Today, very good cleaning efficiencies (up to 100%) have been obtained for particles larger than 100 nm [20]. These performances already open the door to a wide range of applications. Figure 2 shows the laser removal of 385 nm polystyrene particles from silicon surface. Smaller particle removal requires fluences close to the surface damage threshold. To answer the future requirements of the microelectronics industry, studies are performed to reduce the cleaning thresholds.

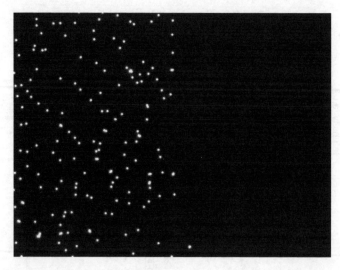

Figure 2: Removal of 385 nm PS particles by XeCl laser irradiation. The right part of the silicon substrate has been irradiated with 10 laser shots at 450 mJ/cm^2.

13.2.1.2. Steam Laser Cleaning (SLC)

Particle removal may also be achieved by the laser induced ejection of a thin liquid film deposited on the surface. This technique, named Steam Laser Cleaning has been investigated since 1987 [21], and an experimental setup has been developed at IBM [22,23].

Many studies have been conducted to determine the physical mechanisms responsible for particle removal in SLC, both experimental [24–27] and theoretical [28–30]. From these studies, it is clear that the laser energy is absorbed by a thin layer of the substrate surface, as for dry laser cleaning, and the heat is transferred from this small volume to the liquid layer. Then, a nucleation process occurs at the interface creating a high density of bubbles which grow and generate an explosive blast wave. This process leads to the ejection of the liquid layer from the irradiated area as well as the particles embedded within this layer with acceleration up to 10^9 m/s^2. The laser-induced explosive vaporization is therefore the main mechanism of steam laser cleaning process.

The optimization of the process requires a good knowledge of the influence of experimental parameters. As for dry laser cleaning, the optical properties of the substrate and the laser wavelength are closely linked through the absorption coefficient. Even if SLC is generally used for very smooth surfaces like silicon wafers, it is interesting to notice that a small surface roughness can improve the nucleation process due to the presence of small absorbing sites, and the particles can play this role. An other key element of this technique is the liquid layer which is generally a mixture of water and alcohol. The film is typically applied to the surface by evaporating the liquid and driving a steam flow over the substrate where the condensation occurs. The film properties influence both the deposition process and the explosive evaporation process. This last point induces a dependence of the laser cleaning threshold on the nature of the liquid. A recent study stated on the influence of the liquid layer thickness on SLC efficiency [24]. The authors demonstrated that a minimum layer thickness (70 nm in their case) is necessary to achieve an efficient particle removal, and above this value, the cleaning threshold depends on the layer thickness. In addition, a thick liquid layer must be avoided to prevent droplet and particle redeposition.

The most interesting feature of steam laser cleaning is the universal threshold of particle removal independently of particle size and nature [31]. These parameters have an influence on adhesion forces, but the minimum fluence required to produce the explosive evaporation process induces ejection forces already higher than the adhesion ones. SLC technique allowed to remove particles with diameter as small as 60 nm from silicon surface [31]. Moreover, the steam laser cleaning threshold is generally lower than the DLC one, and that is very important to prevent substrate damage.

As a consequence, this technique seems more appropriate to the development of industrial applications, because the process efficiency is independent of nature of pollution and the risk of surface damage is reduced. However, this process requires the use of an external component, the liquid, which can be prohibited for some applications, and the setup must be carefully designed to accurately control the liquid film thickness.

13.2.1.3. Laser-Induced Shockwave Cleaning (LSC)

Shockwave generation induces sufficient forces on the particles to overcome the adhesion forces. A method to generate such wave is the laser-induced gas breakdown. When an intense nanosecond laser beam is focused in gas at atmospheric pressure, an ionization process occurs leading to an increase of the temperature and pressure plasma. Its fast expansion generates a strong shockwave, and if the surface is close enough to the focal point, the shockwave can shift the particles. Theoretical studies of this process are not abundant, but it seems that the particle motion begins with a rolling effect followed by the particle detachment [32]. The particles are blown out of the zone located below the laser focal point.

The cleaning efficiency is correlated to the shockwave velocity at the surface which depends on both the laser intensity and the distance between the surface and the focal point, but is independent of optical properties of the substrate. Therefore, this technique may be very powerful for cleaning of non-absorbing substrates like in optical applications. The LSC is a dry and non-contact process, but there are still some risks of surface damage [33]. They are mainly due to thermal and mechanical effects occurring when the focal point is too close to the surface. This technique allowed to remove particles of few hundred of nanometers deposited on fragile surface with complex geometry.

A very interesting adaptation of this process has been studied [34]. It consists in the irradiation, with a UV laser, of the surface just before the shockwave generation. The UV laser irradiation leads both to break the organic bounding between the particles and the surface and to initiate the particle lift-off by surface expansion, like in DLC. The use of this UV prepulse greatly improve the cleaning efficiency, and silica particles of 40 nm diameter have been removed from wafer surface without any damage of the small (120 nm) and fragile patterns [34].

13.2.2. Laser Removal of Superficial Layers

13.2.2.1. Oxide Removal

In many applications, the removal of surface oxides is necessary in order to maintain the requested function of metallic component or to prepare the surface for the next step of manufacturing process. For instance, metallic surfaces must be cleaned before welding because the presence of oxides increases the tendency of brittle behavior of the joint and decreases its mechanical strength. In nuclear technology, it is frequently required to remove the contaminated oxide layer from the reactor vessel without inducing any damage on the steel substrate underneath, in order to ensure a prolonged lifetime and reactor safety.

The two main processes leading to laser oxide removal are thermal and mechanical, and one of them is predominant as a function of the oxide properties. Binding energies and thickness are correlated to the conditions of the oxide formation (pressure, temperature, nature of surrounding gas or liquid, etc.) [35]. The thermal mechanism occurs mainly for thick absorbing oxide layers which have grown on relatively rough surfaces. In this case, each laser pulse, applied with fluence higher than the oxide ablation threshold, removes part of the layer. The thickness of the removed oxide, depends on fluence, pulse duration and wavelength of the laser, physical and optical properties of the oxide, but is typically hundred of nanometers for ablation with nanosecond lasers. The use of short wavelength (i.e. UV excimer lasers) allows to have a lower ablation threshold but also a lower ablation rate than for visible or IR YAG lasers. For both, YAG and excimer lasers, the absorption coefficient of oxides are generally high and the laser energy is absorbed in a thin layer at the oxide surface leading to efficient ablation. Moreover, the short pulse duration of these sources (5–25 ns) limits the thermal diffusion in the bulk.

In some cases, mechanical effects can lead to the complete removal of the oxide layer with few laser shots, and without any damage of the substrate. When the substrate is smooth, the irradiation of the oxidized surfaces generates stresses, which are normal to the oxide–metal interface. The development of the stress field results in the fracture of the interface and the spallation of the surface layer, while the relaxation of the compressive stresses of the thermal oxides that follows enhances their expulsion [36]. Then, no ablation phenomena occurs during this process. It has been demonstrated that the cleaning efficiency is significantly improved when the surface is covered by an electrochemically controlled liquid during laser irradiation. The liquid film confines the absorbed laser energy on the surface and then enhances the thermal and mechanical effects. It has also been proposed that this film induces an electrochemical potential at the surface which strongly modifies the imaginary part of the optical parameter of the oxide and then modifies the absorption length [37]. Another effect of electrochemically controlled liquid confinement is the hydrogen entrapment in the oxide layer during the cathodic reduction process which results in defect formation [38]. In the frame of nuclear decontamination, a comparison between dry process and the use of liquid films exhibited cleaning factors 30 times higher with water film, 85 times higher with nitric acid 0.5 M and 650 times higher with nitric acid 5 M [39].

The oxide layer removal is also a key point in electronic and microelectronic industry. Copper oxide must be removed in electronic device fabrication in order to improve the surface wetability and so achieve a good quality solder joint. Successful oxide removal was achieved with YAG laser (266 nm, 532 nm, 1064 nm) in single-pulse operation, and the ablation mechanisms combined both thermal and mechanical effects [40]. In the frame of the new process development for the next generation of microelectronic components, the laser treatments of surfaces such as silicon or aluminum are studied. The removal of natural oxide layer before the deposition process improves the adhesion and can reduce the electrical losses between the two layers [41].

13.2.2.2. Coating Removal

The ablation mechanisms of coatings deposited on a surface depend on optical properties of the layer at the irradiation wavelength. If the photon energy is higher than the binding energies of coating compounds, the ablation may be photochemically originated by bound breaking as for polymer ablation with ArF laser [42–44]. When absorption coefficient of the coating is high at the laser wavelength, the thermal process is generally predominant. Each laser shot induces the removal of a thin layer of the coating as for oxide removal. Moreover, the ablation threshold of coating is often lower than the damage threshold of the substrate, and irradiations with laser fluences between these two thresholds lead to a very efficient and safe cleaning. When the coating is transparent, or slightly absorbent, the laser energy is absorbed by the substrate and the interaction processes occur at the interface substrate – coating. Mechanisms like heating, volume expansion and shock wave generation lead to the ejection of the coating with a single laser shot. This phenomenon, named spallation, is observed for laser cleaning of graffitis on buildings or paint stripping from metallic surfaces with YAG laser at 1064 nm. Most of the paints are partly transparent in the near infrared range but have a strong absorption coefficient in UV, and the use of excimer laser for this kind of applications leads to a thermal or photochemical ablation. Another technique also very efficient for the coating removal is the backside irradiation. Indeed, if a layer with a strong absorption coefficient is deposited on a transparent substrate, the laser irradiation of the layer through the substrate induces similar interaction processes at the interface and leads to the ejection to the whole thickness of the coating. These single laser shot cleaning processes due to energy absorption at the interface are very efficient, but the confinement of the interaction may induce irreversible damage to the substrate surface, even if the laser fluence is largely below the typical damage threshold of the substrate without confinement. For instance, this effect has been observed for the removal of metallic coating deposited on glass substrates which have been damaged at fluences ten times lower than their damaged thresholds.

Among the multiple applications of laser coating removal, the paint stripping has been intensively investigated. The laser process presents several advantage compared to traditional methods: selective removal of individual layers, suitable for sensitive surfaces, workpiece ready for immediate repainting and no production of secondary wastes. The Urenco Company developed a 2 kW TEA CO_2 laser especially designed for paint stripping of large aircraft [45]. This kind of high power laser emitting in infrared is certainly the most efficient tool for this application, but requires a precise control of the process to prevent any thermal effect on the substrate. Studies performed on graffiti removal from building show that for commercial spray paints the use of UV laser neither causes material deposition nor chemical, thermal or mechanical change of the surface. On the other side, irradiation at 532 nm can induce additional discoloration of the substrate [46].

Laser cleaning process is also very effective for optical surfaces, and especially for large telescope mirrors that require an in situ technique. Photoacoustic stress waves and photothermal vaporization are important at removing particles and thin films of water or oil, and photochemical effect is also efficient for organic pollution removal. A specific study exhibits that UV laser cleaning removes contaminants that standard CO_2 snow cleaning techniques do not remove as well or not at all. For long exposure times, the laser cleaning also restores the thermal emissivity better than the CO_2 snow [47]. UV laser has also proven its efficiency for the removal of organic contamination, like fingerprints, from quartz substrates. In this case, the irradiation from the back of the substrate was more effective [48]. For optical material, the cleanness of the surface is very important not only to maintain the optical properties but also to keep the laser-induced damage threshold as high as possible. Indeed, local damages come from energy absorption by defects, and a careful UV laser cleaning of the surface at low fluence allows to significantly increase the damage threshold for visible or IR laser irradiation.

The laser coating removal also concerns the ejection of liquid films from surfaces. There is a great interest for this process in the field of bacteriological decontamination, especially for the removal of biofilms which have a great role in numerous infectious diseases. Plasma processes are efficient for a various types of decontamination, but microorganisms contained in biofilms are protected by the aqueous medium and resist most of the cleaning processes. The use of UV laser presents the advantage to associate both the photochemical effect to sterilize some bacteria and the mechanical effect to eject the biofilm from the surface. Figure 3 illustrates this mechanical effect by showing the ejection of an entire liquid drop after a KrF laser irradiation without vaporization effect.

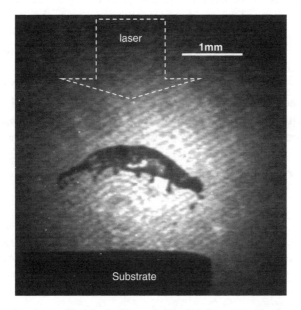

Figure 3: Ejection by mechanical effect of a liquid drop deposited on an aluminum surface. The laser irradiated the whole surface with a fluence of 0.4 J/cm^2 (single shot). The picture has been taken with an ICCD camera by shadow technique 400 μs after the laser irradiation with a gate width of 40 ns. The real picture size is scale.

13.3. Laser Cleaning Applications

Pulsed lasers are well adapted to develop dry cleaning processes which do not generate secondary wastes. This technique is very powerful to remove thin layer of pollution and especially attractive if the cleaning threshold is lower than the surface damage threshold. In this part, we will present the different components of a laser cleaning setup and some specific applications of this process.

13.3.1. Typical Laser Cleaning Setup

Four main parts must be taken into account to define a laser cleaning device: the laser source, the beam delivery system, the suction or blowing of the ejected pollution and the real time control of surface cleanliness [49].

13.3.1.1. Laser Source
As discussed previously, the laser cleaning efficiency is mainly related to the energy density, but two other laser parameters play a leading role in the ablation process and must be considered in the choice of the laser source: the wavelength and the pulse duration. The absorption coefficient is wavelength dependant [50], and choice of a specific wavelength can strongly modify the process efficiency. For example, the use of short wavelength is more appropriate to induce photochemical ablation process. So, surface cleaning can often be done by several sources, but a pertinent choice of the laser is necessary to optimize the process efficiency for a specific application. The thickness of heat-affected zone is proportional to the square root of the pulse duration, and then a short laser pulse is required to reduce the modification of the surface physical properties.

The cleaning velocity V ($cm^2 \cdot s^{-1}$) is proportional to the average power of the laser. It can be estimated by the formula:

$$V = \frac{\alpha P}{FN} = \frac{\alpha Ef}{FN}$$

where P is the laser average power (W), α the attenuation due to the beam delivery system, E the energy per pulse (J), f the laser repetition rate (Hz), F and N are respectively the fluence ($J \cdot cm^{-2}$) and the number of shots required to achieve the surface cleaning.

13.3.1.2. Laser Beam Delivery
Two steps must be considered in the development of beam delivery system: beam shaping and beam manipulation. The first is necessary to obtain a uniform energy repartition on the surface with the fluence required by the application.

Generally, simple optical setup allows to fulfill the requirements. However, some specific applications need to have a very uniform fluence to prevent surface modifications if the damage threshold is too close to the cleaning threshold, for example. The transformation of a Gaussian beam in a top hat beam is sometimes necessary. In these case, more complex optical devices, like beam homogenizer, must be designed to develop the cleaning process.

The beam shaping is sufficient when the elements to clean are small enough to be moved in front of the laser beam. However, for large surface cleaning, it is necessary to consider the beam manipulation. Mobile reflective optics can be used to move the beam without significant losses of energy. However, this technique is not useful for the treatment of complex surfaces and to keep a uniform fluence when the beam optical path length varies a lot during the cleaning procedure. The other solution is to use optical fibers to transport the beam from the output of the laser to the surface. An optical device must be set at the fiber output to obtain a uniform energy repartition with a sufficient fluence to achieve an efficient cleaning. Due to the numerical aperture of the fibers, it is not possible to reach high fluences (10 J/cm^2) with a good uniformity. But the cleaning applications never require such level of energy density which is higher than the damage threshold. Silica fibers are generally used for the transmission of high power laser beams. The losses are very low in the visible spectral range and some IR wavelengths, and systems have been developed for YAG laser beam manipulation with optical fibers [51]. Due to color center formation (typically E′) by high energy photons, it is impossible to efficiently transport high average power of deep UV lasers like KrF (248 nm) or ArF (193 nm). Some studies have been performed to optimize the operating parameters of an XeCl laser (308 nm) in order to reduce the losses in the fibers during beam transportation [52]. At 308 nm, the formation of Non-Bridging Oxygen Hole Centers (NBOHC) is the main reason of UV photon absorption in the fiber. This process is proportional to the peak power density, but it is reversible and after few seconds the transmission goes back to its initial value. Figure 4 presents the evolution of the transmission efficiency as a function of the laser frequency for three input power densities. For a fixed laser average power, the transmission value is higher for high pulse rate frequency and low energy. Furthermore, the transmission of high average power excimer laser must be done through a bundle of fibers to increase the entrance surface of the fiber, and thus, to reduce the fluence. A transmission efficiency of 50% has been obtained at 308 nm (150 W output power) with a 10 meter long bundle of high [OH] silica fibers.

13.3.1.3. Ablated Compound Collection

During laser cleaning process, the pollution generally does not disappear, but it is just ejected perpendicularly to the laser irradiated surface. The ablated compounds can move initially with velocities of about 10^5 cm/s and even under ambient air,

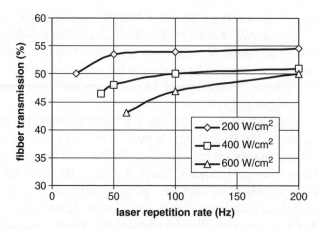

Figure 4: Evolution of the overall fiber transmission efficiency as a function of the laser pulse repetition rate for three input power densities. The source was an XeCl laser (308 nm) with a pulse duration of 25 ns.

expansion distances of few centimeters are observed, depending on the laser energy density. In order to prevent the redeposition of the pollution on the surface, a recovery setup must be used. One can use a blowing system in which a gas flow pushes the ejected pollution outside of the surface. For some specific applications, like nuclear decontamination, the ablation products cannot be disseminated and collection system must be used. In this case, both experimental and modeling studies of the plasma expansion have been performed to improve the collection of the contaminated particles [53,54]. The coaxial nozzle of a cell is set close (few mm) to the surface, and is connected to a vacuum pump through a filter to ensure a sufficient depression in the irradiated zone and to trap the particles. An optical system mounted in the cell focuses the laser beam on the surface through the nozzle. Gas flows in front of the optics to protect them against particle deposition and to drag the particles towards the filter.

13.3.1.4. Real Time Control of Surface Cleanliness
The pollution thickness is never uniform, and, to computer control the cleaning process, it is necessary to have a real time measurement of the surface cleanliness. Three main techniques are used to achieve this control, and they are based on acoustic, electrical or optical analysis. Most of them are efficient when the interaction process induce the formation of ablation plasma. Its formation is correlated with shock and acoustic waves that can be detected by a simple microphone and careful analysis give lot of information on cleaning efficiency [55]. The plasma also generates a high electrical field, and its characteristics depend on ablated compounds.

The measurement of the current generated by this field allows to determine if the ejected species come from the pollution layer or from the surface itself. This technique is particularly sensitive for the control of oxide removal [56]. Emission spectroscopy, Laser Induced Breakdown Spectroscopy (LIBS), can also be used to identify the ablation plasma compounds, and then to determine if pollution is still present at the surface. The measurement of reflectivity variations of a laser probe beam on the irradiated surface has been used to control the oxide removal process [57]. The advantage of this technique is to determine the surface cleanliness before the laser irradiation, and then to better control the number of shots. Light scattering is used to monitor the efficiency of particle removal [58]. This measurement can be done both in air and directly on the surface.

13.3.2. Microelectronic Industry

The design of microelectronic devices is currently presenting structures with higher density and smaller dimensions. As a consequence, the surface contamination problems appearing during manufacturing processes become critical. For instance, the increasing of recording density of the new hard disk drive implies that the distance between the magnetic head slider and the surface is about 10 nm, and the presence of particles larger than few tens of nanometers can damage both the slider and the disk surface. Also, as the semiconductor industry develops processes for the 65 and 45 nm nodes, wafer cleaning faces new challenges for both front-end-of-line (FEOL) and back-end-of-line (BEOL) requirements, including increased fragility of structures and materials, and more stringent requirements for cleanliness and material loss. The surface contamination such as particles, metals, microroughness, watermarks and Si/SiO_2 loss will have a greater impact on the next-generation device yield.

Wet chemical cleaning is still the wafer cleaning process of choice for mainstream semiconductor manufacturing because of its robustness and a history that makes it risk free. Their effects are both etching of the surface and generation of electrostatic repulsive forces. The chemical compounds and the procedure are matched with each type of pollution (metal, polymer, etc.). Some other cleaning processes are based on shock waves generated by vapor/liquid spay or cavitation effects (megasonics). This last technique has been used for many years to clean wafers, creating and imploding bubbles in the fluid to remove particles from the wafer surface. It is often used in association with chemistry. But megasonics has begun to cause problems, even at the 90 nm node, and can cause even more damage at 65 nm and beyond. The forces induced by fast phase change (liquid → gas, solid → gas) can also be used to push the contamination away from the surface. The stringent surface contamination requirements outlined in the International Technology Roadmap for Semiconductors (ITRS) pose new challenges for the surface preparation technology. To meet the ITRS requirements for surface preparation and overcome the posed challenges, new processes and technology will be needed.

Laser cleaning has emerged as a promising technique to achieve efficient cleaning in microelectronic industry. The physics of this process has been previously described (Section 13.2.1), and we will present some specific applications. The Field Emitters Arrays (FEA) are panels of small electron sources which are very promising for vacuum applications in microelectronics. A reliable operation of these devices requires a very homogeneous tip surface, and pollution, like oxidation or contamination with carbon or water, leads to a strong reduction of the emission efficiency and unstable emission behavior. Then, cleaning is of prime importance to allow the development of FEA applications. Laser cleaning has been successfully used for decontamination of these surfaces, yielding a significant increase in emission current [59]. The photochemical decomposition of organic contaminants is the main mechanism of this process, and then it is strongly wavelength dependant. The photon energy of IR and visible light are too low to induce an efficient cleaning, and too deep UV laser irradiation leads to a reduction of emission current. This current is increased by a factor of 5 when the laser cleaning is achieved at 349 nm [60]. The manufacturing of inkjet nozzles for printer cartridge requires the drilling of small holes (typically 40 μm of diameter) on a polyimide flexible circuit. Excimer laser (KrF) drilling is currently used to achieve this operation, and the laser ablation process induces the formation of organic compounds which are deposited around the holes. This contamination surface affects the printing quality and reduces the electrical contact between the devices and the flexible circuit. The two methods currently used to clean these polyimide circuits, chemical cleaning and plasma etching, present some risks of surface damage and significantly increase the processing time. As for FEA cleaning, the laser process is mainly based on photochemical mechanisms, and irradiation at 532 nm with few tens of mJ/cm^2 allows to realize an efficient cleaning of the surface [61].

Laser cleaning has also been successively used to remove contamination from magnetic disk surface without causing any damage [62]. At last, one of the most exciting challenges for the microelectronics industry is the wafer cleaning. In laboratory, particles as small as 100 nm have been removed with dry laser cleaning technique and steam laser procedure allowed to remove particles as small as 60 nm. At the beginning of the fabrication process, FEOL and first steps of BEOL, the surfaces (Al, Cu, Si, etc.) do not yet present fragile structures, and DLC has demonstrated its ability to remove contamination induced by the industrial process, where other cleaning techniques (scrubber, wet strip) failed to clean the surfaces. The photochemical effect due to deep UV irradiation ($\lambda = 193$ nm) associated to the mechanical effect induced by the short laser pulse appear to be very efficient for polymer removal. However, the main challenge is the wafer cleaning during BEOL, when the surface is very delicate, and laser irradiation at fluence higher than 200 mJ/cm^2 often induce an irreversible damage of the structure (Fig. 5). It is then necessary to reduce the cleaning threshold and the SLC which requires a lower fluence to achieve an efficient cleaning could be the right solution.

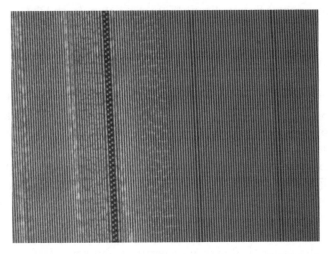

Figure 5: Illustration of laser-induced damage on wafer during the final steps of the fabrication process. The left part has been irradiated by an XeCl laser with a fluence of 300 mJ/cm^2. The smallest structures have a period of 1 μm.

13.3.3. Nuclear Industry

Decontamination techniques used in nuclear facilities are based on chemical (foams, HNO_3, etc.), electrochemical and mechanical processes (ultrasonic cleaning, CO_2 ice blasting, high pressure water, etc.). Their ability to be well adapted for site specific applications depends on a wide variety of parameters such as material, surface, type and composition of contaminant, access, safety, efficiency or operating history of the plant. The laser cleaning has also demonstrated its capacity to fulfill all the requirements for the development of an efficient nuclear decontamination process. Moreover, that is a dry technique which generates only a small amount of secondary wastes, and the reduction of the total waste volume to be stored is one of the major concerns of the nuclear industry.

The few projects of laser cleaning for nuclear decontamination reported in open literature have mainly been performed in United States of America and in France.

Among them, we can mention a group of the Ames Laboratory which used both a KrF excimer laser (248 nm, 27 ns) and a YAG laser (1064 nm, 100 ns) to study the removal of radioactive oxide from metallic surfaces (aluminum, steel, copper, lead). They worked on large surface equipment of nuclear facilities and demonstrated that the efficiency of the ablation process is higher for larger ionization potential of the ambient gas [63,64]. High power infrared lasers have been used at the Argonne National Laboratory for the decontamination of concrete [65]. In this case, the contaminants deeply diffused into the materials, and the decontamination is achieved by the ablation of a thick layer of concrete (several millimeters). Laser cleaning has also been tested for the decontamination of a vapor generator of nuclear facilities [66,67], and COGEMA patented a system for nuclear fuel rod cladding using Nd:YAG laser [68].

Figure 6 presents the decontamination efficiency as function of the number of laser shots for different fluences. For these experiments, an XeCl laser has been used to remove the radioactive oxide layer formed on the stainless steel surface of vapor generator [69]. The thick (few tens of micrometers) oxide layer has been created under high pressure (170 bar) and high temperature (300°C) of water. During the first shots, the unfixed contamination is easily removed, even at low fluence, by mechanical and thermal effects. Then, a thin oxide layer is removed by each laser shot, and the thickness of this layer increases with the fluence. However, it is almost impossible to remove the remaining part of the contamination which diffused into the material. Removal of particles and natural oxide layer has also been performed with a high efficiency (close to 100%) and lower fluences which guarantee not to modify the surface properties.

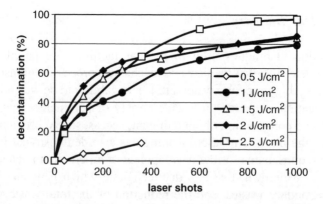

Figure 6: Decontamination efficiency of a vapor generator in nuclear facilities as function of the number of laser shots for different fluences. An XeCl laser was used to perform these experiments.

Pulsed laser cleaning is shown to be an excellent technique for the nuclear decontamination. This process allows a considerable reduction of the radioactive wastes compared to the other techniques, and demonstrated high performance and capacities in removing both particles and oxide layer contamination. Laser decontamination is a dry and clean process, well adapted for large area surface treatment, which can be used for the removal of most nuclear contaminations.

13.3.4. Art Cleaning

Laser cleaning is becoming a major process in restoration of artworks, antiquities and historic building [70–72]. It provides the advantage to selectively remove the pollution layers and to leave the original delicate surface unaffected. However, due to the historical importance of the surfaces to be cleaned, this application requires techniques for the on-line monitoring and in situ control at all stages of the process. The Laser Induced Breakdown Spectroscopy (LIBS) is currently used for this purpose [73]. It is also of prime importance to be able to match the laser parameters with the pollution and substrate characteristics. Lot of scientific studies have been performed during the last ten years to understand the physical processes occurring during laser cleaning of artworks and to investigate the short-and long-term effects of this technique [74].

Zafiropulos gave some results obtained for the cleaning of complex polymeric substrates like painting and the removal of inorganic encrustation of marble or stone surfaces [72]. The main mechanism taking place for the removal of polymerized material from painting is the photochemical one which induces the bound rupture and the fragment ejection. Then, the short wavelengths are preferable to achieve an efficient cleaning, typically 193 nm or 248 nm. However, the laser irradiation can induce discoloration that depends on the nature of the pigments. To prevent this effect, it is necessary to leave the first layer of few micrometers thick covering the painted substrate. This layer acts like a filter and protects the underlying pigments from direct laser exposure [75]. As an example, we can mention the laser cleaning of an eighteenth-century Flemish tempera painting on a wooden panel. The other processes failed due to a very hard varnish deposited on the paint surface during a previous attempt of restoration. A fluence of 0.38 J/cm^2 with a KrF laser (248 nm) allowed to remove this 9 μm thick layer with an ablation rate of 0.25 μm/pulse [72].

In the case of encrustation on surfaces of marble or stone, the role of vaporization and spallation mechanisms is much more significant. The thermal processes can induce an optical phenomenon, named yellowing, which has been studied by many groups. It has been attributed to the light scattering occurring at the voids created by selective vaporization [76] and to the modification of the spectral absorption of the remaining layer [77]. The use of UV lasers is often the unique solution to remove biological encrustation like fungi or lichen. The high absorption coefficients of these materials at short wavelength lead to an amplified mechanical effect associated to a photochemical effect [78]. Moreover, in this last study achieved in the frame of restoration of Parthenon, no yellowing effect has been observed. Specific devices have been developed to perform laser cleaning of historical buildings. A group in Florence (Italy) developed laser sources with variable pulse duration from hundred of nanoseconds to microseconds. Experiments realized at 1064 nm with a 20 μs pulse duration YAG laser on different stone types such as marble, sandstone, limestone exhibit a very efficient cleaning without coloring or discoloring of the patinas. For instance, successful interventions have been made with this prototype on the Musoleum of Theodoric in Ravenna and Palazzo Rucellai in Florence [79]. Laser-based industrial process has been developed by the Laserblast Company and used to clean monuments such as Petit Palais in Paris or city hall in Bruxelles with cleaning velocity higher than 1 m²/h.

13.4. Integration of Laser Cleaning in a Transformation Process

As described previously, laser cleaning can be used for a wide range of applications. It is generally very important not to modify the surface during the pollution removal, as for art cleaning or particle removal from silicon wafer. However, in some cases, laser irradiation can bring much more than a simple cleaning effect. The treatment can modify both chemical and physical characteristics of the surface, and improves some specific properties related to the final application. The insertion of the laser treatment in the transformation process becomes then very attractive.

13.4.1. Surface Pre-treatment: PROTAL® Process

The PROTAL® process, which combines a conventional thermal spraying process (APS, HVOF, ARC, etc.) with a pulsed laser surface preparation in one production step [80], has emerged as a potential technique to replace conventional surface preparation. The nanosecond pulsed laser treatment produces surface ablation eliminating the contaminants (such as oxides, organic particles, absorbed gas, etc.) before the impingement of the first sprayed particles. Promising results have already been reported [81]. Despite little modification of the surface roughness after the laser treatment, a considerable influence on the bond strength or interface toughness is observed. So it was suggested that the laser treatment not only allows the efficient removing of surface pollutants but also promotes physico-chemical bonding at the interface [82].

Previous studies highlighted the laser-induced modification of surface morphology and surface energy. Recent works were focused on surface morphology evolution on different substrates and their relation with interface adhesion after laser irradiation. It must consider that cleaning is not the only one effect affecting the improvement in coating adhesion. Various surface modifications (roughness, composition, etc.) induced by the laser energy may have a great influence. Transient oxidation observed in recent studies must be taken into account [83].

13.4.2. Surface Preparation for Adhesion

13.4.2.1. Polymer Treatment

Polymers are widely used for industrial applications, especially in aeronautic and aerospace, and the fast development of those materials with new properties extend every day their fields of applications. In order to use polymers, treatments must be developed to clean and to prepare the surfaces. One of the most important points which must be addressed is the adherence properties of polymer surfaces to improve the adhesion between polymer and polymer or metal. Plasma processes are often used to perform these treatments with a very good efficiency. However, they are generally carried out at low pressure in specific mixtures and that can generate some technical problems for the treatment of large area surfaces. Laser treatment in ambient condition has been successfully tested in aeronautic industry, especially to improve the polymer–polymer adhesion. Studies investigated the effects of excimer laser irradiation characteristics on surface modifications [84,85]. The treatments were achieved in air below the polymer ablation threshold fluence (typically 20 mJ/cm^2), and the X-rays Photoelectron Spectroscopy (XPS) analysis shows the formation of polar-oxygenated functional groups (C=O, O–C=O). Peel tests realized after the deposition of an Aluminum film on the polymer surface exhibit a significant enhancement of polymer metal adhesion. This improvement is attributed to an increasing number of Al–O–C stable and cohesive complexes formed at the interface polymer–metal. Laser treatment of polymer in air is then a simple process to improve the surface adherence properties without any modification of the bulk. The irradiation conditions (low fluence and small number of shots) allow to achieve large area surface treatments with a low average power excimer laser.

13.4.2.2. Treatment of Metallic Surfaces

Industrial activities such automotive, aerospace or rail industry require the use of metallic parts which generally present a strong surface pollution. Laser process can remove this pollution through the different mechanisms previously described, but can also induce some modifications of the material properties without any specific environment. For example, it has been shown that high-power laser irradiation may modify the surface of metals and alloys to obtain a better resistance to wear and corrosion [86]. Moreover, excimer laser irradiation in air or nitrogen atmosphere can lead to significative surface nitridation, which is known to improve the hardness and corrosion resistance of the surface materials [87]. With respect to other processes such as solvent degreasing, grit blasting, conversion coatings or electrochemical methods, the laser treatment avoids chemical compounds and appears to be a new environment-friendly process.

In order to replace all the mechanical and chemical steps of surface preparation processes, many studies have been performed to understand the surface modifications induced by laser treatments in air. Pulsed laser irradiation of steel surface with high fluences (few J/cm^2) resulted in surface nanostructuring (Figure 7). Indeed, due to the high pressure of the surrounding gas, the plasma remains confined close to the sample surface and the ablated species react with ambient gas to form nanoparticles which are then deposited around the spot area by backward flux (Figure 8). The morphological characteristics of this nanostructured layer is strongly dependant of plasma density and temperature, and then of irradiation conditions [88]. The XPS and Auger Electron Spectroscopy (AES) analysis demonstrate the cleaning effect and show that the deposited layer is composed of Fe^{3+} oxides mainly Fe_2O_3. Incorporation of nitrogen over 1–2 µm thickness and

Figure 7: SEM images of nanostructured layer deposited around the laser spot after irradiation of steel surfaces at 308 nm (a), 532 nm (b) and 1.06 µm (c). The laser irradiation conditions were 10 shots at 10 J/cm^2.

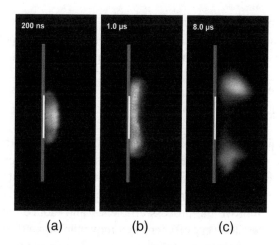

Figure 8: Time resolved ICCD images of the plasma produced laser ablation of steel in air at atmospheric pressure. An XeCl laser fluence was 10 J/cm². The gate of ICCD camera was 30 ns and the time delays relatively to laser pulse are 200 ns (a), 1 μs (b) and 8 μs (c), respectively.

superficial oxidation on 200–400 nm depth have been evidenced by Nuclear Reaction Analyses (NRA). X-ray diffraction at grazing incidence has shown the formation of FeO and Fe_2O_3 oxide phases as well as γ-Fe(N), and ε-Fe_xN for a sufficiently high amount of nitrogen incorporated. In nanosecond regime, the laser wavelength plays a major role in the relative importance of the surface modification processes. Nitrogen incorporation has been found to occur during the interaction between the reactive N of the plasma and the iron molten bath. The nitriding process is promoted in the IR wavelength range, and oxidation which takes place during the cooling step is significantly present in the case of UV treatment [89]. The formation of nanoaggregates in the plasma is also wavelength dependant and mainly occurs for UV irradiation at high fluence [90]. Then, the different morphological and chemical analysis showed that after the laser treatment, the surface could be described as a multilayer structure, namely a nanostructured layer of iron oxide nanoparticles covering the thermal oxide layer and a nitrided volume.

These modifications induce significant changes of the material properties. Both nitriding and Fe_2O_3 layer allow to improve the resistance to corrosion, and then to replace the phosphatation treatment currently applied for this purpose. Nitride formation is also well known to increase the tribological properties of the material [91]. Different peel tests (dry, wet and under corrosive atmosphere) demonstrated better adhesion properties between rubber and metallic surfaces when the latter have been prepared with laser treatments than with the conventional mechano-chemical techniques. This is due to cleaning, nanostructuration of the surface and modification of its chemical properties which favors the adhesion process [92].

13.5. Summary and Outlook

Thermal, mechanical and photochemical mechanisms are at the origin of laser cleaning process. The laser parameters such as wavelength, pulse duration and fluence must be carefully chosen as a function of the contaminants and the substrate in order to optimize the removal mechanisms. A pertinent choice of the wavelength can often strongly improve the process efficiency and reduce the surface damage. For instance, UV lasers are very efficient to remove the organic compounds and the painting layer removal can be sometimes achieved in single pulse operation with IR irradiation. Even if the laser matter interaction processes are rather well known in nanosecond regime, the development of a new specific cleaning application requires a step of optimization of the laser parameters. Indeed, the different thresholds (cleaning, damage, ablation) are strongly dependent on the surface and interface characteristics which change with the conditions of formation of contamination. It is also of prime importance to collect the ablated products in order to prevent their redeposition on the cleaned surface. Moreover, many techniques have been developed in situ to control the cleaning process and then to prevent the surface damage.

Laser cleaning reached then a mature stage and there is a real need of new cleaning techniques to answer the new technological and environmental requirements. Some applications have already crossed the border between the laboratory and the industry, and we can expect that many others will follow. The most important is not to believe that laser cleaning is a single mechanism process. The physic and the efficiency of the process are dependant of the irradiated materials but can be optimized with laser parameters. That implies a careful definition of the laser cleaning setup, but allows to address a wide range of applications [92].

References

[1] Chrisey, D. and Hubler, G., *Pulsed Laser Deposition of Thin Films*, John Wiley & Sons, New York, **1994**.
[2] Bauerle, D., (Ed.), *Laser Processing and Chemistry*, Springer-Verlag, **2000**.
[3] Fogarassy, E. and Lazare, S., *Laser Ablation of Electronic Materials*, North Holland Pub., **1992**.

[4] Gallais, L. and Natoli, J.-Y., *Appl. Opt.*, **2003**, *42*(6), 960–971.

[5] Mittal, K., *Particles on Surfaces*, Vol. 1, Plenum Press, New York, **1988**.

[6] Wu, X., Sacher, E. and Meunier, M., *J. Adhesion*, **1999**, *70*, 167–178.

[7] Arnold, N., *Laser Cleaning*, Luk'yanchuk, B. (Ed.), World Scientific Publishing, **2002**, 51–102.

[8] Lu, Y., Zheng, Y. and Song, W., *Appl. Phys. A*, **1999**, *68*, 569–572.

[9] Luk'yanchuk, B., Zheng, Y. and Lu, Y., *Proc. SPIE*, **2001**, *4423*, 115–126.

[10] Arnold, N., *Appl. Surf. Science*, **2003**, *208–209*, 15–22.

[11] Luk'yanchuk, B., Mosbacher, M., Zheng, Y.W., Münzer, H.-J., Huang, S.M., Bertsch, M., Song, W.D., Wang, Z.B., Lu, Y.F., Dubbers, O., Boneberg, J., Leiderer, P., Hong, M.H. and Chong, T.C., *Laser Cleaning,* Luk'yanchuk, B. (Ed.), World Scientific Publishing, **2002**, 103–178.

[12] Dobler, V., Oltra, R., Boquillon, J.P., Mosbacher, M., Boneberg, J. and Leiderer, P., *Appl. Phys. A* [suppl.], **1999**, *69*, S335–337.

[13] Grojo, D., Cros, A., Delaporte, Ph. and Boyono-Onana, M., *to be published in Appl. Surf. Sci.*, **2006**.

[14] Luk'yanchuk, B., Wang, Z., Song, W. and Hong, M., *Appl. Phys. A*, **2004**, *79*, 747–751.

[15] Mosbacher, M., Münzer, H.-J., Zimmermann, J., Solis, J., Boneberg, J. and Leiderer, P., *Appl. Phys. A*, **2001**, *72*, 41–44.

[16] Pleasants, S., Luk'yanchuk, B. and Kane, D., *Appl. Phys. A*, **2004**, *79*, 1595–1598.

[17] Münzer, H.-J., Mosbacher, M., Bertsch, M., Zimmermann, J., Leiderer, P. and Boneberg, J., *J. of Microscopy*, **2001**, *202*, 129–135.

[18] Zheng, Y., Luk'yanchuk, B., Lu, Y., Song, W. and Mai, Z., *J. Appl. Phys.*, **2001**, *90*(5), 2135–2142.

[19] Lee, J., Watkins, K. and Steen, W., *Appl. Phys. A*, **2000**, *71*, 671–674.

[20] Kane, D., Fernandes, A. and Halfpenny, D., *Laser Cleaning*, Luk'yanchuk, B. (Ed.), World Scientific Publishing, **2002**, 181–228.

[21] Zapka, W., Asch, K., Keyser, J. and Meissner, K., German Patent DE 3721940C2, **1987**.

[22] Zapka, W., Ziemlich, W. and Tam, A., *Appl. Phys. Lett.*, **1991**, *58*(20), 2217–2219.

[23] Tam, A., Leung, W., Zapka, W. and Ziemlich, W., *J. Appl. Phys.*, **1992**, *71*(7), 3515–3523.

[24] Lang, F., Mosbacher, M. and Leiderer, P., *Appl. Phys. A*, **2003**, *77*, 117–123.

[25] Lee, Y., Lu, Y., Chan, D., Low, T. and Zhou, M., *Jpn. J. Appl. Phys.*, **1998**, *37*, 2524–2529.

[26] Yavas, O., Schilling, A., Bischof, J., Boneberg, J. and Leiderer, P., *Appl. Phys. A*, **1997**, *64*(4), 331–339.

[27] Zapka, W., *Laser Cleaning*, Luk'yanchuk, B. (Ed.), World Scientific Publishing, **2002**, 23–48.

[28] Kim, D. and Lee, J., *J. Appl. Phys.*, **2003**, *93*(1), 762–764.

[29] Lu, Y., Song, W. and Low, T., *Mat. Chem. Phys.*, **1998**, *54*(1–3), 181–185.

[30] Wu, X., Sacher, E. and Meunier, M., *J. Appl. Phys.*, **2000**, *87*(8), 3618–3627.

[31] Mosbacher, M., Dobler, V., Boneberg, J. and Leiderer, P., *Appl. Phys. A*, **2000**, *70*, 669–672.

[32] Soltani, M. and Ahmadi, G., *J. Adhesion*, **1994**, *44*, 161–175.

[33] Vanderwood, R. and Cetinkaya, C., *J. Adhesion Sci. Tech.*, **2003**, *17*(1), 129–147.

[34] Lee, J., You, S., Park, J. and Busnaina, A., *Semicond. Int.*, **2003**, *26*(8).

[35] Psyllaki, P. and Oltra, R., *Mat. Sci. and Eng. A*, **2000**, *282*, 145–152.

[36] Oltra, R., Yavas, O., Cruz, F., Boquillon, J.P. and Sartori, C., *Appl. Surf. Sci.*, **1996**, *96–98*, 484–490.

[37] Pasquet, R., Del Coso, R., Boneberg, J., Leiderer, P., Oltra, R. and Boquillon, J.P., *Appl. Phys. A* [suppl.], **1999**, *69*, S727–730.

[38] Oltra, R., Yavas, O. and Kerrec, O., *Surf. and Coat. Tech.*, **1996**, *88*, 157–161.

[39] Dupont, A., Aix – Marseille II University, PhD Thesis, No. 207 94 53, **1994**.

[40] Kearns, A., Fischer, C., Watkins, K.G., Glasmacher, M., Kheyrandish, H., Brown, A., Steen, W.M. and Beahan, P., *Appl. Surf. Sci.*, **1998**, *127–129*, 773–780.

[41] Cheon, L., Kim, D., Lee, K. and Hui, C., *J. Electron. Mater.*, **2005**, *34*(2), 132–136.

[42] Lazare, S. and Granier, V., *Appl. Phy. Lett.*, **1989**, *54*, 862–865.

[43] Sauebrey, R. and Pettit, G., *App. Phy. Lett.*, **1989**, *55*(5), 421–423.

[44] Singleton, D. and Paraskevopoulos, G., *Chem. Phy.*, **1990**, *144*, 415–423.

[45] Schweizer, G. and Werner, L., *Proc. SPIE*, **1995**, *2502*, 57–62.

[46] Gomez, C., Caballero, O., Costella, A., Garcia Moreno, I. and Sastre, R., *2nd workshop on laser cleaning*, Madrid, 13-14 June **2002**.

[47] Kimura, W., Kim, G. and Balick, B., *Astro. Soc. of the Pacific*, **1995**, *107*, 888–895.

[48] Lu, Y., Komuro, S. and Aoyagi, Y., *Jpn. J. Appl. Phys.*, **1994**, *33*, 4691–4696.

[49] Sentis, M., Delaporte, Ph., Marine, W. and Uteza, O., *Quant. Electr.*, **2000**, *30*(6), 495–500.

[50] Dausinger, F. and Shen. J., *ISIJ International*, **1993**, *33*, 925.

[51] Boquillon, J.-P., Restauratorenblätter, Sonderband, Lacona, König E. and Kautek W., (Eds.), **1997**, 103.

[52] Gouillon, A.-S., Gatto, A., Delaporte, Ph., Fontaine, B., Sentis, M. and Uteza, O., *SPIE*, **1998**, *3404*, 295–300.

[53] Le, H., Zeitoun, D., Parisse, J.-D., Sentis, M. and Marine, W., *Physical Review E*, **2000**, *62*(3), 4152–4161.

[54] Itina, T., Hermann, J., Delaporte, Ph. and Sentis, M., *Physical Review E*, **2002**, *66*(6), 066406.

[55] Lee, J. and Watkins, K., *Opt. and Lasers in Engineering*, **2000**, *34*, 429–442.

[56] Kabashin, A., Nikitin, P., Marine, W. and Sentis, M., *Quant. Elect.*, **1998**, *28*, 24–28.

[57] Chaoui, N., Pasquet, P., Solis, J., Afonso, C. and Oltra, R., *Surf. and Coatings Tech.*, **2001**, *150*, 57–63.

[58] Song, W., Hong, M., Lee, S., Lu, Y. and Chong, T., *Appl. Surf. Science*, **2003**, *208–209*, 306–310.

[59] Yavas, O., Suzuki, N., Takai, M., Hosono, A. and Kawabushi, S., *Appl. Phys. Lett.*, **1998**, *72*, 2797–2799.

[60] Takai, M., Suzuki, N. and Yavas, O., *Laser Cleaning*, Luk'yanchuk, B. (Ed.), World Scientific Publishing, **2002**, 417–432.

[61] Gu, J., Low, J., Lim, P. and Gu, P., *SPIE*, **2001**, *4595*, 293–300.

[62] Song, W., Lu, Y., Hong, M. and Low, T., *SPIE*, **1998**, *3550*, 19–26.

[63] Edelson, M., Pang, H. and Ferguson, R., *Proc. of ICALEO Conf.*, **1995**, 768–777.

[64] Ferguson, R., Edelson, M. and Pang, H., Laser ablation system and method of decontaminating surfaces, US Patent No. 5,780,806, **1998**.

[65] Pellin, M., Leong, K. and Savina, M., *EMPS project summaries*, **1998**, *4–5*.

[66] Clar, G., Martin, A. and Cartry, J.P., Patent, US5256848, **1993**.

[67] Delaporte, Ph., Gastaud, M., Marine, W., Sentis, M., Uteza, O., Thouvenot, P., Alcaraz, J.-L., Le Samedy, J.-M. and Blin, D., *Appl. Surf. Sci.*, **2002**, *197–198*, 826–830.

[68] Picco, B. and Marchand, M., Patent FR2774801, **1999**.

[69] Delaporte, Ph., Gastaud, M., Marine, W., Sentis, M., Uteza, O., Thouvenot, P., Alcaraz, J.-L., Le Samedy, J.-M. and Blin, D., *Appl. Surf. Sci.*, **2003**, *208–209*, 298–305.

[70] Chevillot, C. and Watelet, S., *J. of Cultural Heritage*, **2003**, *4*, 27s–32s.

[71] Klein, S., Stratoudaki, T., Marakis, Y., Zafiropulos, V. and Dickmann, K., *Appl. Surf. Sci.*, **2000**, *157*, 1–6.

[72] Zafiropulos, V., *Laser Cleaning*, Luk'yanchuk, B. (Ed.), World Scientific Publishing, **2002**, 345–392.

[73] Maravelaki, P.V., Zafiropulos, V., Kylikoglou, V., Kalaitzaki, M. and Fotakis, C., *Spectrochimica Acta Part B*, **1997**, *52*, 41–53.

[74] Georgiou, S., Zafiropulos, V., Anglos, D., Balas, C., Tornari, V. and Fotakis, C., *Appl. Surf. Sci.*, **1998**, *127–129*, 738–745.

[75] Castillejo, M., Martin, M., Oujja, M., Silva, D., Torres, R., Manousaki A., Zafiropulos, V., Van Den Brink, O.F., Heeren, R.M., Teule, R., Silva, A. and Gouveia, H., *Analytical Chemistry*, **2002**, *74*(18), 4662–4671.

[76] Zafiropulos, V., Balas, C., Manousaki, A., Marakis, Y., Maravelaki-Kalaitzaki, P., Melesanaki, K., Pouli, P., Stratoudaki, T., Klein, S., Hildenhagen, J., Dickmann, K., Luk'Yanchuk, B.S., Mujat, C. and Dogariu, A., *J. of Cultural Heritage*, **2003**, *4*, 249s–256s.

[77] Klein, S., Fekrsanati, F., Hildenhagen, J., Dickmann, K., Uphoff, H., Marakis, Y. and Zafiropulos, V., *Appl. Surf. Sci.*, **2001**, *171*(3–4), 242–251.

[78] Marakis, G., Pouli, P., Zafiropulos, V. and Maravelaki-Kalaitzaki, P., *J. of Cultural Heritage,* **2003**, *4*, 83–91.

[79] Salimbeni, R., Pini, R. and Siano, S., *J. of Cultural Heritage*, **2003**, *4*, 72s–76s.

[80] Coddet, C., Montavon, G., Marchione, T. and Freneaux, O., *Thermal Spray: Meeting the Challenges of the 21st Century,* Vol.1., Coddet, C. (Ed.), ASM International, **1998**, 1321–1325.

[81] Verdier, M., Montavon, G., Costil, S. and Coddet, C., *Thermal spray 2001: New Surface for a New Millennium*, Berndet, C.C. *et al.*, (Eds.), ASM International, Materials Park, Ohio, USA, **2001**, 553–560.

[82] Barbezat, G., Folio, F., Coddet, C. and Montavon, G., *Thermal spray: surface engineering via applied research,* Berndt, C.C. (Ed.), ASM International, Materials Park, OH, **2000**, 57–61.

[83] Dimogerontakis, Th., Oltra, R. and Heintz, O., *Appl. Phys. A,* **2006**, *81*(6), *1173–1179.*

[84] Ardelean, H., Petit, S., Laurens, P., Marcus, P. and Arefi-Khonsari, F., *Appl. Surf. Sci.,* **2005**, *243,* 304–318.

[85] Laurens, P., Ould Bouali, M., Meducin, F. and Sadras, B., *Appl. Surf. Sci.,* **2000**, *154–155,* 211–216.

[86] Yue, T.M., Yu, J.K. and Man, H.C., *Surf. Coat. Technol.,* **2001**, *137,* 65–71.

[87] Schaaf, P., *Progress in Materials Science,* **2002**, *47*(1), 1–161.

[88] Pereira, A., Cros, A., Delaporte, Ph., Georgiou, S., Manousaki, A., Marine, W. and Sentis, M., *Appl. Phys. A,* **2004**, *79,* 1433–1437.

[89] Thomann, A.L., Basillais, A., Wegscheider, M., Leborgne, C., Pereira, A., Delaporte, Ph. and Sentis, M., *Appl. Surf. Sci.,* **2004**, *230,* 350–363.

[90] Pereira, A., Delaporte, Ph., Sentis, M., Marine, W., Thomann, A.L. and Boulmer-Leborgne, C., *J. of Appl. Phys.,* **2005**, *98,* 064902.

[91] Sicard, E., Boulmer-Leborgne, C., Andreazza-Vignolle, C. and Frainais, M., *Appl. Phys. A,* **2001**, *73*(1), 55–60.

[92] Pereira, A., Aix – Marseille II University, PhD Thesis, **2003**.

Index